Functional Coatings

Edited by
Swapan Kumar Ghosh

Related Titles

Streitberger, H.-J., Kreis, W. (Eds.)

Automotive Paints and Coatings

2nd completely revised and enlarged edition

2006

ISBN 3-527-30971-3

Buxbaum, G., Pfaff, G. (Eds.)

Industrial Inorganic Pigments

3rd completely revised and enlarged edition

2005

ISBN 3-527-30363-4

Herbst, W., Hunger, K.

Industrial Organic Pigments

Production, Properties, Applications

2004

ISBN 3-527-30567-9

Caruso, F. (Ed.)

Colloids and Colloid Assemblies

Synthesis, Modification, Organization and Utilization of Colloid Particles

2004

ISBN 3-527-30660-9

Decher, G. / Schlenoff, J. B. (Eds.)

Multilayer Thin Films

Sequential Assembly of Nanocomposite Materials

2002

ISBN 3-527-30440-1

Hunger, K. (Ed.)

Industrial Dyes

Chemistry, Properties, Applications

2002

ISBN 3-527-30426-6

Smith, H. M. (Ed.)

High Performance Pigments

2001

ISBN 3-527-30204-2

Functional Coatings

by Polymer Microencapsulation

Edited by
Swapan Kumar Ghosh

WILEY-VCH

WILEY-VCH Verlag GmbH & Co. KGaA

The Editor

Dr. Swapan Kumar Ghosh
Arcelor Research Industry Gent
OCAS N. V.
John Kennedylaan 3
9060 Zelzate
Belgien

Library of Congress Card No.:
applied for

British Library Cataloguing-in-Publication Data:
A catalogue record for this book is available from the British Library.

**Bibliographic information published by
Die Deutsche Bibliothek**
Die Deutsche Bibliothek lists this publication in the Deutsche Nationalbibliografie; detailed bibliographic data are available in the Internet at <http://dnb.ddb.de>.

Typesetting TypoDesign Hecker GmbH, Leimen
Printing Strauss GmbH, Mörlenbach
Binding Litges & Dopf GmbH, Heppenheim

Printed in the Federal Republic of Germany
Printed on acid-free paper

ISBN-13 978-3-527-31296-2
ISBN-10 3-527-31296-X

Table of Contents

Functional Coatings. Edited by Swapan Kumar Ghosh
Copyright © 2006 WILEY-VCH Verlag GmbH & Co. KGaA, Weinheim
ISBN 3-527-31296-X

Preface

Over the past few decades, coating technology has been transformed into an extensive field of materials research. Today's industry demands additional requirements and functionalities from coatings, in addition to their basic protective and decorative values. Recent advances in polymer science and inorganic chemistry – especially at the nanoscale level – have further enhanced this growth. In this context, microencapsulation, a commercially successful technology which is used mainly in the paper and pharmaceutical industries, provides an additional input to the growing needs of functional coatings. This book aims to review the art of microencapsulation and to provide the readers with a comprehensive and in-depth understanding of the fundamental and applied research into microcapsules containing functional coatings.

The topics detailed in Chapter 1 provide an overview of the different microencapsulation techniques available, and the use of microcapsules in functional coatings. Chapters 2 to 5 describe the different aspects of encapsulating organic and inorganic particles, with the relevant properties and applications. Developments of this technology, based on conducting polymers, are outlined in Chapter 6, while Chapter 7 emphasizes the growth and applications of microencapsulation in the textile industry, highlighting especially the future prospects of smart coatings. Sol-gel technology and electrolytic co-deposition techniques, both of which have attracted much attention during the past few years in the development of functional coatings, are detailed in Chapters 8 and 9, respectively.

I would like to express my gratitude to the contributing authors in making this project a success, and to Dr. Martin Ottmar and Dr. Bettina Bems of Wiley-VCH for their assistance and encouragement in this venture. Finally, I would like to dedicate this book to my wife, Anjana, for her support, encouragement and assistance during its preparation.

Swapan Kumar Ghosh
Spring 2006

Functional Coatings. Edited by Swapan Kumar Ghosh
Copyright © 2006 WILEY-VCH Verlag GmbH & Co. KGaA, Weinheim
ISBN 3-527-31296-X

List of Contributors

Elodie Bourgeat-Lami
Laboratoire de Chimie et Procédés de
Polymérisation
CNRS/CPE
Bât. 308 F, BP 2077
43 Bd. du 11 Novembre 1918
69616 Villeurbanne Cedex
France

Christophe Déjugnat
Max Planck Institute of Colloids and
Interfaces
Wissenschaftspark Golm
14424 Potsdam
Germany

Etienne Duguet
Institut de Chimie de la Matière
Condensée de Bordeaux
CNRS/Université Bordeaux-1
87 Ave. du Docteur Albert Schweitzer
33608 Pessac Cedex
France

Swapan Kumar Ghosh
Arcelor Research Industry Gent
OCAS N.V.
John Kennedylaan 3
9060 Zelzate
Belgium

Serge Hoste
Department of Inorganic and Physical
Chemistry
University of Gent
Krijgslaan 281
9000 Gent
Belgium

Katharina Landfester
Department of Organic Chemistry III
Macromolecular Chemistry and
Organic Materials
University of Ulm
Albert-Einstein-Allee 11
89081 Ulm
Germany

Zhu Liqun
School of Material Science and
Engineering
Beijing University of Aeronautics
and Astronautics
P.O. Box 103
100083, Beijing
China

Dmitry G. Shchukin
Max Planck Institute of Colloids and
Interfaces
Wissenschaftspark Golm
14424 Potsdam
Germany

Functional Coatings. Edited by Swapan Kumar Ghosh
Copyright © 2006 WILEY-VCH Verlag GmbH & Co. KGaA, Weinheim
ISBN 3-527-31296-X

Parshuram G. Shukla
Polymer Science and Engineering
Division
National Chemical Laboratory
Dr. HomiBhaba Road
Pune 411 008
India

Gleb B. Sukhorukov
Max Planck Institute of Colloids and
Interfaces
Wissenschaftspark Golm
14424 Potsdam
Germany

and

IRC at Biomedical Materials
Queen Mary University of London
Mile End Road
London E1 4NS
United Kingdom

Isabel van Driessche
Department of Inorganic and Physical
Chemistry
University of Gent
Krijgslaan 281
9000 Gent
Belgium

Marc van Parys [AWAITING DE-
TAILS] Textile Institute – TO2C
Technical University Gent
Voskenslaan 362
9000 Gent
Belgium

1

Functional Coatings and Microencapsulation: A General Perspective

Swapan Kumar Ghosh

1.1
An Overview of Coatings and Paints

Today, many objects that we come across in our daily lives, including the house in which we live and the materials we use (e.g., toothbrushes, pots and pans, refrigerators, televisions, computers, cars, furniture) all come under the "umbrella" of coated materials. Likewise, fields such as military applications – for example, vehicles, artilleries and invisible radars – and aerospace products such as aircraft, satellites and solar panels all involve the widespread use of coated materials. Clearly, the importance of coatings has increased hugely during the modern era of technology.

Coating is defined as a material (usually a liquid) which is applied onto a surface and appears as either a continuous or discontinuous film after drying. However, the process of application and the resultant dry film is also regarded as coating [1]. Drying of the liquid coating is mostly carried out by evaporative means or curing (cross-linking) by oxidative, thermal or ultraviolet light and other available methods. Paint can be defined as a dispersion that consists of binder(s), volatile components, pigments and additives (catalyst, driers, flow modifiers) [2]. The binder (polymer or resin) is the component that forms the continuous film, adheres to the substrate, and holds the pigments and fillers in the solid film. The volatile component is the solvent that is used for adjusting the viscosity of the formulation for easy application. Depending on their compositions, paints can be divided into three groups: solvent-borne, water-borne and solvent-free (100% solid). Solvent-borne paints consist of resin, additives and pigments that are dissolved or dispersed in organic solvents. Similarly, in water-borne paints the ingredients are dispersed in water. In solvent-free compositions, the paints do not contain any solvent or water and the ingredients are dispersed directly in the resin.

The properties of coating films are determined by the types of binders, pigments and miscellaneous additives used in the formulation. Moreover, types of substrates, substrate pretreatments, application methods and conditions of film formation play additional roles in determining the end properties of the coating. The terms "coating" and "paint" will be used synonymously in this book. In general, collectively or individually, paints, varnishes (transparent solutions) and lacquers (opaque or colored varnishes) are termed as coatings [3].

Coatings occur in both organic and inorganic forms. Inorganic coatings are mainly applied for protective purposes, while organic coatings are mostly used for decorative and functional applications [4]. Organic coatings can be classified as either architectural coatings (house, wall and ceiling coatings) or industrial coatings (appliances, furniture, automobiles, coil coatings) [3]. Although organic and inorganic coatings may be used individually for industrial applications, for specific requirements a combination of both systems – termed hybrid coating – is favored.

1.2
Classification of Coating Properties

Coatings are usually applied as multi-layered systems that are composed of primer and topcoat. However, in some cases – for example automotive coating systems – this may vary from four to six layers. Each coating layer is applied to perform certain specific functions, though its activities are influenced by the other layers in the system. The interactions among different layers and the interfacial phenomenon play an important role in the overall performance of the multi-coat systems [5]. Different properties of coatings are typically associated with specific parts of a coating system (Fig. 1.1) [6].

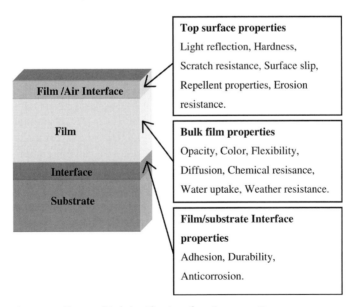

Figure 1.1 Topographical classification of coating properties.

1.3
What are Functional Coatings?

Coatings are mainly applied on surfaces for decorative, protective, or functional purposes, but in most cases it is a combination of these. The term "functional coatings" describes systems which possess, besides the classical properties of a coating (i.e., decoration and protection), an additional functionality [7]. This additional functionality may be diverse, and depend upon the actual application of a coated substrate. Typical examples of functional coatings are self-cleaning [8,9], easy-to-clean (anti-graffiti) [10], antifouling [11], soft feel [12] and antibacterial [13–15]. Although various mechanisms are involved, as well as numerous applications, there is a common feature that is of particular benefit and which satisfies some users' demands. Most coatings (whether inorganic, organic or ceramic) perform critical functions, but as these fields are extensive it is beyond the scope of this book to include all of them at this stage. Thus, the discussion here is limited to coatings with organic binders.

1.4
Types and Application of Functional Coatings

Apart from their special properties, functional coatings must often satisfy additional requirements; for example, nonstick cookware coatings must be resistant to scratching, abrasion and thermal effects. Typical expectations of functional coatings include:

- durability
- reproducibility
- easy application and cost effectiveness
- tailored surface morphology
- environmental friendliness

Functional coatings can be classified as several types depending on their functional characteristics (Fig. 1.2).

Functional coatings perform by means of physical, mechanical, thermal and chemical properties. Chemically active functional coatings perform their activities either at film–substrate interfaces (anticorrosive coatings), in the bulk of the film (fire-retardant or intumescent coatings), or at air–film interfaces (antibacterial, self-cleaning) [16].

Some applications of functional coatings are discussed in the following sections.

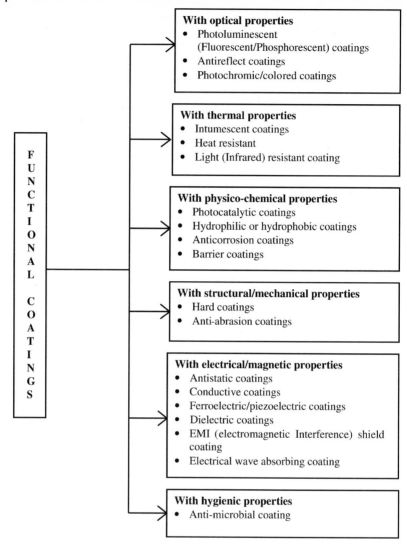

With optical properties
- Photoluminescent (Fluorescent/Phosphorescent) coatings
- Antireflect coatings
- Photochromic/colored coatings

With thermal properties
- Intumescent coatings
- Heat resistant
- Light (Infrared) resistant coating

With physico-chemical properties
- Photocatalytic coatings
- Hydrophilic or hydrophobic coatings
- Anticorrosion coatings
- Barrier coatings

With structural/mechanical properties
- Hard coatings
- Anti-abrasion coatings

With electrical/magnetic properties
- Antistatic coatings
- Conductive coatings
- Ferroelectric/piezoelectric coatings
- Dielectric coatings
- EMI (electromagnetic Interference) shield coating
- Electrical wave absorbing coating

With hygienic properties
- Anti-microbial coating

Figure 1.2 Types of functional coatings.

1.4.1
Anticorrosive Coatings

It is known that, when iron is exposed to a natural atmosphere, then rust is formed. Although the rusting of iron or steel is usually termed as corrosion, the latter is a general term which is used to define the destructive interaction of a material with its environment. Corrosion usually refers to metals, though nonmetallic substrates such as plastics, concrete or wood also deteriorate in the environment. Corrosion causes enormous industrial losses with a depletion of our natural resources. When two areas of a metallic component are exposed to different operational environments, or they differ in their surface structure or composition, an electrical potential is developed. Corrosion is in fact an electrochemical process where the electrical cell is composed of an anode (the corrosion site), an electrolyte (the corrosive medium), and a cathode (part of the metal which is active in the corrosion process but does not itself corrode) [17].

In general, organic coatings are applied onto metallic substrates in order to avoid the detrimental effect of corrosion. The anticorrosive performance of the coating depends upon several parameters, including: adhesion to metal, thickness, permeability, and the different properties of the coating. In most cases, the primer is mainly responsible for protecting the metallic substrate and adhering to other coating layers. In this context, surface preparation is essential in order to provide good adhesion of the primer to the metallic substrate [18]. The mechanisms by which organic coatings offer corrosion protection are summarized as follows.

- Sacrificial means: The use of a sacrificial anode such as zinc to protect steel is a longstanding and well-known industrial practice. The zinc layer on galvanized steel degrades when exposed to an adverse environment, and this protects the underneath surface. Using a similar approach, both inorganic and organic resin-based, zinc-rich coatings have been developed to protect a variety of metal substrates [19,20].
- Barrier effect: In general, polymeric coatings are applied to metallic substrates to provide a barrier against corrosive species. They are not purely impermeable. Moreover, defects or damages in the coating layer provide pathways by which the corrosive species may reach the metal surface, whereupon localized corrosion can occur. Pigments having lamellar or plate-like shapes (e.g., micaceous iron oxide and aluminum flakes) are introduced to polymeric coatings; this not only increases the length of the diffusion paths for the corrosive species but also decreases the permeability of the coating [21]. Other pigments such as stainless steel flakes, glass flakes and mica are also used for this purpose. The orientation of the pigments in the coating must be parallel to the surface, and they should be highly compatible with the matrix resin to provide a good barrier effect. Layered clay platelets such as montmorillonite may also be introduced into organic resin systems to increase the barrier effect towards oxygen and water molecules, thereby enhancing the anticorrosive performance of the coating (Fig. 1.3) [22].
- Inhibition: Traditionally, chromate- and lead-based pigments are the most common compounds used as corrosion inhibitors to formulate anticorrosive primers for metallic substrates. These substances are considered to be toxic and ecologically unsafe, and therefore the search for new alternative anticorrosive pigments is under way. Today, primers containing metallic phosphate, silicate, titanate or molybdate compounds are available commercially. These pigments form a protective oxide layer on the metallic substrates, and often also form anticorrosive complexes with the binder. To reduce the cost, a number of elements and compounds have been combined to develop an effective anticorrosive pigment, including aluminum zinc phosphate, calcium zinc molybdate, zinc molybdate

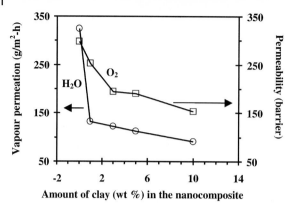

Figure 1.3 Permeability of water (H_2O) and oxygen (O_2) as a function of clay (montmorillonite) content in polystyrene-clay nanocomposite material (adapted from [22]).

phosphate, calcium borosilicate, and strontium phosphosilicate. Organofunctional silanes have emerged recently as alternative chromate treatments for metals due to their environmental friendliness and good anticorrosion properties [23]. The main disadvantage of using silane is that the substrate must bear hydroxyl groups on its surface. Thus, silane technology offers less flexibility compared to the titanate-based chemistry [24]. Another modern development is the use of intrinsically conductive polymers (ICPs) in the corrosion protection of metals (see Chapter 6) [25,26]. A different approach is the use of core-shell materials (e.g., a ferric oxide core with a shell of zinc phosphate or anticorrosive titanium dioxide coated with an organic polymer) to develop anticorrosive primers [27,28]. Self-priming, chromate-free, corrosion-resistant coating compositions have also been investigated [29]. Smart corrosion-inhibiting coatings such as the inclusion of a pH indicator into a paint formulation that can cause color change when corrosion occurs are presently under investigation. Recent developments also include the use of nanoclay that can exchange anticorrosive agents with the corrosive species when needed. Although these innovative research projects have not yet provided any new commercial products, they offer a variety of interesting routes for future developments.

1.4.2
High Thermal-Resistant and Fire-Retardant Coatings

High thermal-resistant coatings are required for a wide variety of metallic substrates that we encounter in everyday life, including nonstick cookware, barbecues and boilers. Fluorine- or silicon-based products are used to obtain a high thermal resistance for the above-mentioned products. Fluorinated coatings are not suitable for high-temperature applications as they degrade above ~300 °C and produce toxic byproducts. Although other binders such as phenolic or epoxy are used to prepare high thermal-resistant coatings, at present silicon-containing coatings dominate the market. Silicon-containing polymers offer better thermal resistance due to the high energy required to cleave silicon bonds compared to carbon bonds in analogous molecules. Recently developed silicon-based coatings are able to resist temperatures of up to 1000 °C. Silicon derivatives such as silicone resins (siloxanes) or inorganic silicates are commonly used for high-temperature applications.

Silicon-containing materials are expensive, however, and consequently copolymers or blends of silicones with acrylate, epoxy or urethanes are very often used to save costs. Recent reports have been made of innovative ways to design thermal-resistant coatings; for example, titanium esters in combination with aluminum flakes have been incorporated into binders that resist temperatures up to 400 °C. Above this temperature "burn off" occurs and a complex coating of titanium-aluminum is formed that deposits on the substrate and enhances thermal resistance up to 800 °C [30].

The devastating nature of fire creates havoc and results in great loss of lives and property. Thus, the need to develop fire-retardant coatings is constantly growing. Although protection against fire by the use of coatings for indefinite periods is impossible, the use of fire-retardant coatings can delay the spread of fire or keep a structure intact against fire, thereby allowing sufficient time for safety measures to be taken. Today, several types of fire retardant are available, including phosphorus-containing, halogen-based and intumescent fire-retardant systems, each with a different principle of operation. Phosphorus-containing compounds function by forming a protective layer either as a glassy surface barrier or by producing char. Halogen- and antimony-based fire retardants are both toxic and ecologically unsafe.

Intumescent coatings form an expanded carbonaceous char which acts as a protective barrier against heat transfer and hinders the diffusion of combustible gases and melted polymer to the site of combustion. These coatings are composed of three components: (i) an inorganic acid (dehydrating agent); (ii) a carbonaceous char-forming material; and (iii) a blowing agent. The performance of the intumescent system depends on the choice of the ingredients and their appropriate combination [31,32]. Nowadays, expandable graphites are available commercially as fire-retardant agents; these contain chemical compounds, including an acid, entrapped between the carbon layers. Upon exposure to higher temperatures, exfoliation of the graphite takes place and this provides an insulating layer to the substrate [33]. A combination of polyurethane and phosphate serves as a well-known fire-retardant intumescent system. One problem associated with these systems results from the solubility of phosphates in water, and this leads to problems of migration. Nonetheless, this difficulty can be avoided by encapsulating phosphates (di-ammonium hydrogen phosphate) within a polyurethane shell and, indeed, the use of microencapsulated fire retardants in polyurethane coatings has shown good fire resistance [34]. Today, silicon- or inorganic hydroxide-based fire-retardant coatings are used in a wide variety of industrial applications [35,36]. Recently, polymer clay (layered silicates) nanocomposites have also been explored for the development of fire-retardant coatings [37,38].

1.4.3
Scratch- and Abrasion-Resistant Coatings

Coatings are susceptible to damage caused by scratch and/or abrasion. Clearly, the consumer prefers to retain the aesthetic appearance of coated materials, and for this reason clear coats used on automobiles must have good scratch and abrasion resistance. An added problem is that scratches may also cause damage to the underlying substrate.

Many companies worldwide have undertaken the challenge of improving the scratch resistance of a coating, without adversely affecting its other properties. Scratch resistance can be obtained by incorporating a greater number of crosslinks in the coating's binder, but unfortunately highly crosslinked (hard) films have poor impact resistance due to less flexibility. A less-crosslinked (softer) film will show better performance with regard to other properties such as anti-fingerprint and impact resistance, but will have less scratch and abrasion resistance. Thus, in order to

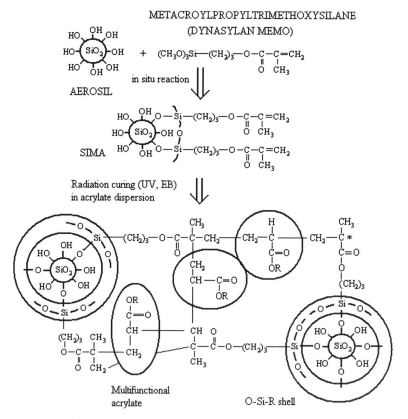

Figure 1.4 Schematic structure of grafted silica nanoparticles and radiation-cured grafted silica/acrylate networks (adapted from reference [41]).

obtain optimal scratch resistance the correct combination of hardness and flexibility is required. In this context, organic-inorganic hybrid films are paving the way for scratch-resistant coating developments. Recent advances in nanotechnology plays an important role in the development of scratch-resistant coatings [39,40]. Gläsel et al. have shown the use of siloxane-encapsulated SiO_2 nanoparticles to develop scratch- and abrasion-resistant coatings [41,42] (Fig. 1.4).

PPG industries have developed scratch-resistant coatings by incorporating SiO_2 nanoparticles into an organic matrix that can migrate to the surface. In this way the scratch resistance is enhanced due to an enrichment of the nanoparticles near the coating surface [43]. Coatings with good abrasion and scratch-resistant properties have also been reported by others [44,45].

1.4.4
Self-Cleaning Coatings

Self-cleaning coatings, as the name suggests, have a special functional property, and today the term Lotus effect® and self-cleaning are synonymous. Although these surfaces can be soiled, manual cleaning is unnecessary and a shower of rain is sufficient to carry out the cleaning process. In 1997, Barthelott and coworkers showed that the self-cleaning property of lotus leaves was due to their specialized surface morphology and hydrophobicity [46] (Fig. 1.5a).

This specialized morphology prevents dirt from forming an intimate contact with the surface, while the high hydrophobicity makes the leaf water-repellent. Consequently, as the water droplets roll onto the leaf surface, they carry along the contaminants (Fig. 1.5b). Since the initial discovery by Barthelott, many groups have attempted to mimic this activity to develop self-cleaning or lotus-effect coatings [47]. Detailed discussions on this concept, the underlying mechanism and the different applications of self-cleaning coating surfaces can be found elsewhere

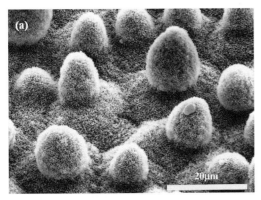 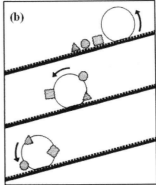

Figure 1.5 (a) Scanning electron micrograph of lotus leaf. (b) Schematic depicting the relationship between surface roughness and self-cleaning (adapted from [46], with kind permission of Springer Science and Business Media.

[9,48,49]. One very recent report noted that lotus leaves may be either hydrophilic or hydrophobic, depending on the contact of water molecules at the leaf surface [50].

During the past few years, self-cleaning coatings using photocatalytic titanium dioxide (TiO_2; especially the anatase crystalline form) have attracted considerable attention both in academic and industrial sectors. When photocatalytic TiO_2 particles are illuminated with an ultraviolet light source (e.g., sunlight), electrons are seen to be promoted from the valence band (VB) to the conduction band (CB) of the particle [51,52]. This creates a region of positive charge (h^+), holes, in the VB and a free electron in the CB. These charge carriers can either recombine or migrate to the surface, while the holes can react with the hydroxyl or adsorbed water molecules on the surface and produce different radicals such as hydroxyl radicals ($OH\cdot$) and hydroperoxy radicals ($HO_2\cdot$). By contrast, the electrons combine with the oxygen and produce superoxide radicals. These photo-produced radicals are powerful oxidizing species and can cause the deterioration of organic contaminants or microbial species on the particle surface. The other beneficial effect of TiO_2 is its super hydrophilic behavior, commonly known as the "water sheathing effect" [53]. This allows contaminants to be easily washed away with water or rainfall if the coatings are applied to external surfaces. Both photocatalysis and hydrophilicity occur simultaneously, despite their underlying mechanisms being of an entirely different nature. The addition of silicon oxide to TiO_2 has also been shown to enhance the overall self-cleaning properties [54].

Photocatalytic TiO_2 particles cannot be incorporated or deposited on the organic coating, as they oxidize the polymer. Recent developments have revealed the use of TiO_2 particles in combination with organic resins [55–57].

1.4.5
Antibacterial Coatings

In today's world, reports of outbreaks of disease in hospitals or problems caused by food poisoning are all-too-frequent occurrences. Microorganisms such as bacteria, fungi or viruses represent potential threats for our modern hygienic lifestyle. Microbial growth on coated substrates may have several adverse consequences, including problems of aesthetics (discoloration of the coating), risks to health and hygiene, malodor, biofilm development or microbial corrosion in the case of metallic substrates. Organic coatings are susceptible to microbial attack, and the properties of the coating and its composition, the presence of nutrients on the surface and the nature of substrates represent the main parameters that determine the types of microorganisms able to colonize the coating. A schematic representation of biofilm formation by microorganisms is shown in Figure 1.6.

The classical biocides function either by inhibiting the growth of bacteria (biostatic) or by killing them (biocidal) (Fig. 1.6). However, new legislations, combined with growing pressure from the environmentalists and the possibility of bacterial mutation have forced coating manufacturers to seek new alternatives. Today, more emphasis is placed on the development of biorepulsive (without killing) antibacterial coatings.

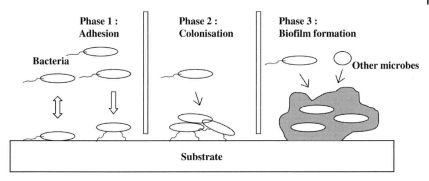

Figure 1.6 Schematic of biofilm formation by microorganisms.

A wide variety of organic or inorganic biocides are available commercially, and these demonstrate a wide variety of biocidal and biostatic mechanisms [58]. For example, biocides containing heavy metal ions function by penetrating the cell wall and inhibiting the bacterium's metabolic enzymes, whereas antimicrobial agents with cationic surfaces cause rupture of the bacterium's cytoplasmic membrane. Examples of organic biocides include polymers, tertiary alkyl amines and organic acids [59,60], while inorganic biocides include silver, zinc oxide (ZnO), copper oxide (CuO), TiO_2, and selenium [61–66]. Microcapsules containing biocides have also been developed in order to increase the longevity and efficiency of antimicrobial coatings [67–69].

1.4.6
Antifouling Coatings

Marine organisms represent a major threat to all objects used within a marine environment, and the unwanted growth or deposition of such organisms being termed as "fouling". Fouling is generally more prominent in coastal waters where ships or boats are either docked or travel at slow speed. Depending upon the types of marine organism involved, fouling is of two main types, namely microfouling and macrofouling.

Microfouling is caused by diatoms and bacteria, whereas *macrofouling* is caused by marine animals (barnacles, tubeworms) and plants (algae). Both biocidal and nonbiocidal coatings are used to prevent foulings. Biocide-based antifouling coatings function by slow leaching of the incorporated biocides into the coating. For reasons of stringent legislation and toxicity, the use of biocides is restricted on a daily basis. For example, tributyl tin (TBT) is a highly efficient marine biocide, but it is no longer used due to its toxicity. It is important that the biocide does not have any adverse effects on marine life while carrying out its antifouling activity. For long-term antifouling effects, either controlled-release or contact-active biocides are required [70].

Recently, a number of antifouling products have been developed using microencapsulation technology [71,72]. For the nonbiocidal approach, polymers with low surface energy are used in order to avoid the adhesion of marine organisms, and silicone elastomers are widely used for this purpose. However, this approach is only effective when the vessels move at relatively high speeds, for example ferries.

The specialized applications of coatings are almost unlimited, and developments of special effect pigments offer new possibilities for the design of functional coatings [73,74]. In this context, infrared (IR)-reflective pigments are in great demand for the development of "cool roof" coatings, while ultraviolet (UV)-resistant coatings have been developed for outdoor applications [75]. Water-borne functional coatings are becoming more popular than the solvent-borne systems due to their eco-friendly behavior [76].

1.5
Microencapsulation

The encapsulation of materials has evolved from examples in Nature, wherein numerous examples exist, ranging from macroscale to nanoscale. Nature envelops materials to protect them from environmental influences; the simplest example on a macroscopic scale is a bird's egg or a seed, while on a microscopic scale the best example is that of a cell along with its contents [77]. The development of microencapsulation began with the preparation of capsules containing dyes; these were incorporated into paper for copying purposes and replaced carbon paper [78]. The pharmaceutical industry has long used microencapsulation for the preparation of capsules containing active ingredients, though as time passed a variety of new technologies have emerged and are being developed in many fields of research. During the past 10 years this approach has been explored widely by the agricultural, food, cosmetic, and textile industries. Microencapsulation provides the possibility of combining the properties of different types of material (e.g., inorganic and organic) – a process which is difficult to achieve using other techniques. Although microencapsulation offers great potential in the coating industry, very little development has been carried out to date in this area. Extended reviews of microencapsulation techniques and processes can be found in references [79–83].

Microencapsulation cannot be defined as a product or as a component of a product. Rather, it is described as a process of enclosing micron-sized particles of solids or droplets of liquids or gasses in an inert shell, which in turn isolates and protects them from the external environment. The inertness is related to the reactivity of the shell with the core material. This technology is mainly used for the purpose of protection, controlled release, and compatibility of the core materials (see Chapter 7).

1.6
Microcapsules

The resultant product of the microencapsulation process is termed a "microcapsule". Such capsules are of micrometer size (>1 μm), and have a spherical or ir-

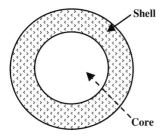

Figure 1.7 Schematic of microcapsule.

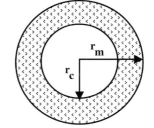

Figure 1.8 Cross-section of an idealized microcapsule.

regular shape. Microcapsules can be divided into two parts, namely the core and the shell. The core (the intrinsic part) contains the active ingredient (e.g., a hardener or a biocide), while the shell (the extrinsic part) protects the core permanently or temporarily from the external atmosphere. A microcapsule is shown schematically in Figure 1.7.

Core materials in microcapsules may exist in the form of either a solid, liquid or gas. The core materials are used most often in the form of a solution, dispersion or emulsion. Compatibility of the core material with the shell is an important criterion for enhancing the efficiency of microencapsulation, and pretreatment of the core material is very often carried out to improve such compatibility. The size of the core material also plays an important role for diffusion, permeability or controlled-release applications. Depending on applications, a wide variety of core materials can be encapsulated, including pigments, dyes, monomers, catalysts, curing agents, flame retardants, plasticizers and nanoparticles.

The abundance of natural and man-made polymers provides a wider scope for the choice of shell material, which may be made permeable, semi-permeable or impermeable. Permeable shells are used for release applications, while semi-permeable capsules are usually impermeable to the core material but permeable to low molecular-weight liquids. Thus, these capsules can be used to absorb substances from the environment and to release them again when brought into another medium. The impermeable shell encloses the core material and protects it from the external environment. Hence, to release the content of the core material the shell must be ruptured by outside pressure, melted, dried out, dissolved in solvent or degraded under the influence of light (see Chapter 7). Release of the core material through the permeable shell is mainly controlled by the thickness of the shell wall and its pore size. The dimension of a microcapsule is an important criterion for industrial applications; the following section will focus on spherical core-shell types of microcapsules (Fig. 1.8).

Assuming that the density of the core (ρ_c) and shell (ρ_s) materials are identical (i.e., $\rho_c = \rho_s$), it is possible to establish the relationship between the shell thickness ($d_s = r_m - r_c$) and the ratio of the weight of the shell material (w_s) to that of the core material (w_c):

$$\frac{w_s}{w_c} = \frac{(4/3)\pi(r_m^3 - r_c^3)\cdot\rho_s}{(4/3)\pi r_c^3\cdot\rho_s} \tag{1}$$

After rearranging, the following equation is obtained:

$$d_s = (r_m - r_c) = \left[\left(\frac{w_s}{w_c} + 1 \right)^{1/3} - 1 \right] r_c \qquad (2)$$

Equation (2) shows a linear relationship between the shell thickness and the capsule diameter when the ratio of $w_c/(w_s+w_c)$ is in the range of 0.50 to 0.95 [83].

1.7
Morphology of Microcapsules

The morphology of microcapsules depends mainly on the core material and the deposition process of the shell. Microcapsules may have regular or irregular shapes and, on the basis of their morphology, can be classified as mononuclear, polynuclear, and matrix types (Fig. 1.9).

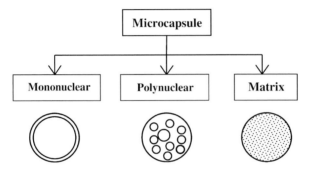

Figure 1.9 Morphology of microcapsules.

Mononuclear (core-shell) microcapsules contain the shell around the core, while polynuclear capsules have many cores enclosed within the shell. In matrix encapsulation, the core material is distributed homogeneously into the shell material. In addition to these three basic morphologies, microcapsules can also be mononuclear with multiple shells, or they may form clusters of microcapsules.

1.8
Benefits of Microencapsulation

Microcapsules have a number of interesting advantages, and the main reasons for microencapsulation can be summarized as follows:

- Protection of unstable, sensitive materials from their environments prior to use.
- Better processability (improving solubility, dispersibility, flowability).

- Self-life enhancement by preventing degradative reactions (oxidation, dehydration).
- Controlled, sustained, or timed release.
- Safe and convenient handling of toxic materials.
- Masking of odor or taste.
- Enzyme and microorganism immobilization.
- Controlled and targeted drug delivery.
- Handling liquids as solids.

1.9
Microencapsulation Techniques

Numerous preparation technologies available for the encapsulation of core material have been reported [81,82,84]. The present discussion focuses on the different microencapsulation techniques that are more relevant to the coating industries, and also provides a comprehensive review of recently developed methods. In general, microencapsulation techniques are divided into two basic groups, namely chemical and physical, with the latter being further subdivided into physico-chemical and physico-mechanical techniques. Some of the important processes used for microencapsulation are summarized in Table 1.1.

Table 1.1 Different techniques used for microencapsulation.

Chemical processes	Physical processes Physico-chemical	Physico-mechanical
• Suspension, dispersion and emulsion polymerization • Polycondensation	• Coacervation • Layer-by-layer (L-B-L) assembly • Sol-gel encapsulation • Supercritical CO_2-assisted microencapsulation	• Spray-drying • Multiple nozzle spraying • Fluid-bed coating • Centrifugal techniques • Vacuum encapsulation • Electrostatic encapsulation

1.9.1
Chemical Methods

In-situ processes such as emulsion, suspension, precipitation or dispersion polymerization and interfacial polycondensations are the most important chemical techniques used for microencapsulation [85–90]. An image of microcapsules with an aqueous core and silicone shell prepared using *in-situ* polymerization is shown in Figure 1.10.

An in-depth discussion on the major *in-situ* polymerization processes is provided in Chapter 4. In addition, encapsulation using the mini emulsion process is discussed in Chapter 2, and interfacial polycondensations processes are described in Chapter 5.

Figure 1.10 Scanning electron micrograph of silicone microcapsules containing an aqueous solution of self-tanning composition
(Courtesy: G. Habar, Microcapsules-Technologies).

1.9.2
Physico-Chemical Processes

1.9.2.1 Coacervation

The first systematic approach of phase separation – that is, partial desolvation of a
homogeneous polymer solution into a polymer-rich phase (coacervate) and the poor
polymer phase (coacervation medium) – was realized by Bungenberg and colleagues
[91,92]. These authors termed such a phase separation phenomenon "coacervation".
The term originated from the Latin ›acervus‹ , meaning "heap". This was the first reported process to be adapted for the industrial production of microcapsules.

Currently, two methods for coacervation are available, namely simple and complex processes. The mechanism of microcapsule formation for both processes is
identical, except for the way in which the phase separation is carried out. In simple
coacervation a desolvation agent is added for phase separation, whereas complex
coacervation involves complexation between two oppositely charged polymers.

Complex coacervation

Complex coacervation is carried out by mixing two oppositely charged polymers in
a solvent (usually water); the process is shown schematically in Figure 1.11.

The three basic steps in complex coacervation are: (i) preparation of the dispersion or emulsion; (ii) encapsulation of the core; and (iii) stabilization of the encapsulated particle. First the core material (usually an oil) is dispersed into a polymer
solution (e.g., a cationic aqueous polymer). The second polymer (anionic, water-
soluble) solution is then added to the prepared dispersion. Deposition of the shell
material onto the core particles occurs when the two polymers form a complex.
This process is triggered by the addition of salt or by changing the pH, temperature
or by dilution of the medium. The shell thickness can be obtained as desired by

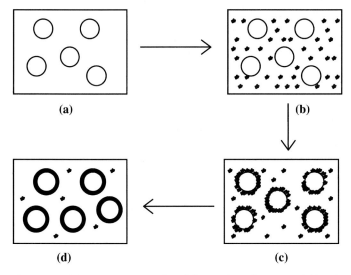

Figure 1.11 Schematic representation of the coacervation process. (a) Core material dispersion in solution of shell polymer; (b) separation of coacervate from solution; (c) coat- ing of core material by microdroplets of coacervate; (d) coalescence of coacervate to form continuous shell around core particles.

controlled addition of the second polymer. Finally, the prepared microcapsules are stabilized by crosslinking, desolvation or thermal treatment.

Complex coacervation is used to produce microcapsules containing fragrant oils, liquid crystals, flavors, dyes or inks as the core material. Porous microcapsules can also be prepared using this technique. When using this technique, certain conditions must be met to avoid agglomeration of the prepared capsules [93]. A micrograph of microcapsules prepared using the coacervation technique is shown in Figure 1.12.

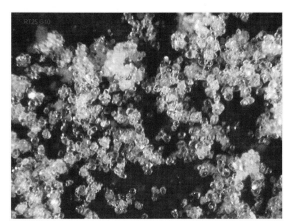

Figure 1.12 Gelatin microcapsules containing a phase-change material prepared by the coacervation method (Courtesy: G. Habar, Microcapsules-Technologies).

1.9.2.2 Encapsulation by Polyelectrolyte Multilayer

Layer by layer (L-B-L) electrostatic assembly of electrically charged particles has attracted much attention due to its enormous potential in multilayered thin film preparations with a wide range of electrical, magnetic and optical properties [94–98]. Polyelectrolyte multilayers are the most widely studied examples of L-B-L assembly, and are prepared by sequentially immersing a substrate in positively and negatively charged polyelectrolyte solutions in a cyclic procedure. However, other charged particles such as nanoparticles, ionic dyes and metal ions are used for preparing L-B-L assembly. Core-shell particles with tailored size and properties are prepared using colloidal particle as the core material that serves as a template onto which multilayers are fabricated. Hollow capsules of organic, inorganic or hybrid particles can be obtained by dissolving the core material. This technique is both versatile and simple, with the multilayer film thickness being controlled precisely by varying the total number of layers deposited; in this way the final properties can be tuned. A detailed discussion of L-B-L assembly for microcapsule preparations is provided in Chapter 3.

1.9.2.3 Polymer Encapsulation by Rapid Expansion of Supercritical Fluids

Supercritical fluids are highly compressed gasses that possess several advantageous properties of both liquids and gases. These fluids have attracted much attention in recent years, the most widely used being supercritical CO_2, alkanes (C_2 to C_4), and nitrous oxide (N_2O). They have low hydrocarbon-like solubility for most solutes and are miscible with common gases such as hydrogen (H_2) and nitrogen (N_2). A small change in temperature or pressure causes a large change in the density of supercritical fluids near the critical point – a property which enhances their use in several industrial applications. Supercritical CO_2 is widely used for its low critical temperature value, in addition to its nontoxic, nonflammable properties; it is also readily available, highly pure and cost-effective. It has found applications in encapsulating active ingredients by polymers. Different core materials such as pesticides, pigments, pharmaceutical ingredients, vitamins, flavors, and dyes are encapsulated using this method [99–101]. A wide variety of shell materials that either dissolve (paraffin wax, acrylates, polyethylene glycol) or do not dissolve (proteins, polysaccharides) in supercritical CO_2 are used for encapsulating core substances. The most widely used methods are as follows:

- Rapid expansion of supercritical solution (RESS)
- Gas anti-solvent (GAS)
- Particles from gas-saturated solution (PGSS)

Rapid expansion of supercritical solution

In this process, supercritical fluid containing the active ingredient and the shell material are maintained at high pressure and then released at atmospheric pressure through a small nozzle. The sudden drop in pressure causes desolvation of the shell material, which is then deposited around the active ingredient (core) and forms a coating layer. The disadvantage of this process is that both the active in-

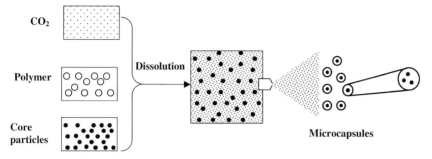

Figure 1.13 Microencapsulation by rapid expansion of supercritical solutions (RESS).

gredient and the shell material must be very soluble in supercritical fluids. In general, very few polymers with low cohesive energy densities (e.g., polydimethylsiloxanes, polymethacrylates) are soluble in supercritical fluids such as CO_2. The solubility of polymers can be enhanced by using co-solvents. In some cases nonsolvents are used; this increases the solubility in supercritical fluids, but the shell materials do not dissolve at atmospheric pressure. A schematic of the microencapsulation process using supercritical CO_2 is shown in Figure 1.13.

Kiyoshi et al. had very recently carried out microencapsulation of TiO_2 nanoparticles with polymer by RESS using ethanol as a nonsolvent for the polymer shell such as polyethylene glycol (PEG), poly(styrene)-b-(poly(methylmethacrylate)-*co*-poly(glycidal methacrylate) copolymer (PS-b-(PMMA-*co*-PGMA) and poly(methyl methacrylate) [102].

Gas anti-solvent (GAS) process

This process is also called supercritical fluid anti-solvent (SAS). Here, supercritical fluid is added to a solution of shell material and the active ingredients and maintained at high pressure. This leads to a volume expansion of the solution that causes supersaturation such that precipitation of the solute occurs. Thus, the solute must be soluble in the liquid solvent, but should not dissolve in the mixture of solvent and supercritical fluid. On the other hand, the liquid solvent must be miscible with the supercritical fluid. This process is unsuitable for the encapsulation of water-soluble ingredients as water has low solubility in supercritical fluids. It is also possible to produce submicron particles using this method.

Particles from a gas-saturated solution (PGSS)

This process is carried out by mixing core and shell materials in supercritical fluid at high pressure. During this process supercritical fluid penetrates the shell material, causing swelling. When the mixture is heated above the glass transition temperature (T_g), the polymer liquefies. Upon releasing the pressure, the shell material is allowed to deposit onto the active ingredient. In this process, the core and shell materials may not be soluble in the supercritical fluid.

Within the pharmaceutical industry, preformed microparticles are often used for the entrapment of active materials using supercritical fluids under pressure. When the pressure is released, the microparticles shrink and return to their original shape and entrap the ingredients.

1.9.3
Physico-Mechanical Processes

1.9.3.1 Co-Extrusion

The co-extrusion process was developed by Southwest Research Institute in the United States, and has found a number of commercial applications. A dual fluid stream of liquid core and shell materials is pumped through concentric tubes and forms droplets under the influence of vibration (Fig. 1.14). The shell is then hardened by chemical crosslinkings, cooling, or solvent evaporation. Different types of extrusion nozzles have been developed in order to optimize the process [103].

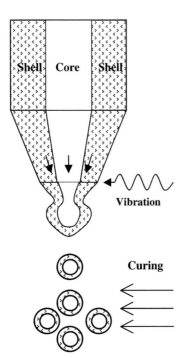

Figure 1.14 Schematic presentation of the coextrusion process.

1.9.3.2 Spray-Drying

Microencapsulation by spray-drying is a low-cost commercial process which is mostly used for the encapsulation of fragrances, oils and flavors. Core particles are dispersed in a polymer solution and sprayed into a hot chamber (Fig. 1.15). The shell material solidifies onto the core particles as the solvent evaporates such that the microcapsules obtained are of polynuclear or matrix type. Very often the en-

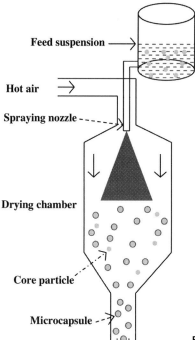

Figure 1.15 Schematic illustrating the process of micro-encapsulation by spray-drying.

capsulated particles are aggregated and the use of large amounts of core material can lead to uncoated particles. However, higher loadings of core particles of up to 50–60% have been reported [104]. Water-soluble polymers are mainly used as shell materials because solvent-borne systems produce unpleasant odors and environmental problems.

1.9.3.3 Fluidized-Bed Technology

With the high demand for encapsulated materials in the global market, fluid-bed coaters have become more popular. They are used for encapsulating solid or porous particles with optimal heat exchange [105]. The liquid coating is sprayed onto the particles and the rapid evaporation helps in the formation of an outer layer on the particles. The thickness and formulations of the coating can be obtained as desired. Different types of fluid-bed coaters include top spray, bottom spray, and tangential spray (Fig. 1.16).

- In the *top spray* system the coating material is sprayed downwards on to the fluid bed such that as the solid or porous particles move to the coating region they become encapsulated. Increased encapsulation efficiency and the prevention of cluster formation is achieved by opposing flows of the coating materials and the particles. Dripping of the coated particles depends on the formulation of the

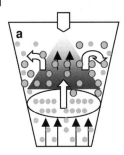 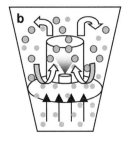 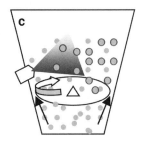

Figure 1.16 Schematics of a fluid-bed coater. (a) Top spray; (b) bottom spray; (c) tangential spray.

coating material. Top spray fluid-bed coaters produce higher yields of encapsulated particles than either bottom or tangential sprays.

- The *bottom spray* is also known as "Wurster's coater" in recognition of its development by Prof. D.E. Wurster [106]. This technique uses a coating chamber that has a cylindrical nozzle and a perforated bottom plate. The cylindrical nozzle is used for spraying the coating material. As the particles move upwards through the perforated bottom plate and pass the nozzle area, they are encapsulated by the coating material. The coating material adheres to the particle surface by evaporation of the solvent or cooling of the encapsulated particle. This process is continued until the desired thickness and weight is obtained. Although it is a time-consuming process, the multilayer coating procedure helps in reducing particle defects.

- The *tangential spray* consists of a rotating disc at the bottom of the coating chamber, with the same diameter as the chamber. During the process the disc is raised to create a gap between the edge of the chamber and the disc. The tangential nozzle is placed above the rotating disc through which the coating material is released. The particles move through the gap into the spraying zone and are encapsulated. As they travel a minimum distance there is a higher yield of encapsulated particles.

1.9.3.4 Spinning Disk

The microencapsulation of suspended core materials using a rotating disc was first developed by Prof. R.E. Sparks [107]. A schematic diagram of the process is shown in Figure 1.17. Suspensions of core particles in liquid shell material are poured into a rotating disc and, due to the spinning action of the disc, the core particles become coated with the shell material. The coated particles, along with the excess shell material, are then cast from the edge of the disc by centrifugal force, after which the shell material is solidified by external means (usually cooling). This technology is rapid, cost-effective, relatively simple and has high production efficiencies. For optimum encapsulation, spherical core particles with diameters of ~100 to 150 μm and rapidly cooling shell materials are required.

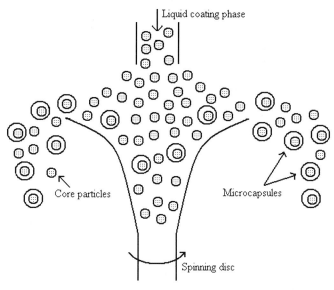

Liquid coating phase

Core particles

Microcapsules

Spinning disc

Figure 1.17 Schematic representation of microcapsule formation by spinning disk.

Although a variety of alternative microencapsulation techniques is available (for details of sol-gel techniques, see Chapter 8), no single method is suitable for encapsulating different types of core material. Ultimately, the best method will depend upon the type of core material, the required particle size, the permeability of the shell wall, and the different properties of the microcapsule, and consequently the process must be custom-tailored in order to provide a satisfactory outcome. An overview of the size of microcapsules obtained by different techniques is provided in Table 1.2.

Table 1.2 Microencapsulation processes with their relative particle size ranges.

Microencapsulation process	Particle size [μm]
Extrusion	250–2500
Spray-drying	5–5000
Fluid bed coating	20–1500
Rotating disk	5–1500
Coacervation	2–1200
Solvent evaporation	0.5–1000
Phase separation	0.5–1000
In-situ polymerization	0.5–1100
Interfacial polymerization	0.5–1000
Miniemulsion	0.1–0.5
Sol-gel encapsulation	2–20
Layer-by-layer (LBL) assembly	0.02–20

1.10
Enhancing Coating Functionalities with Microcapsules

Microcapsules can be used in a wide variety of applications [82,108,109], since the versatility of microencapsulation technologies offers unlimited combinations of core and shell materials for their production. To date, few investigations have been made into possible applications of microcapsules in functional coating developments. Microcapsules are applied onto substrates in various ways. For example, they may be sprayed over an existing coating layer, perhaps to provide immediate release of lubricants or perfumes. The most two common process of applying microcapsules in coatings are either to incorporate them into a coating formulation or by their electrolytic co-deposition with metal ions (Fig. 1.18; see also Chapter 9) [110,111].

The mixing of microcapsules with coating binders require compatibility of the shell material with the binder. Generally, microcapsules are used in coatings for controlled-release applications, but microcapsules containing active ingredients such as biocides can also be trapped inside a coating matrix that will release the contents slowly over time. Another interesting example is to use microcapsules in the development of self-healing coatings [112]. For this, microcapsules containing monomer, crosslinker or catalysts are incorporated into a coating matrix such that, when a coating ruptures, the microcapsules along the rupture break open and release their contents. Subsequently, the monomer polymerizes, crosslinks, and fills the damage, thereby preventing further propagation. An innovative example is the use of microencapsulated phase-change material (PCM) particles in interior coatings for buildings [113,114]. During the day, as the temperature rises, the core material melts and stores heat. During the night, when the temperature falls, the heat stored inside the capsules is released, thereby reducing energy needs. Clearly, for

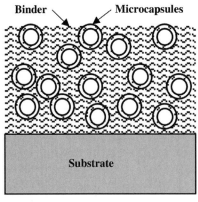

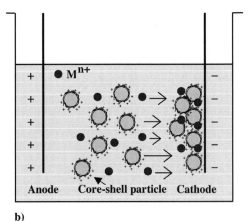

a)   b)

Figure 1.18 Schematic diagram showing pathways for microcapsule incorporation into coatings. (a) Blending of microcapsules with binders; (b) electrolytic co-deposition of microcapsules with metallic ions.

heat management to be effective, the correct quantities of microcapsules must be used in the preparation of such coatings. Other applications include microencapsulated dyes used to formulate color coatings, and foaming agents (e.g., sodium bicarbonate) which can be microencapsulated to generate foams during curing processes. Microcapsules containing perfumes, insecticides, chemicals, and heat- or pressure-sensitive dyes can also be used for functional coating preparations. The use of polymers with different T_g values can be used to create microcapsules that can be added to coatings in order to produced specialized functions, for example vibration damping. Finally, microcapsules containing nanoparticles may be used in the design of functional surfaces with improved physical, optical, mechanical, electrical, or chemical properties.

1.11
Conclusions

Microencapsulation has already been proven as a successful technology for commercial applications in the pharmaceutical and agrochemical industries and, more recently, also in the textile industry. In general, however, the technology remains largely unexplored, notably in the field of functional coatings where the possibilities of obtaining functional surfaces using microcapsules are almost unlimited. The technology allows combinations to be made of the properties of different materials that are difficult or even impossible with other available technologies. The high cost of microencapsulation may often be a prohibitive factor, though in some cases it is justified by the added value of the products. As worldwide demands for functional coatings continue to increase, new, cost-effective microencapsulation technologies will be developed and the technology will remain at the forefront of future research. At present, the industry's major problem is to provide functional coatings that are easy to apply and have long-term stability; consequently, attention will be focused in this area.

Abbreviations

CB	conduction band
GAS	gas anti-solvent
ICP	intrinsically conductive polymer
L-B-L	layer by layer
PCM	phase-change material
PGSS	particles from gas-saturated solution
RESS	rapid expansion of supercritical solution
SAS	supercritical fluid anti-solvent
TBT	tributyl tin
T_g	glass transition temperature
VB	valence band

References

1. Z.W. Wicks, Jr., F.N. Jones, S.P. Pappas, *Organic Coatings: Science and Technology Vol. I*, John Wiley & Sons, Inc., New York, **1992**, Chapter 1.
2. R. Lambourne, in: *Paints and Surface Coatings, Theory and Practice* (Eds. R. Lambourne, T.A. Strivens), 2nd edn. Chem Tech Pub. Inc., **1999**, Chapter 1.
3. J.V. Koleske, in: *Encyclopedia of Analytical Chemistry* (Ed. R.A. Meyers), John Wiley & Sons Ltd., Chichester, **2000**, Chapter 4.
4. A.D. Wilson, J.W. Nicholson, H.J. Prosser, *Surface Coatings-1*, Elsevier Applied Science, New York, **1987**, Chapter 1.
5. C.R. Hegedus, *JCT Research* **2004**, *1(1)*, 5–19.
6. V.V. Verkholantsev, *Eur. Coat. J.* **2003**, 9, 18–25.
7. M. Wulf, A. Wehling, O. Reis, *Macromol. Symp.* **2002**, *187*, 459–467.
8. I. P. Parkin, R. G. Palgrave, *J. Mater. Chem.* **2005**, *15(17)*, 1689–1695.
9. E. Nun, M. Oles, B. Schleich, *Macromol. Symp.* **2002**, *187*, 677–682.
10. M. Kuhr, S. Bauer, U. Rothhaar, D. Wolff, *Thin Solid Films* **2003**, *442 (1–2)*, 107–116.
11. M. Perez, M. Garcia, A. del Amo, G. Blustein, M. Stupak, *Surf. Coat. Int. Part B – Coat. Trans.* **2003**, *86 (4)*, 259–262.
12. L.C. Zhou, B. Koltisko, *Jct. Coatingstech.* **2005**, *2(15)*, 54–60.
13. J.C. Tiller, C.J. Liao, K. Lewis, A.M. Klibanov, *Proc. Natl. Acad. Sci. USA* **2001**, *98(11)*, 5981–5985.
14. K. Johns, *Surf. Coat. Int. Part B: Coat. Trans.* **2003**, *86* (B2), 91–168.
15. http://www.agion-tech.com.
16. V.V. Verkholantsev, *Eur. Coat. J.* **2003**, *10*, 32–37.
17. P.J. Gellings, *Introduction to corrosion prevention and control*, Delft University Press, **1985**.
18. C.I. Elsner, E. Cavalcanti, O. Ferraz, A.R. Di Sarli, *Prog. Org. Coat.* **2003**, *48*, 50–62.
19. N. Kouloumbi, P. Moundoulas, *Pigment & Resin Technol.* **2002**, *31(4)*, 206–215.
20. N. Kouloumbi, P. Pantazopoulou, P. Moundoulas, *Pigment & Resin Technol.* **2003**, *32(2)*, 89–99.
21. L. Hochmannov , *Eur. Coat. J.* **2003**, 9, 26–32.
22. Jui-Ming Yeh, Shir-Joe Liou, Chih-Guang Lin, Yen-Po Chang, Yuan-Hsiang Yu, Chi-Feng Cheng, *J. Appl. Polym. Sci.* **2004**, *92*, 1970–1976.
23. V. Palanivel, W.J. Van Ooij, in: *Silanes and Other Coupling Agents* (Ed. K.L. Mittal), Volume 3, VSP, **2004**, pp. 135–159.
24. S. Monte, *Polymer Paint Color J.* December **2003**.
25. M. Kendig, M. Hon, L. Warren, *Prog. Org. Coat.* **2003**, *47*, 183–189.
26. M. Kraljic, Z. Mandic, Lj. Duic, *Corr. Sci.* **2003**, *45*, 181–198.
27. T. Rentschler, A. Claßen, *Eur. Coat. J.* **2002**, 9, 24–27.
28. A. Kumar, L.D. Stephenson, *Coat. World* **2004**, 24–26 & 28–33.
29. D. Bhattacharya, Worldwide, WO03093380, November 13, **2003**.
30. D.L. Gangotri, A.D. Chaware, *Paint India* September **2004**, 39–42.
31. G. Camino, R. Delobel, in: *Fire Retardance of Polymeric Materials* (Ed. C.A. Wilkie, A.F. Grand) Chapter 7, Marcel Dekker, Inc., New York , **2000**, pp. 217–243.
32. F.J.W.J. Labuschagné, *Metal catalysed intumescence of polyhydroxyl compounds*, PhD Thesis, University of Pretoria, **2003**.
33. K. Rathberger, *Addconworld 2004*, Amsterdam, Rapra Conference Proceedings, Paper 11.
34. S. Girand, S. Bourbigot, M. Rochery, I. Vroman, L. Tighzert, R. Delobel, *Polym. Degrad. Stab.* **2002**, *77(2)*, 285–297.
35. T. Kashiwagi, J.W. Gilman, in: *Fire Retardance of Polymeric Materials* (Ed. C.A. Wilkie, A.F. Grand), Chapter 10, Marcel Dekker, Inc., New York, **2000**, pp. 353–389.
36. W.E. Horn, Jr., in: *Fire Retardance of Polymeric Materials* (Ed. C.A. Wilkie, A.F. Grand), Chapter 9, Marcel Dekker, Inc., New York, **2000**, pp. 285–332.
37. B. Kandola, S. Nazaré, R. Horrocks and various papers, in: *Fire and Polymers, Polym. Mater. Sci. Eng. Preprints*, American Chemical Society, August 22–26, **2004**.
38. M. Zanetti, S. Lomakin, G. Gamino, *Macromol. Mater. Eng.* **2000**, *279*, 1–9.
39. T. Sawitowski, K. Schulte, *Nano and Hybrid Coatings*, PRA Coatings Technology,

24–25 January **2005**, Manchester, UK, Paper no. 8.

40. D.R. Baer, P.E. Burrows, A.A. El-Azab, *Prog. Org. Coat.* **2003**, *47*, 342–356.

41. H.J. Gläsel, F. Bauer, H. Ernst, M. Findeisen, E. Hartmann, H. Langguth, R. Mehnert, R. Schubert, *Macromol. Chem. Phys.* **2000**, *201*, 2765–2770.

42. R. Mehnert, E. Hartmann, H.J. Glasel. S. Rummel, F. Bauer, A. Sobottka, Ch. Elsner, *Mat. -wiss. U. Werkstofftech.* **2001**, *32*, 774–780.

43. G.A. Lawrence, A.K. Barkac, M.A. Chasser, A.S. Desaw, E.M. Hartman, E. Mavis, E.D. Hayes, R.T. Hockswender, L.K. Kuster, A.R. Montague, M. Nakajima, G.K. Olson, S.J. Richardson, J.R. Sadvari, A.D. Simpson, S. Tyebjee, F.T. Wilt, US Patent 6387519, May 14, **2002**.

44. F.A.P. Thomas, J.A.W. Stoks, A. Buegman, US Patent 6291054, September 18, **2001**.

45. http://www.nano-x.de/index.html

46. W. Barthlott, C. Neinhuis, *Planta* **1997**, *202*, 1–8.

47. A. Nakajima, K. Hashimoto, T. Watanabe, *Monatshefte für Chemie* **2001**, *132*, 31–41.

48. R. Blossey, *Nature Materials* **2003**, *2*, 301–306.

49. I.P. Parkin, R.G. Palgrave, *J. Mater. Chem.* **2005**, *15(17)*, 1689–1695.

50. Y.T. Cheng, D. Rodak, *Appl. Phys. Lett.* **2005**. *86* (144101) 1–3.

51. A. Fujishima, K. Hashimoto, T. Watanabe, *TiO$_2$ Photocatalysis: Fundamentals and Applications*. BKC, Tokyo, **1999**, p. 66.

52. T. Watanabe, *Bull. Ceram. Soc. Jpn.* **1996**, *31*, 837–840.

53. A. Fujishima, T.A. Rao, D.A. Tyrk, *J. Photochem. Photobiol. C: Photochem. Rev.* **2000**, *1*, 1–21.

54. K. Guan, *Surf. Coat. Technol.* **2005**, *191*, 155–160.

55. A. Nakajima, K.Hashimoto, T. Watanabe, K. Takai, G. Yamauchi, A. Fujishima, *Langmuir* **2000**, *16*, 7044–7047.

56. N. Niegisch, M. Akarsu, Z. Csögör, M.Ehses, H. Schmidt, *Hygienic Coatings*, Brussels, Belgium, 8–9 July **2002**, paper 20.

57. G. Yamauchi, Y. Riko, Y. Yasuno, T. Shimizu, N. Funakoshi, *Nano and Hybrid Coatings*, The Paint Research Association,

Manchester, UK, 24–25 January **2005**, paper 20.

58. J. Ross, *Polymer Paint Color J.* **2004**, *194 (4479)*, 18–20.

59. D. Thölmann, B. Kossmann, F. Sosna, *Eur. Coat. J.* **2003**, *1(2)*, 16–33.

60. G. Sauvet, S. Dupond, K. Kazmierski, J. Chojnowski, *J. Appl. Polym. Sci.* **2003**, *75*, 1005–1012.

61. A.J. Trogolo, C.F. Rossitto, K.E. Welch, II, World Patent, WO 03/055941, July 10, **2003**.

62. M. Wagener, *Hygienic Coatings & Surfaces*, *PRA Coatings Technology Centre*, Paris, 16–17 March **2005**, paper 14.

63. T. Xu, C.S. Xie, *Prog. Org. Coat.* **2003**, *46*, 297–301.

64. G. Borkow, *Hygienic Coatings & Surfaces*, *PRA Coatings Technology Centre*, Paris, 16–17 March **2005**, paper 20.

65. E.J. Wolfrum, J. Huang, D.M. Blake, P.C. Maness, Z. Huang, J. Fiest, W.A. Jacoby, *Environ. Sci. Technol.* **2002**, *36*, 3412–3419.

66. B. Windsor, *Hygienic Coatings & Surfaces*, *PRA Coatings Technology Centre*, Paris, 16–17 March **2005**, paper 23.

67. J.C. Kim, M.E. Song, E.J. Lee, S.K. Park, M.J. Rang, H.J. Ahn, *J. Dispersion Sci. Technol.* **2001**, *22(6)*, 591–596.

68. M. Edge, N.S. Allen, D. Turner, J. Robinson, K. Seal, *Prog. Org. Coat.* **2001**, *43*, 10–17.

69. S.D. Worley, F. Li, R. Wu, J. Kim, C.I. Wei, J.F. Williams, J.R. Owens, J.D. Wander, A.M. Bergmeyer, M.E. Shirtliff, *Surf. Coat. Inter. Part B: Coat. Trans.* **2003**, *86(4)*, 273–277.

70. M. Thouvenin, J.J. Peron, C. Charreteur, P. Guerin, J.Y. Langlois, K. Valle-Rehel, *Prog. Org. Coat.* **2002**, *44*, 75–83.

71. M. Perez, M. Garcia, B. del Amo, G. Blustein, M. Stupak, *Surf. Coat. Inter. Part B: Coat. Trans.* **2003**, *86(4)*, 259–262.

72. H. Gold, D.R. Levy, M. Temchenko, H.E.T. Mendum, T. Tanaka, T. Enoki, G. Wang, European Patent EP 1406732, April 14, **2004**.

73. G. Wagner, *Eur. Coat. J.* **2002**, *7/8*, 44–47.

74. J.R. White, B. De Poumeyrol, J.M. Halle, R. Stepheson, *J. Mater. Sci.* **2004**, *39*, 3105–3114.

75. M. Pridöhl, S. Heberer, R. Maier, R. Mertsch, G. Michael, European Coatings Conference: Smart Coatings III.,

Berlin, Germany, 7–8 January **2004**, 141–148.

76. A. Sen, R. Pandey, *Paintindia* **2001**, *Golden Jubilee Issue*, 143–156.

77. A.R. Hemsley, P.C. Griffiths, *Philos. Trans. R. Soc. London Ser.* **2000**, *358*, 547–564.

78. L. Schleicher, B.K. Green, US Patent 2730456, **1956**.

79. M.M. Gutcho, *Microcapsules and Microencapsulation Techniques*, Noyes Data Co., New Jersey, USA, **1976**.

80. J.E. Vandegaer, *Microencapsulation: Processes and Applications*, Plenum Press, New York, **1973**.

81. S. Benita, *Microencapsulation: Methods and Industrial applications*, Marcel Dekker, Inc., New York, **1996**.

82. R. Arshady, *Microspheres, Microcapsules and Liposomes*, Citrus Books, London, United Kingdom, **1999**.

83. W. Sliwka, *Angew. Chem. Int. Ed.* **1975**, *14(8)*, 539–550.

84. M.W. Ranney, *Microencapsulation Technology*, Noyes Development Corporation, Park Ridge, 1969, p. 275.

85. J.L. Luna-Xavier, E. Bourgeat-Lami, A. Guyot, *Colloid. Polym. Sci.* **2001**, *279*, 947–958.

86. W.F. Liu, Z.X. Guo, J. Yu, *J. Appl. Polym. Sci.* **2005**, *97(4)*, 1538–1544.

87. M. Okubo, H. Minami, Y. Jing, *J. Appl. Polym. Sci.* **2003**, *89*, 706–710.

88. M.L.Soto-Portas, J.F. Argillier, F. Méchin, N. Zydowicz, *Polym. Int.* **2003**, *52*, 522–527.

89. B.Z. Putlitz, K. Landfester, H. Fischer, M. Antonietti, *Adv. Mater.* **2001**, *13(7)*, 500–503.

90. A.J.P. van Zyl, R.D. Sanderson, D. de Wet-Roos, B. Klumperman, *Macromolecules* **2003**, *36*, 8621–8629.

91. G. Bungenberg de Jong, H. Kruyt, *Prog. Kungl. Ned. Acad. Wetensch.* **1929**, *32*, 849–856.

92. H.G. Bungenberg de Jong, in: *Colloid Science* (Ed. H.R. Kruyt), Elsevier, **1949**, pp. 232–258.

93. J. Okada, A. Kusai, S. Ueda, *J. Microencapsulation* **1985**, *2*, 163–173.

94. G. Decher, *Science* **1997**, *277*, 1232–1237.

95. M. Freemantle, *Chem. Eng. News* **2002**, 80 *(18)*, 44–48.

96. C.S. Peyratout, L. Dähne, *Angew. Chem. Int. Ed.* **2004**, *43*, 3762–3783.

97. G.B. Sukhorukov, A. Fery, M. Brumen, H. Möhwald, *Phys. Chem. Chem. Phys.* **2004**, *6*, 4078–4089.

98. D. Lee, M.F. Rubner, R.E. Cohen, *Chem. Mater.* **2005**, *17*, 1099–1105.

99. R. Ghaderi, *A Supercritical Fluids Extraction Process for the Production of Drug Loaded Biodegradable Microparticles*, PhD Thesis, ACTA Universitatis Upasaliensis, Uppsala, **2000**.

100. H. Liu, M.Z. Yates, *Langmuir* **2002**, *18*, 6066–6070.

101. P. Chambon, E. Cloutet, H. Cramail, *Macromolecules* **2004**, *37*, 5856–5859.

102. K. Matsuyama, K. Mishima, K.I. Hayashi, H. Matsuyama, *J. Nanoparticle Res.* **2003**, *5*, 87–95.

103. J.T. Goodwin, G.R. Somerville, *Chemtech.* **1974**, 623–626.

104. J. Brener, *Perfumer and Flavorist* **1983**, 8, 40–44.

105. K. Lehmann, in: *Microcapsules and Nanoparticles in Medicine and Pharmacy* (Ed. M. Donbrow), CRC Press, Boca Raton, **1992**, pp. 73–97.

106. D.E. Wurster, US Patent 2648609, September 8, **1953**.

107. R.E. Sparks, M. Norbert, US Patent 4675140, June 6, **1987**.

108. G.O. Fanger, *Chemtech.* **1974**, 397–405.

109. H. Yoshizawa, *KONA* **2004**, *22*, 23–31.

110. A. Dietz, M. Jobmann, G. Rafler, *Mat.-wiss. u. Werkstofftech.* **2000**, *31*, 612–615.

111. Z. Liqun, Z. Wei, L. Feng, *J. Mater. Sci.* **2004**, *39*, 495–499.

112. E.N. Brown, S.R. White, N.R. Sottos, *J. Mater. Sci.* **2004**, *39*, 1703–1710.

113. J.F. Su, L.X. Wang, L. Ren, *J. Appl. Polym. Sci.* **2005**, *97(5)*, 1755–1762.

114. M.N.A. Hawlader, M.S. Uddin, H.J. Zhu, *Int. J. Energy Res.* **2002**, *26*, 159–171.

2

Encapsulation Through (Mini)Emulsion Polymerization

Katharina Landfester

2.1
Introduction

The encapsulation of a liquid or a solid, an inorganic or an organic, or of a hydrophobic or hydrophilic material into a polymer shell, is of great importance for many applications. In the case of a liquid, the nanocapsule can act as nanocontainer in order to retain the liquid and therefore to prevent it from leaking into the continuous phase. This can be of great importance for example, in biomedical applications such as controlled drug release. In the case of encapsulating a solid, the solid can be protected against the environment (or the environment against the solid) by the polymer shell. Here, many different materials such as organic and inorganic pigments, magnetite, or other solid nanoparticles can be thought to be encapsulated in a polymer.

The formulation of nanocapsules composed of polymeric material and an encapsulating material is demanding though, in principle, many different approaches can be used to generate nanoparticles. In order to prepare polymer particles, a variety of processes such as microemulsion and/or emulsion polymerization are used, these being based on kinetic control during the preparation. In this way the particles are built from the center to the surface, and the particle structure is governed by kinetic factors. However, this dictate of kinetics may lead to serious disadvantages, including a lack of homogeneity and restrictions in the accessible composition, although in general these must be accepted. Consequently, the formation of nanocapsules using these methods is not straightforward. One possible approach is to take advantage of the potential thermodynamic control for the design of nanoparticles, and to utilize the concept of "nanoreactors", where the essential ingredients for nanoparticle formation, as well as the encapuslating material, are in place at the start of the synthesis. The reaction may then take place in nanoreactors in highly parallel fashion, with the synthesis being carried out in 10^{18} to 10^{20} nanocompartments separated from each other only by a continuous phase. This concept of polymerization within a nanoreactor has been achieved technically, and with a great degree of perfection, in the process of suspension polymerization,

Functional Coatings. Edited by Swapan Kumar Ghosh
Copyright © 2006 WILEY-VCH Verlag GmbH & Co. KGaA, Weinheim
ISBN 3-527-31296-X

where droplets can be created in the micrometer range and then polymerized without any loss or change in particle identity. The suspension principle to produce smaller droplet sizes was further developed by Ugelstad [1], who scaled down the droplet size to several hundred nanometers by shearing the system.

A recent development, where high-shear devices such as ultrasound and high-pressure homogenizers were used to reduce droplet size and the nanoreactor diameter to 30–100 nm, thus allowing the formulation of different types of nanocapsule, forms the main subject of this chapter.

2.2
The Miniemulsion Process

A system where small droplets of high stability in a continuous phase are created by using high shear [2–4] is termed classically as a "miniemulsion". One trick to obtain high stability of the droplets is to add an agent which dissolves in the dispersed phase, but which is highly insoluble in the continuous phase. The small droplets can then be hardened by either a subsequent polymerization or by lowering the temperature (if the dispersed phase is a low-temperature melting material). For a typical oil-in-water miniemulsion, an oil, a hydrophobic agent (or several agents), an emulsifier and water are homogenized by high shear (Fig. 2.1) to produce homogeneous and monodisperse droplets in the size range of 30 to 500 nm [5]. However, in the first step of the miniemulsion process, small stable

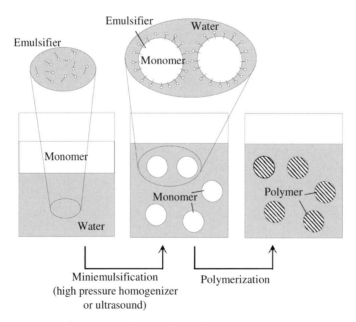

Figure 2.1 The principle of miniemulsion polymerization.

droplets of 30 to 500 nm size are formed by shearing a system containing the dispersed phase, the continuous phase, a surfactant, and an osmotic pressure agent. In a second step, these droplets are polymerized without changing their identity.

Based on the principle of miniemulsion, it is now possible to prepare new nanoparticles which could not be created by using heterophase processes. In particular, this includes the encapsulation of different liquid and solid materials, and this topic will form the main content of this chapter.

When creating a miniemulsion, the step of homogenization is of great importance as fairly monodisperse, small droplets must be produced. The homogenization can be achieved using an ultrasonifier (for the miniemulsification of small quantities in a laboratory-scale batch process) or a high-pressure homogenizer (for larger-scale processes). At the start of the homogenization, polydispersity of the droplets is still quite high, but by constant fusion and fission processes induced by the high shear, the size and polydispersity is decreased until the miniemulsion reaches a steady state [6]. Originally, the homogenization process could be followed using different methods, for example by measurements of turbidity and/or surface tension. A constant value indicating steady state is achieved using either approach. The surface tension may reach high values, indicating that coverage of the droplets by surfactant molecules is very low. In fact, incomplete coverage of droplets by surfactant molecules is an important characteristic of miniemulsions and shows that the surfactant is very efficiently used. It has been observed that the coverage of surfactant is dependent upon droplet size – the smaller the droplets, the higher must be the coverage in order to obtain stable droplets. The exact droplet size can be adjusted selectively by the type and amount of surfactant used for the stabilization. Anionic and cationic surfactants allow the formation of monodisperse droplets between about 30 and 200 nm, whereas nonionic oligomeric or polymeric surfactants are suited to the formation of droplets between about 100 and 800 nm.

Such minidroplets were previously regarded as a rather unstable dispersion state of matter for two growth mechanisms for the droplets, namely growth by Ostwald ripening and growth by collisions (and subsequent coalescence). The suppression of both processes is necessary in order to formulate a stable miniemulsion. Coalescence can be controlled by the effective use of a surfactant, while Ostwald ripening can be suppressed efficiently by the addition of a hydrophobic agent to the dispersed phase. This agent cannot diffuse from one droplet to the other and is trapped in each droplet; this leads to the production of an osmotic pressure inside the droplets which counteracts the Laplace pressure. The effectiveness of the hydrophobe increases with decreasing water solubility in the continuous phase.

This mechanism was used previously to stabilize fluoroalkane droplets by adding the ultrahydrophobe perfluorodimorphineopropane, the results being the production of an effective and stable blood substitute [7]. A variety of molecules can be used as hydrophobes, although they must be chosen such that they add useful properties to the final product; for example, the molecule may be a dye, a plasticizer, or a cross-linker, while for biomedical applications this component might also be a fluorescent marker or a drug.

The addition of an ultrahydrophobe does not completely block droplet growth (due to a still finite solubility, the existence of droplet collisions, and surfactant-assisted transport), but remarkably the growth is greatly slowed. The final state to be expected is determined by the balance of osmotic pressure and Laplace pressure. Since immediately after miniemulsification the Laplace pressure is usually larger than the osmotic pressure, the miniemulsion tends to grow on the timescale of days to weeks before reaching an equilibrium state. Due to the timescale involved, this growth usually is not relevant to synthetic applications, but it may be handled in thermodynamic fashion. This can be achieved either by increasing the amount of osmotic agent, by increasing the particle size, or by adding a second dose of surfactant after the dispersion (this will lower both the surface tension and the related Laplace pressure)[5].

The extraordinarily high droplet stability against exchange processes can be demonstrated in a very illustrative manner by the formation of a nickel murexid complex inside the droplets in inverse miniemulsion systems [8,9]. Here, a miniemulsion with droplets containing a $FeCl_3$ solution, and another miniemulsion containing a $K_4[Fe(CN)_6]$ solution, are mixed. The mixed miniemulsion remains colorless for weeks (Fig. 2.2), indicating that the droplets with different species remain separated as colloidal entities on the time scale of most chemical reactions. Repetition of the same experiment with two microemulsions or micellar solutions would lead to an immediate reaction because of unblocked droplet exchange. In miniemulsions, the exchange can be stimulated by mechanical energy, such as ultrasound used to prepare the original miniemulsions. In this case, both fusion and fission processes are induced, and with increasing ultrasound application the miniemulsion indeed turns blue (Fig. 2.2b).

In the inverse situation the droplet size throughout the miniemulsification process achieves an equilibrium state (steady-state miniemulsion); this is characterized by a dynamic rate equilibrium between fusion and fission of the droplets, and can be determined turbidimetrically. A high stability of the droplets *after* high-

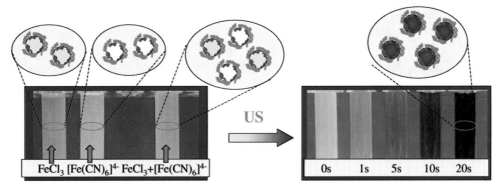

$FeCl_3$ $[Fe(CN)_6]^{4-}$ $FeCl_3+[Fe(CN)_6]^{4-}$ US 0s 1s 5s 10s 20s

Figure 2.2 Illustration of the high stability of miniemulsion droplets. Two separate inverse miniemulsions are prepared, one containing droplets with $FeCl_3$ solution droplets, the oth-er with $K_4[Fe(CN)_6]$. The droplets are stable. Only during ultrasonication do fusion and fission processes lead to a reaction between the components.

shear treatment, however, is obtained with the osmotic agent. The type of lipophobe has no influence on the stability of the inverse miniemulsion. In contrast to regular miniemulsions, the droplet size is dependent upon the amount of osmotic agent used [10]. It appears that in inverse miniemulsions, the droplets experience shortly after miniemulsification a real zero-effective pressure situation (the osmotic pressure counterbalances the Laplace pressure), and this makes them very stable. This is attributed (speculatively) to the different stabilization mechanism and mutual particle potentials, which make pressure equilibration close to the ultrasonication process possible. As a consequence, it is believed that inverse miniemulsions are not critically stabilized, but rather are fully stable systems.

For inverse miniemulsions the surfactant is used in a very efficient manner, at least in comparison to inverse microemulsions [11,12] or inverse suspensions [13], which are used for subsequent polymerization processes. Again, surface coverage of the inverse miniemulsion droplets with surfactant is incomplete and empty inverse micelles are absent, but this is important if any interpretation of the reaction mechanism is required.

The fusion/fission mechanism of minidroplet formation also results in a typical triangular relationship between the amount of surfactant, the resulting particle size, and the surface coverage. With increasing amounts of the surfactant, the particle size decreases. Subsequently, the smaller the particles, the higher their coverage by surfactant. For inverse miniemulsions, these relationships also depend upon the amount of hydrophobe present.

2.3
Different Polymers for the Formation of Nanocapsules in Miniemulsions

In principle, the process of miniemulsion allows all types of monomers to be used in the formation of particles, including those not miscible with the continuous phase. In case of prevailing droplet nucleation or the start of the polymer reaction in the droplet phase, each miniemulsion droplet can indeed be treated as a small nanoreactor. This enables a whole variety of polymerization reactions that lead to nanoparticles (much broader than in emulsion polymerization) as well as to the synthesis of nanoparticle hybrids, which were not previously accessible. All such polymers can be used in the formation of nanocapsules, and details of the different types of polymerization are presented in the following sections.

2.3.1
Radical Polymerization

Styrene has been described as a model monomer for radical homopolymerization of hydrophobic monomers in many reports. The polymerization of acrylates and methacrylates is also well known. It could be also shown that the miniemulsion process also easily allows polymerization of the ultrahydrophobic monomer lauryl methacrylate without any carrier materials as are necessary in emulsion polymerization [4]. Not only hydrophobic, but also fluorinated monomers, were applied for the synthesis of latexes in the size range of 100 to 250 nm by employing rather low doses of protonated surfactants [14]. At temperatures above its melting point, the hydrophobic monomer vinylnaphthalene also forms miniemulsions, with the droplets capable of being polymerized as polymer particles with a refractive index of 1.6818, one of the highest known for polymers. The polymerization of more hydrophilic monomers is also possible, as shown for methyl methacrylate (MMA) and vinyl acetate [15–17]. In the case of monomers with a pronounced water solubility, nucleation in water should be efficiently suppressed in order to avoid secondary nucleation in the water phase. This can be achieved, for example, by using an oil-soluble initiator and the polymerization of acrylonitrile, or by adding a termination agent to the continuous phase. A typical calorimetric curve of MMA polymerization using a hydrophobic initiator shows a rapid conversion. PVC latex particles consisting of two size populations can be generated in a miniemulsion polymerization. The mechanism for the formation of two discrete particle families relies upon the polymerization of two distinct types of droplet [18].

It has been shown that the principle of aqueous miniemulsions could be transferred to nonaqueous media [19]. Here, polar solvents, such as formamide or glycol replace water as the continuous phase, and hydrophobic monomers are miniemulsified with a hydrophobic agent, which stabilizes the droplets against molecular diffusion processes.

In the case of inverse systems, hydrophilic monomers such as hydroxyethyl acrylate, acrylamide, and acrylic acid were miniemulsified in nonpolar media such as cyclohexane or hexadecane [10,19].

The miniemulsion is also well suited to the preparation of copolymers. Here, a mixture of hydrophobic monomers can be used, for example a mixture of styrene and MMA [20] or styrene and butyl acrylate [21], MMA and *p*-methylstyrene, vinyl hexanoate, or vinyl 2-ethylhexanoate [22]. Fluorinated monomers can also be copolymerized with monomers such as MMA and styrene [14]. Miniemulsification of such mixed monomer species allows efficient copolymerization reactions to be performed with standard hydrophobic and hydrophilic monomers in a common heterophase situation, resulting either in core-shell latexes or in statistical copolymers.

The polymerization process of two monomers with different polarities in similar ratios is difficult due to solubility problems. Using the miniemulsion process, it was possible to start from very different spatial monomer distributions, and this resulted in very different amphiphilic copolymers in dispersion [23]. The monomer, which is insoluble in the continuous phase, is miniemulsified in order to form stable and small droplets with a small amount of surfactant. The monomer with the opposite hydrophilicity dissolves in the continuous phase (but not in the droplets). The formation of acrylamide/MMA (AAm/MMA) and acrylamide/styrene (AAm/Sty) copolymers was chosen as examples of the miniemulsion process. In all cases, the syntheses were carried out in water as well as in cyclohexane as the continuous phase. If the synthesis is performed in water, the hydrophobic monomer with a low water solubility (styrene or methyl methacrylate) forms mainly monomer droplets, whereas the hydrophilic monomer acrylamide with a high water solubility is mainly dissolved in the water phase. In the case of inverse miniemulsion, the hydrophilic monomer is expected to form the droplets, whereas the hydrophobic monomer is dissolved in the continuous phase.

2.3.2
Polyaddition

The existence of stable, isolated nanodroplets, in which chemical reactions may depend (though not necessarily so) on droplet exchange (the so-called "nanoreactors") enables the miniemulsion process to be applied over a much broader range.

In contrast to the process of creating a secondary dispersion (as for the preparation of polyurethanes and epoxide resins), it was shown that the miniemulsion polymerization process would allow monomeric components to be mixed together, and that the polyaddition and polycondensation reactions could be performed *after* miniemulsification in the miniemulsified state [24].

The principle of miniemulsion polymerization to polyadditions of epoxy-resins was successfully transferred to mixtures of different epoxides with varying diamines, dithiols, or diols heated to 60 °C to form the respective polymers [24]. One requirement to formulate miniemulsions is that both components of the polyaddition reaction have a relatively low water solubility (at least one with solubility $<10^{-5}$ g L^{-1}).

One group showed that polyurethane latexes could be prepared by direct miniemulsification of a monomer mixture of diisocyanate and diol in an aqueous surfactant solution, followed by heating [25]. This was somewhat surprising, as suppression of the polymerization by side reactions might be expected between the highly reactive diisocyanates and the continuous phase water. Thus, it is important that the reaction between diisocyanate and diol occurs more slowly than the time needed for the miniemulsification step, while the side reaction of the diisocyanate with water in the dispersed state must be slower than the reaction with the diol. In this case the functional groups (isocyanate to alcohol groups) were utilized in a 1:1 molar ratio.

It has been shown that the use of hydrophobic organo-tin catalysts, a solvent of the polyaddition medium or an off-stoichiometric ratio of the isocyanate and the diol component allows a considerable increase in the molecular weight of the polymer. It was also shown that hybrid polymer nanoparticles based on polystyrene/polyurethane (PS/PU) or poly(butyl acrylate)/polyurethane (PBA/PU) can be synthesized using a one-pot procedure [26].

Polyaddition was also performed using chitosan as a stabilizer (for further details, see Section 2.5) [27].

2.3.3
Polycondensation

Even in the presence of large amounts of water, the miniemulsion process permits the synthesis of hydrophobic polyesters in a very simple manner and at very low temperatures in order to obtain stable polyester dispersions. The influence of several parameters on the esterification yield has been studied. On the one hand, any modification of the dispersed phase such as the hydrophobicity of the components, viscosity, and the reactant nature results in different yields. With increasing hydrophobicity of the monomers or decreasing viscosity, the yield increases. On the other hand, any modification of the surrounding environment of the droplets such as the interface nature, the ionic strength and the interface area, has no influence on the equilibrium. From a thermodynamic point of view, this polymerization presents the characteristics of a bulk or solution polymerization. Independently of the dispersion state in the range studied, the equilibrium is the same as in bulk or in solution polymerization with an organic phase saturated with water. It is however very unlikely that the reactions occur exclusively in the core of the particle, but in order to provide an answer to this question it would be necessary to conduct a kinetic study related to the interface area.

Polyesterification and radical polymerization can also be combined in a one-pot procedure for the synthesis of well-defined polyester/PS particles of size range 100 nm. Because of the shortened reaction time, the molecular weight of the polyester is relatively low, highlighting the need to identify an effective esterification catalyst that would permit polyester synthesis at low temperature and with a short reaction time.

Although such polyesters are of only low molecular weight, they can be used to great effect as precursors for further reactions (e.g., for polyurethanes).

2.3.4
Anionic Polymerization

For the anionic polymerization of phenyl glycidyl ether (PGE) in miniemulsion, Maitre et al. used didodecyldimethylammonium hydroxide as an "inisurf"; this acts simultaneously as a surfactant and an anionic initiator by means of its hydroxy counterion [28]. As shown by ^{1}H-NMR and FTIR spectroscopy, genuine α,ω-dihydroxylated polyether chains were produced. Moreover, the average molecular weight could be increased by varying the initiator concentration, the type and concentration of surfactants, or by adding an alcohol as co-stabilizer. With increasing conversion, although the polymer chain length increased, the molecular size remained small, with a critical degree of polymerization (DP_{max}) of 8.

2.3.5
Metal-Catalyzed Polymerization Reactions

Ethylene can be polymerized using an aqueous miniemulsion [29] consisting of an organo-transition metal catalyst at ethylene pressures of 10–30 bar and temperatures of 45–80 °C; the resultant large particles have a typical diameter of ca. 600 nm [30]. A maximal productivity of 2520 kg PE g^{-1} atom active metal was achieved, which represents about 60% of the productivity of the same catalyst when used in ethylene suspension polymerization in the organic phase.

2.3.6
Enzymatic Polymerization

Direct enzymatic polymerization of miniemulsions consisting of lactone nanodroplets represents a new and convenient pathway for the synthesis of biodegradable polymer nanoparticles. Moreover, the chemical composition and molecular weight can be varied over a certain range. Oligoesters completely end-capped by an alkene or diene group can also be prepared using this technique. Typically, these building blocks extend polyester application as they allow improved biodegradability to be extended both to siloxane and resin chemistry [31].

2.4
Nanocapsule Particles by Miniemulsion Technologies

2.4.1
Encapsulation of a Liquid: The Formation of Nanocapsules

In the past, control of the morphology of latex particles has been an important area in polymer science, but today's technology has advanced such that a variety of structured particles are now accessible, including core-shell, micro-domain, and inter-penetrating network latexes. Synthetic methods leading to polymer particles having cavities and voids have also undergone extensively investigation. In this chapter, attention is focused on the special case of a hollow sphere or nanocapsule structure. Hollow latex particles can serve as synthetic pigments, which contribute to the opacity of architectural coatings by scattering light [32]. Additionally, these particles can enhance the gloss of paper coatings by influencing the mechanical properties of these high-pigment volume concentration formulations during calendering of the coating [33,34]. The potential value of polyalkylcyanoacrylate nanocapsules has been discussed for a variety of pharmaceutical applications, such as a medium for the controlled release of calcitonin [35], for peroral administration of insulin [36], or as a part of blood substitutes [37].

Nanocapsules are generally considered as spherical, hollow structures with an average diameter <1 μm [38–40]. Typically, the capsule consists of a polymeric wall with a thickness in the nanometer region, filled with an oil which can dissolve

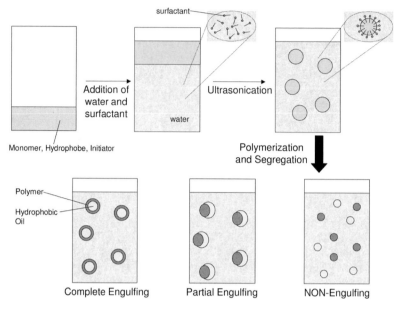

Figure 2.3 Schematic view of the encapsulation reaction.

lipophilic agents. To enable a stable dispersion, the capsule surface is stabilized by surface charges or by adsorption of an amphiphile.

Until now, the formation of capsules of size ≥1 μm has predominantly been described, though for many applications – especially in medicine and high-resolution electronic inks – smaller capsules of 50 to 300 nm attract much more interest. The approach to synthesizing nanocapsules as described below is based on the principle of miniemulsion using the differences of interfacial tension and the phase separation process during polymerization to obtain a nanocapsule morphology.

A miniemulsion polymerization is described that yields an encapsulation of a nonsolvent hydrocarbon by the polymer being formed. Using this process, it is possible to prepare latex particles having voids with facile control of the particle diameter, void fraction, and structure. The process initially involves polymerizing a monomer in a dispersed hydrocarbon-monomer mixture which phase-separates during the polymerization. This phase-separated polymer subsequently serves as a locus for polymerization. The morphology of the demixing structure is determined by the type of surfactant chosen, the polarity of the monomer, and the choice of hydrophobe which modify either the polymer–hydrocarbon or water–polymer interface and determine the degree of engulfing. The principle of the encapsulation process is shown schematically in Figure 2.3.

Pioneer studies related to the theoretical prediction of three-phase interactions in shear and electrical fields were published by Torza and Mason [41]. These authors proposed that the resulting equilibrium configuration of two immiscible liquid drops, designated phase-1 and -3, suspended in a third immiscible liquid, phase-2, and brought into contact could be readily predicted from the interfacial tensions σ_{ij} and spreading coefficients $S_i = \sigma_{jk} - (\sigma_{ij} + \sigma_{ik})$. In terms of the convention $\sigma_{12} > \sigma_{23}$ ($S_1 < 0$), phase-1 is completely engulfed by phase-3 when $S_2 < 0$ and $S_3 > 0$, no engulfing occurs when $S_2 > 0$ and $S_3 < 0$; and S_1, S_2, $S_3 < 0$ leads to partial engulfing and formation of two-phase droplets with three interfaces, the shapes of which can be calculated. The mechanism of engulfing was established with the aid of high-speed cinematography and shown to involve two competitive processes: penetration and spreading.

For simplicity, it was assumed that the final equilibrium state is determined solely by three interfacial tensions σ_{12}, σ_{13}, and σ_{23}. The equilibrium state of three phases of equal density will be that which has a minimum free surface energy $G_s = \Sigma \sigma_{ij} A_{ij}$, where A_{ij} is the area of the ij interface in the configuration. A simple analysis which yields the same results as that based on minimizing G_s can be made from the consideration of the three spreading coefficients:

$$S_i = \sigma_{jk} - (\sigma_{ij} + \sigma_{ik}). \tag{1}$$

If one adopts the convention of designating phase-1 to be that for which $\sigma_{12} > \sigma_{23}$, it follows that $S_1 < 0$. An inspection of Eq. (1) shows that there are only three possible sets of values of S_i corresponding to the three different equilibrium configurations illustrated in Figure 2.4 which are, respectively, complete engulfing, partial engulfing, and non-engulfing.

The description of the expected particle morphology is a system with a complex parameter field. Recognizing the dramatic effect that common emulsifiers have on the interfacial tension between water and organic liquids or solids, it is not surprising to find that the preferred particle morphology reacts sensitively on the chemical natures of the emulsifier, the polymer and the oil.

It is clear that the development of the final morphology in polymer microparticles involves the mobility or diffusion of at least two molecular species influenced by some driving force to attain the phase-separated structure. The ease of movement

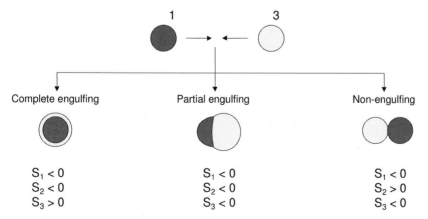

Figure 2.4 Possible equilibrium configurations corresponding to the three sets of relationships for S_i. The medium is phase-2.

may be related to the phase viscosity, but in this approach the main emphasis is placed on the driving force, which is the Gibbs' free energy change of the process.

An analysis of the thermodynamics of two-stage particle formation has been developed by Berg et al. [42,43] in which the system was considered simply in terms of the free energy changes at the interface of a three-phase system (i.e., in their case polymers 1, 2, and water) based on $G = \Sigma \sigma_{ij} A_{ij}$, where G is the Gibbs' free energy of the system. According to this analysis, each particular morphological configuration will have a different value for G, and the arrangement with the minimal free energy will be the one which is favored thermodynamically. Sundberg et al. [44] also discussed that the influence of the surfactant and the nature of the incompatible polymers on polymer particle morphology is seen through their individual and collective effects upon the interfacial tensions. Several apparently different morphologies (hemispherical, sandwich, multiple lobes) have been found to coexist at the same time within a single emulsion, suggesting that they may be simply different states of phase separation and not thermodynamically stable, unique morphologies.

Chen and coworkers [45] developed a thermodynamically based model to describe the free energy differences between different possible particle structures.

For the preparation of polymer particles in aqueous media consisting of two phases, the interfacial energies can for example be influenced by the following parameters:

- the differences in the hydrophilicity of the monomers and the polymers and the solubility of the monomers and polymers in the aqueous phase [46, 47]
- the compatibility of the formed polymers [48]
- type and amount of initiator [49]
- temperature [50]

The formation of nanocapsules was achieved by a variety of approaches. One of the earliest processes for making hollow latex particles was developed in the research laboratories of the Rohm and Haas Company [51–54]. Their concept involved making a structured particle with a carboxylated core polymer and one or more outer shells. Ionization of the carboxylated core with base under the appropriate temperature conditions expands the core by osmotic swelling to produce hollow particles with water and polyelectrolyte in the interior. In addition to this approach, a number of alternative processes have also been patented that are complex in terms of process stages and chemistry [55–57].

McDonald et al. found that the modification of an emulsion polymerization with a water-miscible alcohol and a hydrocarbon nonsolvent for the polymer can influence the morphology and enables the formation of monodisperse particles with a hollow structure or diffuse microvoids [58]. Both kinetic and thermodynamic aspects of the polymerization dictate particle morphology. Complete encapsulation of the hydrocarbon occurs, provided that a low molecular-weight polymer is formed initially in the process. Monodisperse hollow particles with diameters ranging from 0.2 to 1 μm were obtainable, and void fractions as high as 50% are feasible.

Another well-established method is the preparation of alkylcyanoacrylate nanocapsules, the special choice of monomer yielding in thinner capsule walls and generally in a more reproducible capsule structure. The sizes of capsules prepared in the described manner depend on the concentrations of the oil and the monomer components [59]. Capsules from polyalkylcyanoacrylates synthesized by interfacial polymerization were initially described by Florence et al. [60]. In the meantime, various techniques for the preparation of polyalkylcyanoacrylates nanocapsules have been studied [40,61,62].

Berg et al. reported the preparation and evaluation of microcapsules formed by the polymerization of methyl methacrylate in the presence of an oil–water macroemulsion. The oil phase was composed of an alkane (e.g., decane or hexadecane), and the oil–water emulsions were stabilized by a variety of emulsifiers [42,43]. Both oil- and water-soluble initiators were used, and the monomer was introduced by either dissolving in the oil or feeding it through the water phase. The authors view this system to be an opportunity to study morphological characteristics of polymeric microparticles in the 1 to 100 μm size range.

Okubo et al. examined the penetration/release behavior of various solvents into/from the interior of micron-sized monodisperse cross-linked polystyrene/polydivinylbenzene composite particles [63]. The hollow particles were produced by the seeded polymerization utilizing the dynamic swelling method [64]. Itou et al. prepared crosslinked hollow polymer particles of submicron size by means of a seeded emulsion polymerization [65]. The morphology of the particles depends on the composition of divinylbenzene and methyl methacrylate.

Nanocapsules prepared by the polymerization of rather complex entities were made by Stewart and Lui [66], who stabilized block copolymer vesicles made of polyisoprene-*b*-poly(2-cinnamoylethyl methacrylate) diblock copolymers by UV crosslinking. A similar approach using block copolymer building blocks was used by Meier and colleagues, who generated nanocapsules by crosslinking polymerization of ABA triblock copolymer vesicles [67], the size of which can be controlled in the range of 50 nm up to 500 nm. Due to their crosslinked structure, both the Liu and the Meier nanocapsules are shape-persistent, even after their isolation from aqueous solution.

Feldheim et al. used gold particles as templates for the synthesis of hollow polypyrrole capsules [68]. Etching of the gold leaves a structurally intact hollow polymer capsule with a shell thickness governed by polymerization time (5 to 100 nm) and a hollow core diameter dictated by the diameter of the template particle (5 to 200 nm). Microencapsulation of peptides and proteins is achieved by preparing microcapsules using a double emulsion technique [69]. For the induced phase separation method the aqueous drug solution was intensively mixed with the organic polymer solution while an aqueous surfactant solution is added slowly to the oil–water emulsion. The obtained water–oil–water emulsion was stirred under partial vacuum conditions until the organic solvent was removed and the microcapsule built up.

Polymerization in miniemulsion can also be performed in the presence of an oil, which is inert to the polymerization process. During polymerization, the oil and polymer can demix, and many different structures such as an oil droplet encapsulated by a polymer shell, sponge-like architectures or dotted oil droplets can be formed. The formation of such structures is known from classical emulsion polymerization, but is usually kinetically controlled [70–72]. The synthesis of hollow polymer nanocapsules as a convenient one-step process using the miniemulsion polymerization, however, has the advantage of being thermodynamically controlled [73].

Chemical control of the expected particle morphology for an encapsulation process is a system with a complex parameter field. The particle morphology reacts sensitively on the chemical nature of the emulsifier, the polymer and the oil, as well as on additives such as an employed additional hydrophobe, the initiator, or possible functional comonomers. It is clear that development of the final morphology in polymer microparticles also involves the mobility or diffusion of at least two molecular species influenced by some driving force to attain the phase-separated structure. The ease of movement may be related to the phase viscosity, but in this approach the main emphasis is laid on the driving force, which is the Gibbs' free energy change of the process.

It was found that the nanocapsules are formed in a miniemulsion process by a variety of monomers in the presence of larger amounts of a hydrophobic oil. The hydrophobic oil and monomer form a common miniemulsion before polymerization, whereas the polymer is immiscible with the oil and phase-separates throughout polymerization to form particles with a morphology consisting of a hollow polymer structure surrounding the oil. The differences in the hydrophilicity of the oil and the polymer turned out to be the driving force for the formation of nanocapsules.

In the case of poly(methyl methacrylate) (PMMA) and hexadecane as a model oil to be encapsulated, the pronounced differences in hydrophilicity are suitable for direct nanocapsule formation. PMMA is regarded as rather polar (but is not water-soluble), whereas hexadecane is very hydrophobic so that the spreading coefficients are of the right order to stabilize a structure in which a hexadecane droplet core is encapsulated in a PMMA shell surrounded by water [73]. In the state of miniemulsion, the monomer and the hexadecane are miscible, but phase separation occurs during the polymerization process due to the immiscibility of hexadecane and PMMA. Miniemulsions were obtained by mixing the monomer MMA and hexadecane at varying ratios together with the hydrophobic, oil-soluble initiator 2,2'-azo-bis-isobutyronitrile (AIBN) and miniemulsifying the mixture in an aqueous solution of sodium dodecyl sulfate (SDS). After polymerization, the polymer capsules were obtained as shown in Figure 2.5a. Nanocapsules with an increased shell stability can be obtained by using up to 10 wt.% ethylene glycol dimethacrylate (EGDMA) as crosslinking agent. It is a fortunate experimental situation that the particle size in this reaction does not change with the amount of the anionic surfactant SDS or the nonionic surfactant Lutensol AT50 (a hexadecyl-modified poly(ethylene glycol)). This means that with increasing surfactant load, the surface coverage also increases, and the interfacial tension at the droplet–water interface decreases. In this way the influence of the systematic variation of one of the interface tensions on the particle morphology was examined, and a continuous morphological change towards engulfed structures was identified (Fig. 2.5b) [73].

a)

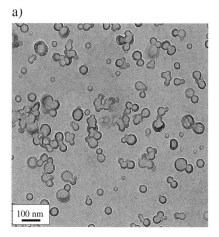

b)

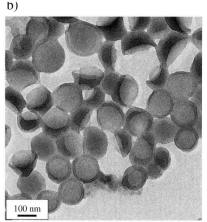

Figure 2.5 (a) Transmission electron micrographs of nanocapsules with a methyl methacrylate to hexadecane ratio of 1:1.

(b) Coexistence of nanocapsules and capped particles in the case of using the surfactant Lutensol AT50.

In the case of styrene as monomer and hexadecane as model oil, the cohesion energy density of the polymer phase is closer to that of the oil, and therefore the structure of the final particles depends much more on those parameters which critically influence the interfacial tensions. A variety of different morphologies in the styrene–hexadecane system can be obtained by changing the spreading parameter. This was done by changing the monomer concentration and the type and amount of surfactant, as well as the initiator and the functional comonomer.

The best results were obtained by using the block copolymer surfactant SE3030 (poly[styrene-*b*-ethylene oxide] together with the nonionic initiator PEGA200, which supports interface stabilization and improves the structural perfection (Fig. 2.6a) of the polystyrene capsule morphology.

Another very powerful approach to improve the perfection of the capsules is the addition of a comonomer to the oil phase. Depending on the polarity of the monomer, it will enter one of the two interfaces (polymer–water) or (polymer–oil) and reduce the corresponding interfacial tensions and spreading coefficients. It was shown that the very hydrophobic comonomer lauryl methacrylate, which is expected to minimize the interfacial tension between styrene and the hexadecane phase, has no significant effect on the resulting morphology of the particles; this means that this interfacial energy is of minor importance as it is already quite low. On the other hand, the slightly more hydrophilic MMA and the very hydrophilic acrylic acid (AA) affect the interfacial tension of styrene to water, and here, a pronounced influence on the morphology was found. The influence depends on the partitioning coefficient: for MMA, about 50 wt.% of monomer was needed to create only close-to-perfect capsules, whereas already 1 wt.% of AA was sufficient in order to saturate the capsule surface with carboxylic groups, and hollow shell structures with constant capsule thickness were found (Fig. 2.6b). There is, however, a

a)   b)

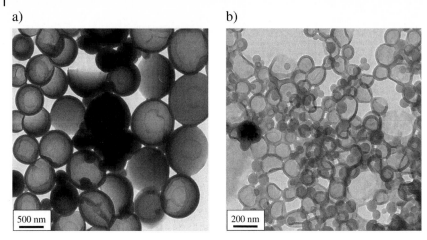

Figure 2.6 Transmission electron micrographs of polystyrene/hexadecane. (a) With SE3030 as surfactant and PEGA200 as initiator; (b) with AA as comonomer.

minor fraction of small homogeneous polymer latexes, which we attribute to secondary nucleation due to the high content of water-soluble acrylic acid.

The influence of surfactant concentration on particle size and stability of nanocapsules with liquid cores, synthesized by an *in-situ* miniemulsion polymerization process, was investigated by van Zyl et al. [74].

Park et al. [75] reported on a synthesis by miniemulsion polymerization of nanocapsules consisting of an encapsulated phase change material in polystyrene. By using differential scanning calorimetry (DSC), it could be shown that phase-change material coated with polystyrene exhibited thermal energy storage and release behavior, and the latent heat was found to be a maximum 145 J g^{-1}. It was noted that the nanocapsules showed a good potential as a thermal energy storage medium.

The use of biocompatible systems is proposed by Rajot et al. [76], who produced nonionic hydrophobic drugs such as indomethacin encapsulated in poly(vinyl acetate). In addition, oligocaprolactone macromonomers obtained by anionic coordinated ring-opening polymerization, benzyl benzoate, or triglycerides from fatty acids were used as the hydrophobe in order to obtain biocompatible systems.

Nanocapsules can also be prepared by using the Ouzo effect [77], which utilizes spontaneous droplet formation and subsequent polymer precipitation or synthesis. It is a preparation of metastable liquid dispersions by homogeneous liquid–liquid nucleation, and is based primarily on a recent study conducted by Vitale and Katz [78]. This spontaneous emulsification occurs upon pouring, into water, a mixture of a totally water-miscible solvent and a hydrophobic oil (and optionally some water), thus generating long-lived small droplets, which are formed even though no surfactant is present.

Baek et al. [79] used the phase-inversion emulsification technique in order to obtain epoxy resin microcapsules containing the flame retardant triphenyl phosphate with excellent physical properties and network structure. This microencapsulation process was adopted for the protection of triphenyl phosphate evaporation and wetting of polymer composite during the polymer blend processing. For the synthesis, triphenyl phosphate, epoxy resin and mixed surfactants were emulsified to oil in water (O/W) by the phase inversion technology and then conducted on the crosslinking of epoxy resin by *in-situ* polymerization. The capsule size and size distribution was controlled by mixed surfactant ratio, concentration and T-6-P-phosphatase (TPP) contents. The formation and thermal property of the capsules were measured by DSC and thermogravimetric analysis.

2.4.2
Nanocapsules with a Shell Obtained by Surface Coating of Miniemulsions with Inorganic Nanoparticles and Crystalline Building Blocks

In many cases, the gas permeation or chemical sensitivity of polymer capsules is still too high to be efficient for encapsulation. Here, the employment of crystalline inorganic materials, such as clay sheets of 1.5 nm thickness, can be recommended. As these clay sheets are fixed like scales onto the soft, liquid miniemulsion droplet, the resulting objects are termed "armored latexes" [80]. Since clays carry a negative surface charge, miniemulsions stabilized with cationic sulfonium-surfactants represented a convenient way to generate those armored latexes or crystalline nanocapsules. Due to their high stability also against changes in the chemical environment, it is possible to use miniemulsion droplets themselves, with polymerized latex particles also being used as templates for such a complexation process. As a result, the liquid droplets or polymer particles become completely covered with clay plates, thereby preventing film formation or coalescence. A synthetic monodisperse model clay with small lateral extensions was employed; consequently, the liquid droplets or polymer particles become covered with clay plates, which is also macroscopically visible by the absence of film formation or coalescence. However, complexation with the clay plates alone was not always sufficient to prohibit the release of low-T_g polymer (e.g., PBA) or liquid material. In order to further glue together the single clay sheets, silicic acid was used as a "mortar". Thus, the stability of the shells can be increased by a condensation reaction with silicic acid, which reacts with itself, as well as with residual surface OH-groups of the clay. In this case, film formation is indeed completely suppressed, as shown by transmission electron microscopy (TEM) of a dried sample with PBA as template (Fig. 2.7a). Figure 2.7a shows intact nanocapsules filled with PBA and of about the same size as found previously in solution. In order to receive an emptied, purely inorganic shell, PMMA latex particles were used as templates. Intense illumination with the electron beam causes PMMA to depolymerize to MMA, which uses or creates small pores in the shell to evaporate under the high-vacuum conditions used in TEM. Figure 2.7b shows such emptied capsules as obtained by electron degradation. Here, the inorganic structure has not disintegrated, and consequently the colloidal "laying of bricks" had been successful. These "armored" droplets and latex particles may be of great interest for the production of pressure-sensitive adhe-

a)

b)

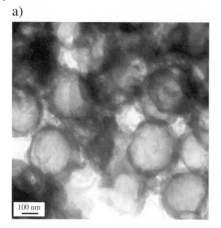
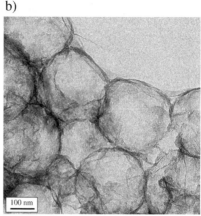

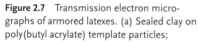

Figure 2.7 Transmission electron micrographs of armored latexes. (a) Sealed clay on poly(butyl acrylate) template particles; (b) sealed clay on poly(methyl methacrylate) (PMMA) template particles after removal of PMMA in the core.

sives or as a new type of filler with unconventional chemical and mechanical performance.

Switching from highly hydrophilic clays towards other inorganic nanoparticles (e.g., colloidal silica) leads, during the interplay with polymerization in miniemulsions, to a potential structural complexity that covers the whole range from embedded particles (as in the case of the calcium carbonate and carbon blacks) to surface-bound inorganic layers (as in the case of the clays). When conducting basic research, these are ideal systems to analyze complex structure formation processes in emulsions, as the original droplet shows a structure that is essentially established by molecular forces and local energy considerations, and is (ideally) just solidified into a polymer structure.

It was discussed earlier that the structure created by the ternary system oil–water–nanoparticle follows the laws of spreading thermodynamics, as they hold for ternary immiscible emulsions (oil 1–oil 2–water) [41,73,81]. The only difference is that the interfacial area and the curvature of the solid nanoparticle must remain constant; hence, an additional boundary condition is added. When the inorganic nanoparticles possess, beside charges, also a certain hydrophobic character, they become enriched at the oil–water interface. This forms the physical basis of the stabilizing power of special inorganic nanostructures, the so-called "Pickering stabilizers" [82–84], and means that the surface energy of the oil–nanoparticle–water system must be lower than the sum of the binary combinations oil–water and water–nanoparticle to enable superstructure formation to occur. Since all three terms can be adjusted by the choice of monomer and the potential addition of surfactants, this spans a composition diagram in which a variety of morphologies may occur. Silica nanoparticles are ideal as model nanoparticles for the systematic examination of compositional phase behavior as they are easy to obtain and to control with respect to their surface structure and interacting forces. The latter is achieved either by variation of pH, which alters the surface charge density, or by the adsorption of cationic organic components, which changes the polarity of the objects.

It was shown that silica nanoparticles in the absence of any surfactant could act as a Pickering stabilizer for a miniemulsion process [85]. The high quality and small overall particle size obtained only under alkaline conditions (pH = 10) and in the presence of the basic comonomer 4-vinylpyridine (which acts as an aminic coupler) [86,87] is shown in Figure 2.8. The particle size depends on the amount of silica in the expected manner: the higher the silica content, the smaller the resulting stable hybrid structures (see Fig. 2.8c and 2.8d). Comparatively small compound particles in the diameter range 120 to 220 nm and with rather narrow size distribution were obtained [85], which underscores the high stabilization power of the silica particles as Pickering stabilizers. This underlines the surface activity of the

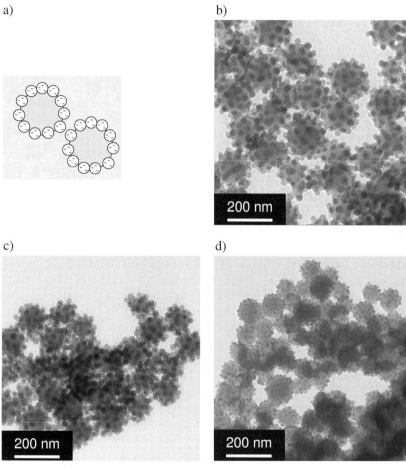

a)

b)

200 nm

c)

200 nm

d)

200 nm

Figure 2.8 (a) Scheme for using silica particles as stabilizer for monomer droplets of miniemulsions (Pickering stabilization).

(b–d) Latexes with different monomer:silica ratios; (b) 1:0.32; (c) 1:0.72; (d) 1:1.08.

silica nanoparticles under the applied conditions and the kinetic-free, equilibrium-type structure of a miniemulsion latex particle. As these systems are free of low molecular-weight surfactants and all chemical byproducts which might act as a surfactant, the measured surface tension reached 71.4 mN m^{-1}, which is almost the value of pure water.

The addition of a surfactant (nonionic, anionic, or cationic) to the same system resulted in a more complex group of structures [85]. Nonionic surfactants are preferentially bound to the silica nanoparticles due to a preferential interaction between the silica and the ethylene oxide chains [88], which screens any interaction with the monomer mixture. The addition of SDS also leads to electrostatic repulsion and competition between surfactant and silica nanoparticles. The most pronounced morphology changes are observed with the cationic surfactant cetyltrimethylammonium chloride (CTMA-Cl). Due to charge coupling as well as to induced dipole interaction, this surfactant binds strongly to silica over the whole pH range. The surfactant is mainly adsorbed by the silica, but under standard conditions there is insufficient CTMA in the mixture to counterbalance the negative charges of the system at pH 10. In the presence of 4-vinylpyridine, the strong acid–base interaction provides an additional stability. Thus, at pH 10 a "hedgehog" morphology is found with a small overall diameter of 90 nm.

At higher CTMA concentrations exceeding the amount adsorbed onto the silica, a different morphology was found. Starting from a calculated surface coverage of 75%, the silica particles become incorporated into the droplets, and stable hybrid structures are obtained. The hybrids now have a "raspberry" morphology, but are somewhat heterogeneous with respect to loading with silica [85].

Zhang et al. [89] showed also that titania particles can be used for the coating process. These authors prepared hybrid microballs with polystyrene cores coated by titania nanoparticles. Acrylic acid was used as a comonomer in order to promote locating titania nanoparticles on the polymer's surface. The morphology of hybrid particles was examined using TEM, while the infra-red spectra of hybrid nanoparticles showed there to be an interaction between titania nanoparticles and the polymer shell.

2.4.3
Encapsulation of Pigments by Direct Miniemulsification

For the encapsulation of pigments by miniemulsification, two different approaches can be used. In both cases, the pigment–polymer interface as well as the polymer–water interface must be carefully adjusted chemically in order to obtain encapsulation as a thermodynamically favored system. The design of the interfaces is mainly dictated by the use of two surfactant systems, which govern the interfacial tensions, as well as by the use of appropriate functional comonomers, initiators, or termination agents. The sum of all the interface energies must be minimized.

For the successful incorporation of a pigment into the latex particles, both type and amount of surfactant systems must be adjusted to yield monomer particles, which have the appropriate size and chemistry to incorporate the pigment by its lat-

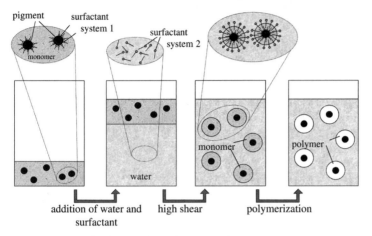

Figure 2.9 Principle of encapsulation by miniemulsion polymerization.

eral dimension and surface chemistry. For the preparation of the miniemulsions, two steps must be controlled (Fig. 2.9). First, the already hydrophobic or hydrophobized particulate pigment with a size up to 100 nm must be dispersed in the monomer phase. Hydrophilic pigments require a hydrophobic surface to be dispersed into the hydrophobic monomer phase, which is usually promoted by a surfactant system I with low hydrophilic-lipophilic balance (HLB) value. This common mixture is then miniemulsified in the water phase, employing a surfactant system II with high HLB, which has a greater tendency to stabilize the monomer (polymer)–water interface.

Erdem et al. described the encapsulation of TiO_2 particles via miniemulsion in the two steps. First, TiO_2 was dispersed in the monomer using the OLOA 370 (polybutene-succinimide) as stabilizer [90]. This phase was then dispersed in an aqueous solution to form stable, submicron droplets [91]. The presence of TiO_2 particles within the droplets limited the droplet size. Complete encapsulation of all TiO_2 in the colloidal particles was not achieved, and the encapsulation of 83% of the TiO_2 in 73% of the polymer was reported. The amount of encapsulated material was also very low, such that a TiO_2:styrene weight ratio of 3:97 could not be exceeded [92,93].

Oh and Kim [94] also encapsulated TiO_2 using a convenient one-step miniemulsion polymerization, but together with leuco dyes and a developer. The morphology of the resultant polymer dye capsules with an average size of 30 to 500 nm was examined using scanning electron microscopy (SEM). This new photobleacheable nanocomposite was developed as a reusable recording media. The leuco dye was originally colorless, but acquired color when it reacted with the developer. The color of the recorded part, by coating the dye capsule solution onto white paper, was persistent, but the recorded mark was bleached upon exposure to UV light. Capsule morphology was not detected in the bleached region as examined by SEM, indicating that TiO$_2$ in the film acted as a photocatalyst to decompose the capsule structure. The bleaching was irreversible so that the recording media was reusable. The bleaching efficiency in photobleaching was shown to exceed 75%.

Nanoparticulate hydrophilic CaCO$_3$ was effectively coated with a layer of stearic acid as surfactant system I prior to dispersing the pigments into the oil phase [95]. The COOH groups act as good linker groups to the CaCO$_3$, and the tendency of the stearic acid to pass to the second polymer–water interface was found to be low. A CaCO$_3$ level of 5 wt.%, based on monomer, could be completely encapsulated into polystyrene particles [95]. The weight limit was limited due to the fact that, at this concentration, each polymer particle already contained one CaCO$_3$ colloid, which was encapsulated in the middle of the latex particle (Fig. 2.10).

Calcium carbonate was also encapsulated by Seul et al. [96], who investigated the optimum conditions for encapsulation with methyl methacrylate. Moreover, it was also shown that other pigments could be used for the encapsulation process [97]. For example, aqueous dispersions of polystyrene latexes encapsulating a copper phthalocyanine blue pigment were formulated using the miniemulsion polymerization technique. The organic pigment was first suspended into the monomer phase, and the resulting oily suspension subsequently converted into stable miniemulsion droplets followed by a polymerization step.

Silica nanoparticles could also be encapsulated with an epoxy resin to produce water-borne nanocomposite dispersions by using the phase-inversion emulsification technique [98]. Microscopy results indicated that all the silica nanoparticles were encapsulated within the composites and uniformly dispersed therein.

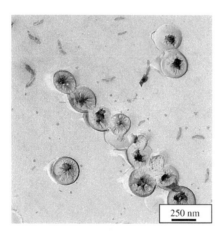

250 nm

Figure 2.10 Encapsulation of calcium carbonate by the miniemulsion process.

2.4.4
Encapsulation by Co-Miniemulsion

2.4.4.1 Nanocapsules Containing Carbon Black

Since carbon black is a rather hydrophobic pigment (depending on the preparation conditions), its encapsulation in latexes by direct dispersion of the pigment powder in the monomer phase prior to emulsification is possible [95]. Here, full encapsulation of non-agglomerated carbon particles can be provided by an appropriate choice of hydrophobe. In this case, the hydrophobe not only acts as the stabilizing agent against Ostwald ripening for the miniemulsion process, but also mediates to the monomer phase by partial adsorption. This direct dispersion only allows the incorporation of 8 wt.% carbon black, however, as the carbon is still highly agglomerated in the monomer. At higher levels, the carbon cluster will break the miniemulsion, and less-defined systems with encapsulation rates less than 100%, which also contained pure polymer latexes, were obtained.

In order to increase the amount of encapsulated carbon to 80 wt.%, another approach was developed [99] whereby both monomer and carbon black were independently dispersed in water using SDS as a surfactant and subsequently mixed in

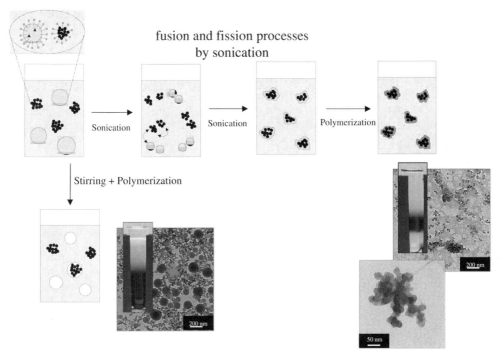

Figure 2.11 Principle of co-miniemulsion where both components must be dispersed independently in water and mixed afterwards. The controlled fission/fusion process in the miniemulsification realized by high-energy ultrasound or high-pressure homogenization destroys all aggregates and liquid droplets, and only hybrid particles being composed of carbon black and monomer remain due to their higher stability.

any ratio between the monomer and carbon. The mixture was then cosonicated, whereupon the controlled fission–fusion process characteristic of miniemulsification destroyed all aggregates and liquid droplets. As a consequence, only hybrid particles composed of carbon black and monomer remained due to their higher stability [99]. This controlled droplet fission and heteroaggregation process can be achieved with either high-energy ultrasound or high-pressure homogenization.

TEM and ultracentrifugation studies showed that this process results in effective encapsulation of the carbon with almost complete yield (Fig. 2.11): only rather small hybrid particles, but no free carbon or empty polymer particles, were found. It must be stated that the hybrid particles with high carbon content do not possess a spherical shape, but rather adopt the typical fractal structure of carbon clusters, coated with a thin (but homogeneous) polymer film. The thickness of the monomer film depends on the amount of monomer, and the exchange of monomer between different surface layers is – as in miniemulsion polymerization – suppressed by the presence of an ultrahydrophobe.

Therefore, the process is best described as a polymerization in an adsorbed monomer layer created and stabilized as a miniemulsion ("ad-miniemulsion polymerization"). The process is shown schematically in Figure 2.11.

2.4.4.2 Nanocapsules Containing Magnetite

Magnetic fluids are stable dispersions of ultrafine magnetic particles or encapsulated magnetic particles in an organic or aqueous carrier medium. The stabilization of these particles can be achieved by surfactants which hinder the particles from flocculation or sedimentation. In ideal cases, the particles also keep their stability under the exposition of magnetic fields.

Magnetic polymer nanoparticles, which are usually dispersed in a carrier liquid, can be tailor-made depending on the final application [100,101]. Several types of magnetic polymer nanoparticles have been produced from both natural and synthetic polymers, with the intention of incorporating groups onto the surface or of treating their surfaces to perform, for instance, selective separations. In particular, magnetic nanoparticles with or without polymer encapsulation can be used for magnetic drug targeting [102,103], tissue engineering, magnetic resonance imaging [104], and hyperthermia [105].

In earlier reports, magnetic fluids were produced by grinding magnetite with heptane or long-chain hydrocarbons and a grinding agent (e.g., oleic acid) [106]. Later, magnetic fluids were produced by the precipitation of an aqueous Fe^{3+}/Fe^{2+} solution with a base, coating these particles with an adsorbed layer of oleic acid and then dispersing them in a non-aqueous fluid [107]. Both processes result in tiny magnetite particles, a surfactant coating these magnetite particles, and a non-aqueous liquid carrier in which the hydrophobic magnetite particles are dispersed. Clearly, the latter process is more feasible in the production of more homogeneous magnetite particles.

Other applications of magnetic fluids rely on water as the continuous phase. Kelley [108] produced an aqueous magnetic material suspension by the conversion of iron compounds to magnetic iron oxide in the aqueous medium under controlled pH conditions in the presence of a petroleum sulfonate dispersant. Shimoiizaka et al. [109] developed a water-based magnetic fluid from oleic acid-coated magnetite particles dispersed by an anionic or nonionic surfactant solution which was suitable to form a second surfactant layer.

Polymer-covered magnetic particles can also be produced by the precipitation *in situ* of magnetic materials in the presence of polymer which acts as a stabilizer. In this way, magnetic polymer nanoparticles are produced in presence of the water-soluble dextran [110], poly(ethylene imine) [111], poly(vinyl alcohol) [112], poly(ethylene glycol) [113], sodium poly(oxyalkylene di-phosphonates) [114], and amylose starch [115]. In all cases, the magnetic particles are surrounded by a hydrophilic polymer shell.

Another method of producing magnetic polymer particles involves the separate synthesis of magnetic particles and polymer particles, followed by their mixing to enable either physical or chemical adsorption of the polymer onto the magnetic material. The polymer material can be produced in different ways, including emulsion or precipitation polymerization [116].

It is also possible to use a strategy comprising polymerization in heterophase in the presence of magnetic particles. The magnetic material (preferably surfactant-coated) is embedded into a polymer using processes such as suspension, emulsion, or precipitation polymerization. Magnetic particles were encapsulated in hydrophilic polyglutaraldehyde by suspension polymerization, which resulted in particles of average diameter 100 nm [111]. Magnetite-containing nanoparticles of 150–200 nm were also synthesized by seed precipitation polymerization of methacrylic acid and hydroxyethyl methacrylate in the presence of magnetite particles containing tris(hydroxy methyl) aminomethane hydroxide in ethyl acetate medium [117]. Polymethacrylate/poly(hydroxymethacrylate)-coated magnetite particles could be also prepared by a single inverse microemulsion process, leading to particles with a narrow size distribution, but with a magnetite content of only 3.3 wt.% [118].

Daniel and coworkers [119] obtained magnetic polymer particles by dispersing a magnetic material in an organic phase which consisted of an organosoluble initiator, vinyl aromatic monomers and/or a water-insoluble compound. The mixture was emulsified in water by using an emulsifier, after which polymerization occurred to produce polymer particles with magnetite contents between 0.5 and 35 wt.% with respect to the polymer. However, the resulting particle size distribution was rather broad (between 30 and 5000 nm). Charmot and Vidil [120] used a similar method to produce magnetizable composite microspheres of a hydrophobic crosslinked vinylaromatic polymer, but obtained a mixture of magnetizable particles and non-magnetizable blank microspheres. Wormuth [121] and Xu et al. [122] used the inverse miniemulsion process [19] to encapsulate magnetic particles by a hydrophilic polymer.

Magnetic polymer nanoparticles should fulfill certain criteria in order to fit further biomedical applications: these include a lack of sedimentation, a uniform size and size distribution, high and uniform magnetic contents, superparamagnetic behavior, a lack of toxicity, no iron leakage, high selectivity in case the particles are used for hyperthermia purposes, and sufficient heat generation at lower frequencies to enhance selective heating [105]. Thus, magnetite particles which are encapsulated homogeneously in a hydrophobic polymer and prevent water-soluble components from contacting the magnetite particles are of major interest. Several reasons have been proposed for the use of polystyrene as a hydrophobic encapsulation material in biomedical applications [123]; namely, it is inexpensive, it is hydrophobic and allows the physical adsorption of antibodies or proteins, and it can also be functionalized (e.g., by COOH groups), thus enabling the covalent binding of antibodies, proteins, or cells.

Magnetite, however, is hydrophilic and relies on an effective surface treatment before the encapsulation process. This is preferentially carried out by the adsorption of a secondary surfactant onto the magnetite surface, as this does not interfere with the primary surfactant system needed to stabilize the polymer particles.

Magnetite particles were encapsulated in a polystyrene matrix by a miniemulsion polymerization process (see Fig. 2.9) [124]. In a first step, a stable dispersion of magnetite particles in styrene is required; for this, an effective surfactant system must be used in order to render the particles hydrophobic and to prevent aggregation. Oleoyl sarcosine acid and more efficient oleic acid as first surfactant system to handle the interface magnetite/styrene oleic acid, and magnetite particles down to 10 nm were effectively stabilized also in styrene and related monomer mixtures. This lipophilic dispersions could be miniemulsified in water by using SDS as a second surfactant system forming a stable emulsion. It is important to note that the simple magnetite particles remain well dispersed in the monomer droplets, as shown by electron microscopy. This means that the acid stays at the magnetite–styrene interface and is not redistributed towards the monomer–water interface, an important prerequisite for keeping such double dispersions stable.

As hexadecane was added to the monomer phase as an ultrahydrophobe to prevent Ostwald ripening, the monomeric double miniemulsion was kinetically stable. Polymerization was then started by raising the temperature. The final dispersion was free of coagulum and stable. The brown color did not change during the polymerization, proving that the radical polymerization process did not significantly interfere with the oxidation state of the magnetite colloid. A simple test with a magnet showed the dispersion to be a ferrofluid (i.e., it is magnetic). Thermogravimetric measurements showed the magnetite loading in the final particles to be 20 wt.%, indicating that no change of composition by selective loss of inorganic material occurred throughout the reaction. TEM of polystyrene particles with encapsulated magnetite is shown in Figure 2.12; typical particles had diameters of ca. 100 nm.

This reaction still showed some imperfections, however, with the distribution of magnetite between the particles and within each particle being somewhat heterogeneous. This was presumably due to interaction between the magnetite moieties and a related size- and content-specific destabilization of the miniemulsion droplets, while the magnetite content of the polystyrene matrix was limited to ca. 15 wt.%. The influence of pH appeared crucial, and was attributed to both the pH dependence of the surface charge of magnetite and to protonation of the oleoyl sarcosine acid and the coupled interface energy.

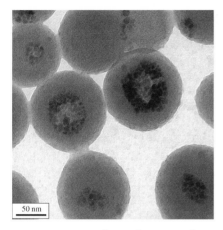

Figure 2.12 Encapsulation of Fe_3O_4 in polystyrene particles.

In order to be able to increase the magnetite content in the polymer particles, it was necessary to develop another process. A limitation of the inorganic material, which is to encapsulate, was also observed in the case of carbon black encapsulation. Here, another approach was developed for the encapsulation of high amounts of carbon black [99]. Both monomer and carbon black were independently dispersed in water using SDS as a surfactant and subsequently mixed in any ratio between the monomer and carbon. The mixture was then cosonicated, with the controlled fission/fusion process characteristic for miniemulsification destroying all aggregates and liquid droplets; consequently, only hybrid particles composed of carbon black and monomer remained due to their higher stability [99]. In order to obtain homogeneous encapsulation with a high magnetite content, a three-step process was developed [125] (Fig. 2.13):

- Step 1: hydrophobized magnetite particles of diameter ca. 10 nm were synthesized in a classical co-precipitation procedure.
- Step 2: the magnetite particles were transformed to magnetite aggregates of ca. 40–200 nm in water by using a miniemulsion process.
- Step 3: the magnetite aggregates were encapsulated with a monomer by an ad-miniemulsification process and, after polymerization, highly magnetite-loaded particles were obtained.

Magnetite particles of mean diameter 10 nm were obtained in a co-precipitation process [107] by quickly adding a concentrated ammonium solution to a solution of Fe^{2+}/Fe^{3+} with a molar ratio of 3:2 which allows the compensation of the oxidation of some iron II to iron III during the co-precipitation in an open vessel.

By adding oleic acid at temperatures above its melting temperature, the magnetic particles were hydrophobized. After evaporation of the water and washing out of the nonadsorbed oleic acid, a dry powder was obtained. The hydrophobized mag-

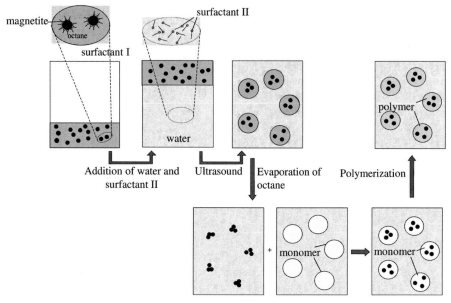

Figure 2.13 Formulation of polymer-coated magnetite particles with a high magnetite ratio. In the first step, hydrophobized magnetite particles are produced and, in a second step, transferred to magnetite aggregates in water by using the miniemulsion process. In a third step, the principle of co-miniemulsion is used. The controlled fission/fusion process in the miniemulsification realized by high-energy ultrasound or high-pressure homogenization destroys all aggregates and liquid droplets, and only hybrid particles composed of magnetite and monomer remain due to their higher stability.

netite particles could be easily dispersed in octane; the typical size of the oleic acid-coated particles in octane was ca. 20 nm (Fig. 2.14).

In a next step, the hydrophobic magnetite particles in octane as dispersion medium are reformulated to stable water-based magnetic fluids for the encapsulation

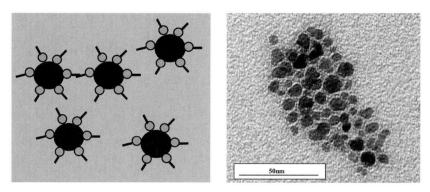

Figure 2.14 Oleic acid-coated magnetite particles in octane.

process. For this, a miniemulsion process was carried out to produce magnetite/octane-in-water dispersions. For the formulation of stable miniemulsions, the addition of a hydrophobic agent is required to provide an osmotic stabilization. However, the magnetite/octane-in-water miniemulsions with and without hexadecane both show a high stability, indicating that oleic acid not only acts as coating agent but can also replace the osmotic pressure agent.

After evaporation of the octane, a water-based ferrofluid consisting of oleic acid-coated aggregated magnetite dispersed in a water phase is obtained (Fig. 2.15). In other words, the magnetite aggregates must have a surfactant double layer: the first layer is oleic acid, which provides a hydrophobicity of the particles for later encapsulation; the second layer, being SDS, promotes the stabilization in water.

In the third step of the synthesis route, a monomer miniemulsion with 20 wt.% dispersed phase is prepared and added to the water-based ferrofluid containing the magnetite aggregates, as obtained above. This mixture is then cosonicated, and the controlled fission/fusion process which is characteristic for miniemulsification is expected to destroy all aggregates and liquid droplets. As a consequence, only hybrid particles composed of magnetite and monomer should remain, presumably since this species shows the highest stability. Polymerization of the monomer is then started for all the samples presented here by adding an initiator.

This structural or composition homogeneity is observed by electron microscopy. As shown in Figure 2.16, full encapsulation of the magnetite particles is obtained.

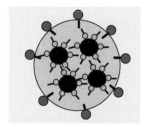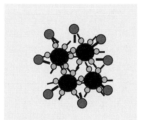

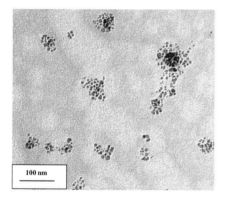

Figure 2.15 Magnetite aggregates obtained after a miniemulsion process in water.

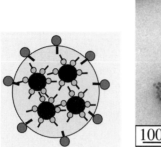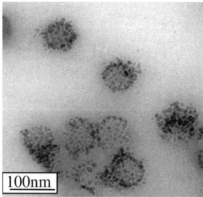

Figure 2.16 Transmission electron micrograph of magnetic polystyrene particles.

The small magnetite particles were found to be well separated, which means that each is presumably completely coated with a thin layer oleic acid and the entire aggregate is then covered with a layer of polymer. However, due to the presence of oleic acid, the polymer formed is rather soft.

By increasing the aggregate size, the density distribution becomes larger, indicating that homogeneous encapsulation of larger particles is more difficult. This is due to statistical reasons during the fusion/fission process: the magnetite aggregates should not be too large because only a few large aggregates may then face many styrene droplets. It should also be noted that magnetite particles of 10 nm are too small to be efficiently encapsulated. A situation where a similar number of magnetite aggregates and styrene droplets exists, and this therefore is favored for a homogeneous encapsulation process.

2.5
Nanocapsules with Non-Radical Polymerizations in Miniemulsion

As shown above, the miniemulsion polymerization is not limited to polymers obtained by radical polymerization. In the following section some examples will be provided for specific polymers.

Polyaddition in miniemulsion was also performed using chitosan as stabilizer with two biocompatible costabilizers, Jeffamine D2000 and Gluadin. A linking diepoxide in the presence of an inert oil results, via an interface reaction, in thin – but rather stable – nanocapsules. Since both water- and oil-soluble aminic costabilizers can be used, these experiments show the way to a great variety of capsules with different chemical structure. Such capsules are expected to be biocompatible and biodegradable, and may find applications in drug delivery [27].

Chitosan is a biodegradable, nontoxic and naturally occurring polymer of β-(1-4)-2-amino-2-deoxy-D-glucopyranose. It is prepared by the partial alkaline deacetyla-

a)   b)

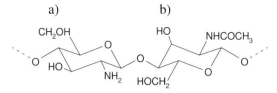

Figure 2.17 Chemical structure of chitosan consisting of:
(a) deacetylated and (b) actetylated units.

tion of chitin, a main structural component of the cuticles of insects, mollusks, and crustaceans [126]. The copolymer structure is depicted in Figure 2.17.

The amphiphilic biopolymer is insoluble in water at pH 7, but becomes soluble and positively charged in acidic media and can therefore be used either as a flocculating agent [127,128] or as a biosurfactant [129]. This polyelectrolyte has been successfully used to stabilize polymer nanoparticles of pMMA [130] and of poly(butyl cyanoacrylate) [131]. In these cases, the chitosan is thought to be grafted onto the particles by a hydrogen abstraction mechanism. These positively charged nanoparticles can be used as ideal candidates for the purification of proteins from a crude biological mixture [130]. If the pH is maintained above 7, the proteins carry a net negative charge and can therefore develop electrostatic interactions with the particles. These nanoparticles are also well suited as site-specific drug carriers [131].

Chitosan was also successfully used in the preparation of stable miniemulsions [27]. The surface tension values of the final dispersions prepared by the low molecular-weight chitosan were ca. 66 mN m^{-1}, indicating that in this case all the chitosan is mainly adsorbed onto the particles. As chitosan bears amine functions, it can be grafted onto the particles via a hydrogen abstraction mechanism [130,131]. However, chitosan alone as a stabilizer for nanometer-sized droplets is insufficient, and a large amount of coagulate is formed which might originated from the fact that the chitosan cannot protect the final polymer particles against collision. Therefore, the emulsification was supported by additional small amounts of other low molecular-weight surfactants after the miniemulsification process, or synthetic polymers were added to the oil phase.

An enhancement of stabilization was obtained by using CTMA-Cl, which allows high solid contents and the synthesis of very small particles with diameters <100 nm. However, CTMA-Cl was chosen as a cross-test to highlight the deficiencies of the system, since its use is prohibited by demands of biocompatibility and biodegradability.

The synthetic biocompatible polymer, Jeffamine D2000, which was shown earlier to have interfacial properties [24], was added to the monomer phase to ensure cationic stabilization of the "weak spots" during the polymerization process. In this case, high solid contents were obtained when adding only small amounts of Jeffamine (0.5 to 2.5 wt.% with respect to the monomer phase), and very small and monodisperse latexes in the size range ca. 100 to 200 nm without any coagulate could easily be synthesized. It is interesting to note that when the Jeffamine content exceeded 1%, the particle size saturated at a similar value of 100 nm. It is expected that the primary particle size would be stabilized throughout the process.

Chitosan was shown to be a biocompatible [132] and biodegradable cationic polyelectrolyte that could be used to prepare capsules for pharmaceutical and biomedical fields [133,134]. The capsules can be prepared in different ways. The first method is the coacervation technique, where a solution of chitosan is blown into a nonsolvent and the polymer precipitates at the surface of the droplets, thereby forming capsules [135]. The second method is the coating process [136], where the oil phase containing the product is blown into the chitosan solution, causing the polymer to precipitate at the surface of the droplets. The surface properties of chitosan were used to produce capsules by crosslinking chitosan at the interface [137–139], but the formed capsules were often quite large (from 500 nm to several micrometers) and the size distribution was wide.

Nanocapsules with diameters down to 100 nm and a cationic chitosan surface layer were also successfully generated using miniemulsion procedures [140]. By using hybridization and stabilization chemistry, it was possible to crosslink chitosan and its cosurfactant Jeffamine D2000 with diepoxides to capsule structures. Here, the amphiphilic hybrid copolymer can be built up *in situ* around the material to be encapsulated, as long as it does not interfere with the polyaddition process in miniemulsion. Previously synthesized chitosan capsules were quite large and not well defined, though the interfacial hybridization reaction in miniemulsions is expected to extend the accessible size range towards the nanocapsule region. A scheme of this interface reaction around the inert droplet is shown in Figure 2.18 [140].

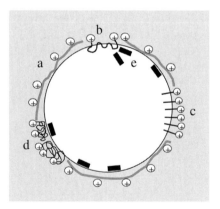

Figure 2.18 Schematic view of the formation of nanocapsules by interfacial reaction. Chitosan acts as a reactive biocompatible stabilizer from the water phase forming patches on the interface (a); a costabilizer such as oligomeric diamines (b) (e.g., Jeffamine D2000) or low molecular-weight cationic surfactants (c) (e.g., CTMA molecules) can improve the surface layer structure from the inside of the droplets and the coupled stabilization efficiency. A stabilization from the water phase can also be provided by a water-soluble, but amphiphilic protein (e.g., Gluadin) (d). A diepoxide (e) can additionally be used as stabilizing crosslinking agent [140].

Toluene containing the diepoxide, Epikote E828, and the Jeffamine was dispersed in the chitosan solution. The starting oil-phase products are miscible, and a stable miniemulsion can be formed. During the crosslinking reaction between the chitosan and the epoxide, phase separation between the toluene and the polymer product occurs, and in the case of appropriate spreading coefficients, as reported earlier, capsules are formed [73].

Free toluene could not be detected after the reaction and could not be separated by centrifugation, indicating that it was indeed encapsulated. As expected, the capsules in the presence of Jeffamine were small (126 nm) and well defined, but were rather large in its absence. It was not possible to obtain transmission electron micrographs of these samples, as the material constituting the shell is a low-T_g polymer and degrades in the electron beam. However, using diaminododecane as the amine component resulted in a polymer with higher T_g and a greater stability against electron degradation, whereupon the shells of nanocapsules could be depicted by TEM as empty hulls (Fig. 2.19) as the toluene was evaporated in the vacuum of the microscope.

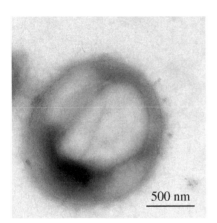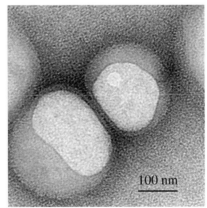

Figure 2.19 Transmission electron micrographs of the capsule preparation with chitosan [140].

A crosslinking/hybridization reaction was achieved by including another water-soluble, amphiphilic, biocompatible and more easily biodegradable amine, namely Gluadin APG. This is a partially hydrolyzed wheat gluten protein with a molecular weight of ca. 5000 g mol^{-1}. Toluene containing Epikote and Jeffamine D2000 was dispersed in a solution of chitosan and Gluadin using the conventional miniemulsion process. It is important to note that, in this case, the costabilizer was water-soluble and approached the droplet surface from the water phase. After the reaction, again no free toluene was found in the latexes, indicating efficient encapsulation. From NMR experiments it was estimated that about 90% of the Gluadin had reacted. Hence, the majority of the Gluadin had bound to the miniemulsion droplets, reacted with the oil-soluble diepoxy derivative, and bridged to the chitosan [140]. It is speculated that the chitosan, due to its rather stiff polysaccharide backbone, shows rather flat adsorption and omits larger unstabilized "patches" which, for reasons of packing, cannot be filled by the chitosan itself. This is why the addition of a second, more flexible component, such as CTMA-Cl (ideally, but not biocompatible), Jeffamine D2000 or Gluadin has such a profitable influence. Similar effects are known from microemulsions and are termed the "tree-grass" principle [141]. Problems of Gluadin arranging with chitosan at the droplet interface were clear from the data set where both Jeffamine D2000 and Gluadin had to be added to the reaction in order to obtain a Gaussian distribution of the nanocapsules without the formation of larger ones. Here, the particle size was dependent upon the amount of Gluadin added, as would be expected for linear costabilizer efficiency [140].

An additional point here is that polymer reactions on biopolymers occur at the relatively high internal surface areas of miniemulsions in a preoriented state (by the gradient of cohesion energy). Therefore, reaction in miniemulsions also allows both hydrophilic and hydrophobic modification of chitosan with rather high efficiency – for example, the otherwise rather complicated coupling or grafting with a polypeptide, as delineated by the coupling with Gluadin [140].

Butyl cyanoacrylate is also used quite often in the preparation of nanocapsules. In this case, the interfacial polymerization in a water-in-oil emulsion leads to nanocapsules [142,143] which can be used for the delivery of oligonucleotides [144,145] or for the subcutaneous and peroral delivery of octreotide, a long-acting somatostatin analogue [146].

2.6
Conclusions

The main aim of this chapter was to summarize and combine recent progress in the field of nanocapsules obtained by the miniemulsion process. It was shown that the use of high-shear, appropriate surfactants, and the addition of a hydrophobe in order to suppress the influence of Ostwald ripening represent key factors in the formation of small and stable droplets in miniemulsions. The strength of the miniemulsion process was shown to be the formation of polymeric nanocapsules consisting of polymers with encapsulated material. These structures are barely accessible using other types of heterophase polymerization. The formation of hybrid materials by the encapsulation of organic or inorganic solid materials or liquids are examples representing the wide applicability of miniemulsions in technologically relevant situations. The choice of polymer for the shell material was also shown to be highly versatile, and polymers obtained by radical polymerization, polyaddition, and enzymatic polymerization can all be used effectively in the encapsulation process. Clearly, the field of miniemulsion in the preparation of nanocapsules is an expanding area in polymer and materials science, with numerous additional possibilities both for fundamental research and applications.

Abbreviations

AA	acrylic acid
AAm	acrylamide
AIBN	2,2'-azo-bis-isobutyronitrile
CTMA-Cl	cetyltrimethylammonium chloride
DP	degree of polymerization
DSC	differential scanning calorimetry
EGDMA	ethylene glycol dimethacrylate
FTIR	Fourier transform infra-red
HLB	hydrophilic-lipophilic balance
MMA	methyl methacrylate
PBA	poly(butyl acrylate)
PGE	phenyl glycidyl ether
PMMA	poly(methyl methacrylate)
PS	polystyrene
PU	polyurethane
SDS	sodium dodecyl sulfate
SEM	scanning electron microscopy
Sty	styrene
TEM	transmission electron microscopy
T_g	glass transition temperature
TPP	T-6-P-phosphatase

References

1. J. Ugelstad, M.S. El-Aasser, J.W. Vanderhoff, *J. Polym. Sci., Polym. Lett. Ed.* **1973**, *11*, 503.
2. P.J. Blythe, E.D. Sudol, M.S. El-Aasser, *Macromol. Symp.* **2000**, *150*, 179.
3. F.J. Schork, G.W. Poehlein, S. Wang, J. Reimers, J. Rodrigues, C. Samer, *Colloids Surf. A: Physicochem. Eng. Asp.* **1999**, *153*, 39.
4. K. Landfester, *Macromol. Rapid Comm.* **2001**, *22*, 896.
5. K. Landfester, in: *Colloid Chemistry II* (Ed. M. Antonietti), Springer, Heidelberg, **2003**, p. 75.
6. K. Landfester, N. Bechthold, F. Tiarks, M. Antonietti, *Macromolecules* **1999**, *32*, 5222.
7. K.C. Lowe, *Art. Cells, Blood Subs., Immob. Biotech.* **2000**, *28*, 25.
8. K. Landfester, *Adv. Mater.* **2001**, *10*, 765.
9. M. Antonietti, K. Landfester, *Prog. Polym. Sci.* **2002**, *27*, 689.
10. M. Willert, Ph.D. Thesis, Universität Potsdam, **2001**.
11. F. Candau, in: *Polymerization in Organized Media* (Ed. E.C. Paleos), Gordon and Breach Science Publisher, Philadelphia, **1992**, p. 215.
12. F. Candau, *Macromol. Symp.* **1995**, *92*, 169.
13. D.J. Hunkeler, J. Hernandez-Barajas, in: *Polymeric Materials Encyclopedia, Vol. 9* (Ed. J.C. Salamone), CRC Press, New York, **1996**, p. 3322.
14. K. Landfester, R. Rothe, M. Antonietti, *Macromolecules* **2002**, *35*, 1658.
15. J. Delgado, M.S. EL-Aasser, C.A. Silibi, J.W. Vanderhoff, *J. Polym. Sci., Polym. Chem. Ed.* **1990**, *28*, 777.
16. S. Wang, F.J. Schork, *J. Appl. Polym. Sci.* **1994**, *54*, 2157.
17. X.Q. Wu, F.J. Schork, *Indust. Eng. Chem. Res.* **2000**, *39*, 2855.

18. R.C. Dewald, L.H. Hart, W.F. Carroll, Jr., *J. Polym. Sci., Polym. Chem. Ed.* **1984**, *22*, 2923.

19. K. Landfester, M. Willert, M. Antonietti, *Macromolecules* **2000**, *33*, 2370.

20. V.S. Rodriguez, M.S. El-Aasser, J.M. Asua, C.A. Silibi, *J. Polym. Sci., Polym. Chem. Ed.* **1989**, *27*, 3659.

21. H. Huang, H. Zhang, J. Li, S. Cheng, F. Hu, B. Tan, *J. Appl. Polym. Sci.* **1998**, *68*, 2029.

22. J. Reimers, F.J. Schork, *Polym. Reaction Eng.* **1996**, *4*, 135.

23. M. Willert, K. Landfester, *Macromol. Chem. Phys.* **2002**, *203*, 825.

24. K. Landfester, F. Tiarks, H.-P. Hentze, M. Antonietti, *Macromol. Chem. Phys.* **2000**, *201*, 1.

25. F. Tiarks, K. Landfester, M. Antonietti, *J. Polym. Sci., Polym. Chem. Ed.* **2001**, *39*, 2520.

26. M. Barrere, K. Landfester, *Macromolecules* **2003**, *36*, 5119.

27. E. Marie, K. Landfester, M. Antonietti, *Biomacromolecules* **2002**, *3*, 475.

28. C. Maitre, F. Ganachaud, O. Ferreira, J. F. Lutz, Y. Paintoux, P. Hemery, *Macromolecules* **2000**, *33*, 7730.

29. F.M. Bauers, R. Thomann, S. Mecking, *J. Am. Chem. Soc.* **2003**, *125*, 8838.

30. A. Tomov, J.P. Broyer, R. Spitz, *Macromol. Symp.* **2000**, *150*, 53.

31. A. Taden, M. Antonietti, K. Landfester, *Macromol. Rapid Comm.* **2003**, *24*, 512.

32. J. Strauss, *Surf. Coat. Aust.* **1987**, *24*, 6.

33. A. Kowalski, M. Vogel, R.M. Blankenship, US Patent 4,427,836, **1984**.

34. T. Kaji, P. Kami, *Gikyoshi* **1992**, *46*, 271.

35. C. Tasset, N. Barrete, S. Thysman, J.M. Ketelegers, D. Lemoine, V. Preat, *J. Controlled Release* **1995**, *33*, 23.

36. C. Damge, C. Michel, M. Aprahamian, P. Couvreuer, J. Devissaguet, *J. Controlled Release* **1990**, *13*, 233.

37. T.M.S. Chang, *Eur. J. Pharm. Biopharm.* **1998**, *45*, 3.

38. M. Gallardo, G. Couarraze, B. Denizot, J. Treupel, P. Couvreur, F. Puisieux, *Int. J. Pharm.* **1993**, *100*, 55.

39. F. Chouinard, F. W. Kann, J.-C. Leroux, C. Foucher, V. Lenaerts, *Int. J. Pharm.* **1991**, *72*, 211.

40. N. Al-Khouri-Fallouh, L. Roblot-Treupel, H. Fessi, J.P. Devissaguet, F. Puisieux, *Int. J. Pharm.* **1986**, *28*, 125.

41. S. Torza, S.G. Mason, *J. Coll. Interf. Sci.* **1970**, *33*, 6783.

42. J. Berg, D. Sundberg, B. Kronberg, *Polym. Mater. Sci. Eng.* **1986**, *54*, 367.

43. J. Berg, D. Sundberg, B.J. Kronberg, *Microencapsulation* **1989**, *3*, 327.

44. D.C. Sundberg, A.P. Casassa, J. Pantazopoulos, M.R. Muscato, *J. Appl. Polym. Sci.* **1990**, *41*, 1425.

45. Y.-C. Chen, V. Dimonie, M.S. El-Aasser, *J. Appl. Polym. Sci.* **1991**, *42*, 1049.

46. M. Okubo, A. Yamada, T. Matsumoto, *J. Polym. Sci., Polym. Chem. Ed.* **1980**, *16*, 3219.

47. S. Muroi, H. Hashimoto, K. Hosoi, *J. Polym. Sci., Polym. Chem. Ed.* **1984**, *22*, 1365.

48. M. Okubo, K. Kanaida, T. Matsumoto, *Colloid Polym. Sci.* **1987**, *265*, 876.

49. M. Okubo, Y. Katsuta, T. Matsumoto, *J. Polym. Sci., Poly. Lett. Ed.* **1982**, *20*, 45.

50. S. Lee, A. Rudin, *J. Polym. Sci., Polym. Chem. Ed.* **1992**, *30*, 2211.

51. A. Kowalski, R. Blankenship, US Patent 4,468,498, **1984**.

52. A. Kowalski, M. Vogel, US Patent 4,469,825, **1984**.

53. A. Kowalski, M. Vogel, US Patent 4,880,842, **1989**.

54. R.M. Blankenship, A. Kowalski, Patent 4,594,363, **1986**.

55. K.K. Nippon-Zeon, Japanese Patent 052779409 A, **1993**.

56. K.K. Nippon-Zeon, Japanese Patent 07021011, **1995**.

57. M. Kaino, Y. Takagishi, H. Toda, US Patent, 5,360,827, **1994**.

58. C.J. McDonald, K.J. Bouck, A.B. Chaput, C.J. Stevens, *Macromolecules* **2000**, *33*, 1593.

59. M. Wohlgemuth, W. Mächtle, C.C. Mayer, *J. Microencapsulation* **2000**, *17*, 437.

60. A.T. Florence, T.L. Whateley, D.A. Wood, *J. Pharm. Pharmacol.* **1979**, *31*, 422.

61. M.S. El-Samaligy, P. Rohdewald, H.A. Mahmoud, *J. Pharm. Pharmacol.* **1986**, 216.

62. F. Lescure, C. Zimmer, D. Roy, P. Couvreur, *J. Colloid Interface Sci.* **1992**, *154*, 77.

63. M. Okubo, H. Minami, Y. Ynamoto, *Coll. Surf. A* **1999**, *153*, 405.

64. M. Okubo, M. Shiozaki, M. Tsujihiro, Y. Tsukuda, *Colloid Polym. Sci.* **1991**, *269*, 222.

65. N. Itou, T. Masukawa, I. Ozaki, M. Hattori, K. Kasai, *Coll. Surf. A* **1999**, *153*, 31.

66. S. Stewart, G.J. Liu, *Chem. Mater.* **1999**, *11*, 1048.

67. C. Nardin, T. Hirt, J. Leukel, W. Meier, *Langmuir* **2000**, *16*, 1035.

68. S.M. Marinakos, J.P. Novak, L.C. Brousseau, A.B. House, E.M. Edeki, J.C. Feldhaus, D.L. Feldheim, *J. Am. Chem. Soc.* **1999**, *121*, 8518.

69. G.E. Hildebrand, J.W. Tack, *Int. J. Pharm.* **2000**, *196*, 173.

70. A. Rudin, *Macromolec. Symp.* **1995**, *92*, 53.

71. K. Landfester, C. Boeffel, M. Lambla, H. W. Spiess, *Macromolecules* **1996**, *29*, 5972.

72. K. Landfester, H.W. Spiess, *Acta Polym.* **1998**, *49*, 451.

73. F. Tiarks, K. Landfester, M. Antonietti, *Langmuir* **2001**, *17*, 908.

74. A.J.P. van Zyl, D. de Wet-Roos, R.D. Sanderson, B. Klumperman, *Eur. Polymer J.* **2004**, *40*, 2717.

75. S.J. Park, K.S. Kim, S.K. Hong, *Polymer-Korea* **2005**, *29*, 8.

76. I. Rajot, S. Bone, C. Graillat, T. Hamaide, *Macromolecules* **2003**, *36*, 7484.

77. F. Ganachaud, J.L. Katz, *Chemphyschem* **2005**, *6*, 209.

78. S.A. Vitale, J.L. Katz, *Langmuir* **2003**, *19*, 4105.

79. K.H. Baek, J.Y. Lee, S.H. Hong, J.H. Kim, *Polymer-Korea* **2004**, *28*, 404.

80. B.Z. Putlitz, K. Landfester, H. Fischer, M. Antonietti, *Adv. Mater.* **2001**, *13*, 500.

81. J. Berg, D. Sundberg, B. Kronberg, *J. Microencapsulation* **1989**, *6*, 327.

82. W. Ramsden, *Proc. Roy. Soc. London* **1903**, *72*, 156.

83. S.U. Pickering, *J. Chem. Soc. Commun.* **1907**, *91*, 2001.

84. T.R. Briggs, *Ind. Eng. Chem. Prod. Res. Dev.* **1921**, *13*, 1008.

85. F. Tiarks, K. Landfester, M. Antonietti, *Langmuir* **2001**, *17*, 5775.

86. C. Barthet, A.J. Hickey, D.B. Cairns, S.P. Armes, *Adv. Mater.* **1999**, *11*, 408.

87. M.J. Percy, C. Barthet, J.C. Lobb, M.A. Khan, S.F. Lascelles, M. Vamvakaki, S.P. Armes, *Langmuir* **2000**, *16*, 6913.

88. G.J.D.A. Soler-Illia, C. Sanchez, *New J. Chem.* **2000**, *24*, 493.

89. M. Zhang, G. Gao, C. Q. Li, F. Q. Liu, *Langmuir* **2004**, *20*, 1420.

90. B. Erdem, E.D. Sudol, V.L. Dimonie, M.S. El-Aasser, *J. Polym. Sci., Polym. Chem.* **2000**, *38*, 4419.

91. B. Erdem, E.D. Sudol, V.L. Dimonie, M.S. El-Aasser, *J Polym. Sci., Polym. Chem. Ed.* **2000**, *38*, 4431.

92. B. Erdem, E.D. Sudol, V.L. Dimonie, M.S. El-Aasser, *Macromol. Symp.* **2000**, *155*, 181.

93. B. Erdem, E.D. Sudol, V.L. Dimonie, M.S. El-Aasser, *J Polym. Sci., Polym. Chem. Ed.* **2000**, *38*, 4441.

94. H. Oh, E. Kim, in: *On the Convergence of Bio-Information-, Environmental-, Energy-, Space- and Nano-Technologies*, Parts 1 and 2, Vol. 277-279, Trans Tech Publications Ltd, Zurich-Uetikon, **2005**, pp. 1029.

95. N. Bechthold, F. Tiarks, M. Willert, K. Landfester, M. Antonietti, *Macromol. Symp.* **2000**, *151*, 549.

96. S.D. Seul, S.R. Lee, Y.H. Kim, *J. Polym. Sci. Part A-Polym. Chem.* **2004**, *42*, 4063.

97. S. Lelu, C. Novat, C. Graillat, A. Guyot, E. Bourgeat-Lami, *Polymer Int.* **2003**, *52*, 542.

98. Z.Z. Yang, D. Qiu, J. Li, *Macromol. Rapid Commun.* **2002**, *23*, 479.

99. F. Tiarks, K. Landfester, M. Antonietti, *Macromol. Chem. Phys.* **2001**, *202*, 51.

100. C. Bergemann, D. Müller-Schulte, J. Oster, L.A. Brassard, A.S. Lübbe, *J. Magn. Magn. Mater.* **1999**, *194*, 45.

101. J. Roger, J.N. Pons, R. Massart, A. Halbreich, J.C. Bacri, *Eur. Phys. J. Appl. Phys.* **1999**, *5*, 321.

102. E. Viroonchatapan, M. Ueno, H. Sato, I. Adachi, H. Nagae, K. Tazawa, I. Horikoshi, *Pharmaceut. Res.* **1995**, *12*, 1176.

103. A.S. Lübbe, C. Bergemann, W. Huhnt, T. Fricke, H. Riess, J.W. Brock, D. Huhn, *Cancer Res.* **1996**, *56*, 4694.

104. D.K. Kim, Y. Zhang, W. Voit, K.V. Rao, J. Kehr, B. Bjelke, M. Muhammed, *Scripta Mater.* **2001**, *44*, 1713.

105. M. Mitsumori, M. Hiraoka, T. Shibata, Y. Okuno, Y. Nagata, Y. Nishimura, M. Abe, M. Hasegawa, H. Nagae, Y. Ebisawa, *Hepato-gastroenterol.* **1996**, *43*, 1431.

106. S.S. Papell, US Patent No. 3215572, **1965**.

107. G.W. Reimers, S.E. Khalafalla, US Patent No. 3843540, **1974**.

108. J.R. Kelley, US Patent No. 4019994, **1977**.

109. J. Shimoiizaka, K. Nakatsuka, T. Fujita, A. Kounosu, in: *Proceedings of the international symposium on fine particles processing, Vol. 2* (Ed. P. Somasundaran), New York, **1980**.

110. R.S. Molday, US Patent No. 4452773, **1984**.

111. A. Rembaum, US Patent No. 4,267,234, **1981**.

112. J. Lee, T. Isobe, M. Senna, *Colloid Surf. A* **1996**, *109*, 121.

113. M. Suzuki, M. Shinkai, M. Kamihira, T. Kobayashi, *Biotechnol. Appl. Biochem.* **1995**, *21*, 335.

114. I. Dumazet-Bonnamour, P.L. Perchec, *Colloid Surf. A* **2000**, *173*, 61.

115. V. Veiga, D.H. Ryan, E. Sourty, F. Llanes, R.H. Marchessault, *Carbohydr. Polym.* **2000**, 353.

116. F. Sauzedde, A. Elaïssari, C. Pichot, *Colloid Polym. Sci.* **1999**, *277*, 846.

117. V.S. Zaitsev, D.S. Filimonov, I.A. Presnyakov, R.J. Gambino, B. Chu, *J. Colloid Interf. Sci.* **1999**, *212*, 49.

118. P.A. Dresco, V.S. Zaitsev, R.J. Gambino, B. Chu, *Langmuir* **1999**, *15*, 1945.

119. J.-C. Daniel, J.-L. Schuppiser, M. Tricot, US Patent No. 4358388, **1982**.

120. D. Charmot, C. Vidil, US Patent No. 5356713, **1994**.

121. K. Wormuth, *J. Colloid Interf. Sci.* **2001**, *241*, 366.

122. Z.Z. Xu, C.C. Wang, W.L. Yang, Y.H. Deng, S.K. Fu, *J. Magn. Magn. Mater.* **2004**, *277*, 136.

123. J.M. Singer, in: *NATO ASI Series E, Vol. 138* (Eds. M.S. El-Aasser, R.M. Fitch), N. Nijhoff Publ., Dordrecht, The Netherlands, **1987**, p. 371.

124. D. Hoffmann, K. Landfester, M. Antonietti, *Magnetohydrodynamics* **2001**, *37*, 217.

125. L.P. Ramirez, K. Landfester, *Macromol. Chem. Phys.* **2003**, *204*, 22.

126. G.A.F. Roberts, *Chitin Chemistry*, Macmillan Press, London, **1992**.

127. M. Ashmore, J. Hearn, *Langmuir* **2000**, *16*, 4906.

128. M. Ashmore, J. Hearn, F. Karpowicz, *Langmuir* **2001**, *17*, 1069.

129. L.F.D. Blanco, M.S. Rodriguez, P.C. Schultz, E. Agullo, *Colloid Polym. Sci.* **1999**, *277*, 1087.

130. C.S. Chern, C.K. Lee, C.C. Ho, *J. Polym. Sci., Polym. Chem.* **1999**, *37*, 1489.

131. S.C. Yang, H.X. Ge, Y. Hu, X. Q. Jiang, C.Z. Yang, *Colloid Polym. Sci.* **2000**, *278*, 285.

132. M.L. Weiner, *Advances in Chitin and Chitosan*, Elsevier Science Publishers Ltd., London, **1993**.

133. M.N.V.R. Kumar, N. Kumar, *Drug Dev. Ind. Pharm.* **2001**, *27*, 1.

134. K.D. Yao, T. Peng, Y.J. Yin, M.X. Xu, *J. M. S.-Rev. Macromol. Chem. Phys.* **1995**, *C35*, 155.

135. R.H. Chen, M.L. Tsaih, *J. Appl. Polym. Sci.* **1997**, 161.

136. P. Calvo, C. Remunan-Lopez, J.L. Vila-Jato, M.J. Alonso, *Colloid Polym. Sci.* **1997**, *275*, 46.

137. T.K. Saha, K. Jono, H. Ichikawa, Y. Fukumori, *Chem. Pharm. Bull.* **1998**, *46*, 537.

138. C.K. Yeom, S.B. Oh, J.W. Rhim, J.M. Lee, *J. Appl. Polym. Sci.* **2000**, *78*, 1645.

139. I. Genta, P. Perugini, B. Conti, F. Pavanetto, *Int. J. Pharm.* **1997**, *152*, 237.

140. E. Marie, R. Rothe, M. Antonietti, K. Landfester, *Macromolecules* **2003**, *36*, 3967.

141. M. Antonietti, R. Basten, S. Lohmann, *Macromol. Chem. Phys.* **1995**, *196*, 441.

142. C. Limouzin, A. Caviggia, F. Ganachaud, P. Hemery, *Macromolecules* **2003**, *36*, 667.

143. S. Li, Z. Y. Qian, X.L. Liu, X.B. Liu, *Acta Polymerica Sinica* **2003**, 679.

144. G. Lambert, E. Fattal, H. Pinto-Alphandary, A. Gulik, P. Couvreur, *Int. J. Pharmaceutics* **2001**, *214*, 13.

145. G. Lambert, E. Fattal, H. Pinto-Alphandary, A. Gulik, P. Couvreur, *Pharm. Res.* **2000**, *17*, 707.

146. C. Damge, J. Vonderscher, P. Marbach, M. Pinget, *J. Pharm. Pharmacol.* **1997**, *49*, 949.

3

Microcapsules through Layer-by-Layer Assembly Technique

Christophe Déjugnat, Dmitry G. Shchukin, and Gleb B. Sukhorukov

3.1
Introduction

Self-assembly of oppositely charged polyelectrolytes is a powerful tool for the fabrication of multilayer flat thin films by the Layer-by-Layer (LbL) technique [1,2]. During the past few years, many studies have been conducted by applying the LbL technique for the covering of colloidal particles (Fig. 3.1). Subsequent dissolution of the sacrificial core leads to polyelectrolyte microcapsules [3].

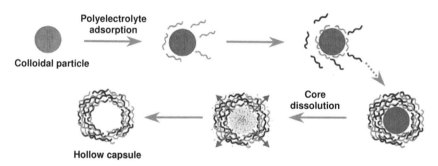

Figure 3.1 Schematic illustration of the polyelectrolyte deposition process and of subsequent core decomposition.

These hollow microcapsules have been extensively studied (permeability of the wall [4], mechanical properties [5,6], etc.) during the past few years, and different cores as well as different polyelectrolytes were used [7,8].

In this chapter we will focus on two emerging, rapidly growing points in the field of microcapsules: (i) stimuli-responsive capsules; and (ii) composite polyelectrolyte/inorganic capsules. Stimuli-responsive polymers are sensitive to specific changes in their environment and can undergo reversible sharp physical or chemical modifications in response to such changes. They have been widely used for

controlled drug delivery, in tissue engineering, in the fabrication of sensors and actuators, and for chromatography [9]. Stimuli-responsive polyelectrolytes can also
be used to design responsive microcontainers. In parallel, the use of microcapsules
as microreactors in the synthesis of inorganic nanoparticles is currently under development. The introduction of such inorganic material into the capsule structure
provides the capsules with new specific properties and potential applications in the
field of catalysis, for example.

3.2
Responsive Multilayers on Planar Substrates

The pH-sensitivity of polyelectrolyte multilayers (PEM) composed of weak polyelectrolytes has been widely investigated. The linear charge density along the polymer chains depends on the pH, and affects the interactions between the polyelectrolytes. For example, multilayers in which the weak polycation poly(allylamine hy-
drochloride) (PAH) was associated to the strong polyanion poly(sodium styrene
sulfonate) (PSS) undergo reversible swelling and porosity transitions, upon changing the pH [10]. The thickness and stability of the films were always determined by
the pH of the dipping solution. Other pH-sensitive films have been prepared, containing always at least one weak polyelectrolyte, such as poly(carboxylic acid)
[11–14]. The reversible formation of hydrogen bonds, depending on the pH, was also used to control the properties of films made from poly(acrylic acid) (PAA) and
poly(acrylamide) (PAAM) [15]. Other systems have been investigated, such as
poly(methacrylic acid) (PMA)/poly(vinylpyrrolidone) (PVPON) [16] or PAA/
poly(ethylene oxide) (PEO), especially for the preparation of highly ion-conductive
multilayers [17]. The well-known thermoresponsive poly(N-isopropylacrylamide)
(PNIPAM) was used in temperature-responsive thin films. In this way, Caruso et
al. [18] combined PNIPAM and PAA, via hydrogen bond stabilization, to produce
"smart" films with a tunable release of adsorbed substances as a function of the
temperature. Moreover, most of the biocompatible polyelectrolytes are weak polymers and can be used for pH-sensitive PEM with potential biological properties,
such as the control of cell adhesion [19]. Hyaluronic acid (HA) has been used in
combination with PAH or poly(lysine) (PLL) for pH-controlled loading and the release of small hydrophilic molecules [20,21]. As many weak polycations and polyanions are available commercially, or are easy to synthesize, it is possible to design a
broad variety of pH-responding systems along a wide pH range in aqueous solutions.

Other sensitive PEMs have been described which respond to different stimuli.
Photosensitive multilayers were prepared using polyelectrolytes functionalized
with photoisomerizable azobenzene chromophores. Barrett et al. have prepared
sensors, the mechanical properties of which change under light irradiation [22].
Solvent-responsive polymer brushes have been prepared by anchoring two types of
polymers (one hydrophobic and one hydrophilic) on a wafer surface. As a function

of the polarity of the solvent in which the system was immersed, the soluble chains extended while the insoluble ones collapsed [23].

3.3
Microcapsules

The use of stimuli-responsive polyelectrolytes as constituents of thinfilms on microcapsules leads to the production of responsive microcontainers that may be used for controlled drug delivery, or as micropumps, sensors, and actuators. The stimuli-responsive capsules, which contain sensitive polymers either as shell constituents or in their interior, are the subject of an intensively growing area of interest.

3.3.1
Switchable Permeability of Hollow Microcapsules:
A Tool for Macromolecule Encapsulation

Hollow polyelectrolyte microcapsules usually exhibit selective permeability: the shell is permeable to small molecules such as dyes, but remains impermeable to polymers [24]. However, as in the case of flat films, the polyelectrolyte multilayers can undergo porosity transitions when some parameters that influence the degree of association of the polymer chains are varied. The main application of this reversible permeability is the encapsulation of macromolecules and, furthermore, their controlled release.

3.3.1.1 Salt-Sensitive Permeability of Microcapsules
The most extensively studied (PSS/PAH) capsules, which are templated on melamine-formaldehyde (MF) spherical microparticles, are impermeable to macromolecules with a molecular weight greater than 4 kg mol^{-1} [24]. Ibarz et al. observed that a 0.1 M NaCl solution can render (PSS/PAH)$_4$ capsules permeable to macromolecules, such as PAH (70 kg mol^{-1}) [25]. The critical concentration of NaCl required to open the pores was found to be about 0.01 M. This salt-induced permeability can be explained by means of screening the electrostatic interactions within the multilayers. Washing with water closed the pores – due to the decrease in ionic strength – and allowed entrapment of the polymer. The PAH can be further released by a similar treatment with salt. However, the filling is low (3 mM), and this might be a limitation for this type of smart encapsulation. In addition, the use of MF templates for the fabrication of the capsules may complicate the system because residual MF oligomers produced during acidic degradation of the core form a gel-like structure inside the capsule [26]. This matrix can strongly influence the loading of the capsule due to electrostatic interactions (the positive MF oligomers can limit the penetration of the polycation PAH) and adsorption.

The salt-sensitivity of microcapsules was also used recently by Gao et al. for the encapsulation of dextran inside capsules made from PSS and poly(diallyl dimethyl ammonium chloride) (PDADMAC) templated on MF particles [27]. In contrast to (PSS/PAH) capsules, these shells were initially permeable to high molecular-weight dextran (2000 kg mol⁻¹). This higher permeability reflects the strong mechanical stress endured by the capsules during the core dissolution, primarily due to the osmotic pressure (formation of charged MF oligomers) and the rigidity of the (PSS/PDADMAC) film (stiffer and less flexible than the (PSS/PAH) version). The addition of salt reduced the permeability due to a partial annealing of the pores, and this caused the capsules to shrink. This process was used to seal capsules in the presence of the polymer to be encapsulated. However, washing the loaded capsules with pure water solution caused an immediate deformation and swelling of the structures, accompanied by release of the dextran.

3.3.1.2 Reversible pH-Induced Permeability

The permeability of hollow $(PSS/PAH)_4$ microcapsules templated on MF cores for labeled dextran was studied as a function of the pH [28]. The capsules are in an "open state" for pH below 6.5, and are closed at pH values above 7.5 (Fig. 3.2). Such behavior might be explained by the pH-response of PAH. The capsules were prepared at pH ca. 7.5, at which the PAH is not completely deprotonated. The neutral amino groups do not participate electrostatically to the polyelectrolyte complexes. When reducing the pH, those groups became protonated and created a local excess of positive charges; the electrostatic repulsions may destabilize the multilayers and form defects or pores in the shells. This corresponds to the "open state" observed at low pH. Increasing of pH deprotonated the excess ammonium groups, reduced the electrostatic repulsion, and finally led to closing of the pores (capsules in "closed state"). This opening-closing process was conducted in the presence of dif-

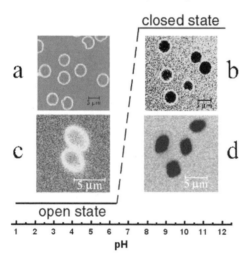

Figure 3.2 Encapsulation of FITC-dextran (75 kg mol⁻¹) into $(PSS/PAH)_4$ microcapsules prepared on melamine-formaldehyde (MF) particles (a,b) and $CdCO_3$ crystals (c,d). pH = 3.5 for (a, c) and pH = 10 for (b,d). (Reproduced with permission from [29]; © Elsevier B.V., 2002.)

ferent polymers, including dextrans and bovine serum albumin (BSA), all of which were finally encapsulated inside the microcapsules.

The presence of MF gel inside the capsules can participate in the formation of pores in acidic conditions (partial degradation and release of residual MF material). In addition, this gel can promote the accumulation of polymers by adsorption. The experiment was then reproduced with (PSS/PAH) capsules templated on fully decomposable $CdCO_3$ cores, in order to eliminate a possible effect of the MF material [29]. The results were reproducible and encapsulation was successful using true capsules, showing that the (PSS/PAH) pair is a usable pH-responsive system.

(PAH/PSS) capsules templated on various cores were studied as a function of the pH [30]. In this case, the PAH was adsorbed in a fully charged state and therefore no change was observed when immersing the capsules in acidic solutions, in contrast to the system described above. However, under basic conditions a swelling of the capsules was observed. Capsules templated on inorganic particles ($MnCO_3$, $CaCO_3$) immediately dissolved when the pH reached values above 11.5, with an almost uncharged PAH. The remaining negative charges of the PSS become uncompensated, leading to strong electrostatic repulsions between the polymeric chains and therefore to an expansion of the capsule (Fig. 3.3a). Stabilization of the swollen state was achieved when using polystyrene (PS) particles as templates, and the capsules did not dissolve immediately. Tetrahydrofuran (THF), which is used for the core dissolution, may partially dehydrate the polyelectrolyte multilayers, making them more compact and more stable. Reducing the pH led to shrinkage of the capsules to a slightly smaller size than the initial one, due to possible re-

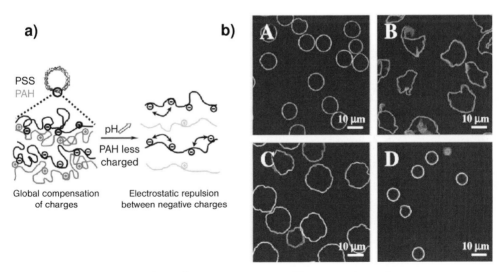

Figure 3.3 (a) (PAH/PSS) capsules swell in alkaline medium due to electrostatic repulsions between negative charges of PSS. (b) pH-induced swelling and shrinking of these capsules (A) in water; (B) at 1 sec after 0.1 M NaOH addition; (C) at 1 min after NaOH addition; and (D) after 0.1 M HCl addition. (Reproduced with permission from [30]; © American Chemical Society, 2004.)

arrangement of the polymeric network (Fig. 3.3b). Several swelling–shrinking cycles could be achieved, thereby providing the reversible character of this pH-responsive system.

In the swollen state, these capsules were permeable to macromolecules, and this was used for the pH-controlled encapsulation of PAA (Fig. 3.4A), when later reducing the pH [31]. The release of encapsulated polymer is achieved by swelling the filled capsules in alkaline medium, which allows the amount of entrapped material to be measured. This amount can be controlled, and is dependent upon on the initial concentration of the surrounding PAA (Fig. 3.4B).

Polyelectrolyte (PE) complexes can also be stabilized using weakly soluble polyelectrolytes in their uncharged form. Microcapsules containing only weak polyelectrolytes were prepared using PAH and PMA, templated on $CaCO_3$ particles

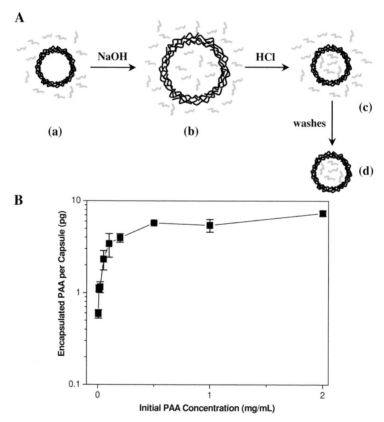

Figure 3.4 (A) General scheme for the encapsulation. Hollow capsules impermeable to poly(acrylic acid) (PAA) (a) are swollen in alkaline medium and the polymer penetrates inside the capsules (b). Reduction of the pH makes the capsules shrink, with entrapment of the polyanion (c), and subsequent washes lead to loaded capsules (d). (B) Quantities of encapsulated PAA per capsule as a function of polymer concentration during the encapsulation.

[32]. The capsules show a swelling in acid conditions, related to a near-complete protonation of the carboxylic groups in the PMA chains and repulsion between the remaining PAH chains. The process was shown to be completely reversible, mainly due to the accurate definition of the system, making it completely tunable.

3.3.1.3 Solvent-Induced Encapsulation

Organic solvents can also induce permeation of macromolecules through the polyelectrolyte shell. Using a 1:1 ethanol:water mixture, Lvov et al. encapsulated urease (5 nm-diameter globules) in (PSS/PAH) capsules templated on MF particles. The pores were closed by treatment with water when washing out the ethanol [33]. Ethanol might partially remove the hydration water between the polyelectrolytes, leading to a segregation of the polyion network and the formation of pores. Encapsulated urease retained a certain activity although it was reduced compared with free urease in bulk solution.

3.3.1.4 Temperature-Sensitive Hollow Capsules

In order to obtain temperature-responsive capsules, the thermosensitive PNIPAM can be used. This polymer is soluble in water if the temperature does not exceed 32 °C, and precipitates in aqueous solutions at higher temperatures. PNIPAM was introduced during the synthesis of both anionic and cationic block-co-polymers, and the LbL technique was then applied on MF particles leading to hollow capsules after dissolution of the core [34]. Upon annealing, the capsules shrank and the permeability of the shell was decreased, as measured by fluorescence recovery after photobleaching (FRAP) measurements using fluorescent 6-carboxyfluorescein and confocal microscopy. When cooling, the capsules became more permeable with time, but they regained only about 30% of their initial permeability after two weeks at 20 °C. These capsules did not indicate any clear thermoreversibility, but this may be a limitation of the system.

3.3.2
Microcapsules Loaded with "Smart" Polymers

These systems are based on the sensitivity of encapsulated stimuli-responsive polymers. In this case, the response of the material may also affect the microcapsule itself, depending on the interactions between the loaded substance and the polymeric microcontainer.

Encapsulated PAA, introduced inside (PAH/PSS) capsules by changing their permeability with pH, remains sensitive to divalent ions such as Ca^{2+} and the formation of an insoluble complex in the interior of the capsules was observed (Fig. 3.5) when these loaded capsules were immersed in an aqueous Ca^{2+} solution [31].

The complexation is reversible as the addition of acid dissolves the complex with release of ions and re-solubilization of the polymer inside the capsule.

Another way to introduce polymers inside the capsules was described by Dähne et al. [35] with the so-called "ship in a bottle" synthesis, applied to PSS. The resulting loaded capsules became swollen under basic conditions [36]. The diameter decreased

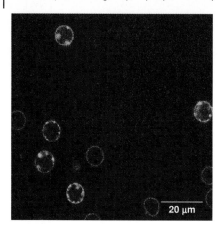

Figure 3.5 Confocal laser scanning micrograph of reversible precipitates of Ca^{2+}/PAA complex formed inside capsules filled with poly(acrylic acid) (PAA).

by treating the swollen capsules with acid, but this system did not show good reversibility, for two reasons. First, the PSS was released almost completely during the first swelling due to increased permeability of the capsule wall. Second, the use of MF is again critical because PSS and MF oligomers can interact strongly. During the release of PSS, much of the MF oligomers may be released at the same time, and consequently the composition and the structure of the capsule may be modified. In addition, it has been shown that hollow (PSS/PAH) capsules themselves are sensitive to pH, depending on their preparation mode. The response of the capsule cannot be neglected, and may compete with the response of the smart encapsulated polymer.

Encapsulation of PNIPAM inside non-temperature-sensitive (PSS/PAH) capsules templated on PS cores was achieved by the same "ship in a bottle" synthesis [37], providing temperature-responsive microcapsules. Encapsulated PNIPAM collapsed when the temperature of the system increased above 34 °C, forming discrete particles inside the capsule, without modification of the shell (Fig. 3.6). This prop-

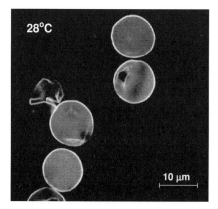 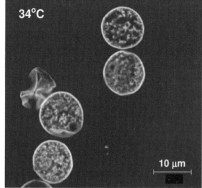

Figure 3.6 Reversible temperature-induced precipitation of entrapped labeled poly(N-isopropylacrylamide) (PNIPAM) inside (PAH/PSS)$_6$ polyelectrolyte microcapsules templated on polystyrene cores. (Reproduced from [37].)

erty was completely reversible, and below the critical temperature the polymer was again soluble, homogeneously filling the interior of the capsule. This situation was very interesting because the polarity of the polymer changes during this transition: in this way, hydrophilic substances adsorbed onto the polymer in its soluble state should be released from the capsule during collapse of the PNIPAM. The resulting system could be used as a drug carrier with controlled release of the drug by increasing the temperature.

3.4
Composite Polyelectrolyte/Inorganic Microcapsules

3.4.1
Synthesis

3.4.1.1 Growing Inorganic Particles Using Encapsulated Polyelectrolytes
Nanoengineering of a polyelectrolyte shell becomes possible by applying the LbL approach, which affords facile, cheap and environmentally friendly processing of both simple and complex polyelectrolyte/inorganic nanostructures. Therefore, by choosing the correct assembling conditions, a polyelectrolyte film can be fabricated with desired properties that are advantageous for synthesizing composite nanomaterials inside. Multilayer thin films of adsorbed polyelectrolytes were utilized as nanoreactors for both metallic (Ag, Ni, Pd, Cu) and semiconductor nanoparticles (PbS) [38–43]. Polyelectrolyte multilayers with a controlled content of free carboxylic acid-binding groups were fabricated with weak polyelectrolytes (PAH and PAA). These groups were used to bind various inorganic ions (Ag^+, $Pd(NH_3)_4^{2+}$, $PdCl_4^{2-}$, Cu^{2+}), which were then converted into nanoparticles by a reducing agent. Spatial control (on the nanoscale) over the growth of the nanoparticles was achieved by the use of multilayer heterostructures containing bilayer blocks that are not able to bind inorganic ions [44]. The resulting composite polyelectrolyte/inorganic films are highly uniform with controlled supramolecular organization and molecular environment of the nanoparticles.

Selective permeability of the capsule wall can lead to the establishment of a concentration gradient even for permeable solutes. Indeed, the presence of charged polymers exclusively either inside or outside the capsule contributes according to Donnan equilibrium in the distribution of all ions for which the shell is permeable. As a result, the small ions – including H^+ and OH^- – may have a concentration gradient across the capsule wall, which can reach 4–5 pH units [45].

A decreased pH inside the capsules loaded with PSS opens the possibility of precipitating tungstate and molybdate ions, which undergo chemical polymerization (polycondensation) in acidic media forming (within the shell) different polytungstates (polymolybdates) with interesting catalytic, electrocatalytic, and electrochromic properties [46]. PSS-loaded capsules were added to a $NaWO_4$ solution and kept for 24 hours [47]. This long period (12–24 h) is required to complete the reaction due to diffusion or equilibrium limitations possibly to be found in the capsule microreactor. Polytungstate nanoparticles in the polyelectrolyte capsules lead to the appearance of Raman signals at 390 cm^{-1}, 670 cm^{-1}, and 940 cm^{-1} corresponding to the vibration of W–O and W–O–W bonds. Dried capsules containing polytungstate nanoparticles have a rough surface, reproducing the bulky shape of the initial polyelectrolyte capsules in solution. However, their diameter is two-fold smaller than that of the initial capsules, indicating partial shrinkage and densification of the capsule shell.

In contrast, cationic PAH, with OH$^-$ as counterions, creates an alkaline pH value in the capsules [48]. The presence of PAH in 0.1 M monomer chain concentration within the capsule interior provokes a pH gradient from 6 outside the capsule to 9 inside. The alkaline pH in the capsule interior is a major force to selectively deposit nanosized Fe$_2$O$_3$, TiO$_2$ gel or hollow spheres, magnetic ferrites, and Fe$_3$O$_4$ exclusively inside the capsules from the corresponding inorganic salts [49–51]. A typical scanning electron microscopy (SEM) image of polyelectrolyte capsules filled with Fe$_3$O$_4$ is shown in Figure 3.7B. Inorganic nanomaterial in the capsule interior prevents capsule collapse during drying, maintaining the original bulky shape. The deposition of the magnetic ferrites and magnetite occurs on the inner side of the PAH/PSS shell (see TEM image, Fig. 3.7A) where PAH molecules, which comprise the first layer of the PAH/PSS shell, are at their highest concentration.

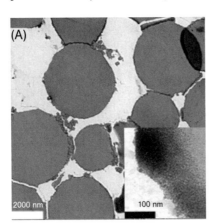 

Figure 3.7 TEM (A) and SEM (B) images of polyelectrolyte capsules modified by magnetic nanomaterials. (Reproduced with permission from [50].)

3.4.1.2 Fabrication of Nanoparticles by Ion-Exchange Reactions

A second approach to reinforce the shell of polyelectrolyte capsules with nanoparticulate inorganic material is that of inorganic nanosynthesis by ion-exchange reactions. Exposing PAH-loaded capsules to a solution of other anions (e.g., 0.1 M H$_3$PO$_4$ or 0.1 M HF) leads to the rapid substitution of OH$^-$ anions to PO$_4^{3-}$ or F$^-$ and the formation of PAH/PO$_4^{3-}$-PAH/PSS or PAH/F$^-$-PAH/PSS capsules. Thus, by using such a simple procedure, PAH/PSS capsules containing one of the reagents inside can be obtained.

Adding F^--loaded capsules to the water solution containing Y^{3+} ions results in the formation of YF_3 only inside the capsules, without any traces of precipitate in the surrounding media [52]. One yttrium compound which precipitates inside the capsules is weakly crystallized YF_3, with traces of $Y(OH)_3$ as minor component. Formation of the latter can be explained by the hydrolysis of yttrium salt in the presence of PAH molecules. The crystallite size for YF_3 inside the capsules, derived from X-ray data, is ca. 7 nm, which is less than that of YF_3 precipitated from aqueous solution (50–100 nm). The thickness and particle size of the YF_3 layer depends greatly on the concentration of yttrium salt in solution. For high concentrations ($10^{-3} \div 5 \times 10^{-5}$ M [Y^{3+}]), a continuous 50- to 100-nm layer comprised of 7- to 10-nm particles is observed, while separate agglomerates or individual particles attached to the inner capsule wall are formed at [Y^{3+}] $<10^{-6}$ M. In being dried, the folds and creases that spread from one YF_3-loaded capsule to another are observed, demonstrating partial shrinkage of the capsule shell. Moreover, part of the polyelectrolyte shell was exfoliated, exhibiting an inner YF_3 layer (Fig. 3.8).

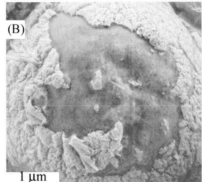

(A)

(B)

5 µm

1 µm

Figure 3.8 SEM images of polyelectrolyte capsules filled with YF_3 nanoparticles. (Reproduced with permission from [52].)

PAH/PO_4^{3-}-PAH/PSS capsules were employed for the biomimetic synthesis of calcium hydroxyapatite, $Ca_{10}(PO_4)_6(OH)_2$, inside polyelectrolyte capsules [53]. Transmission electron microscopy (TEM) analysis established preferable formation of the hydroxyapatite nanoparticles on the inner side of the PAH/PSS shell, and this resulted in empty hydroxyapatite spheres. The thickness of the $Ca_{10}(PO_4)_6(OH)_2$ layer is 100–120 nm, and is composed of 12- to 16-nm particles. The hydroxyapatite particles formed have shape and surface morphology which is different from the particles synthesized by common methods in solution. Other special properties of hydroxyapatite composite hollow shells, including surface acidity, catalytic and biological activity, as well as bone-repairing effects, can also be expected.

viously observed for "initial" PAH/PSS capsules that did not contain any inorganic material. The drying of dextran-containing YF_3/PAH capsules does not influence the controlled-release properties (Fig. 3.10). In the dried stage, fluorescence from the captured dextran molecules was observed, while further redispersion of dried inorganic/organic composite capsules in acidic and alkali media showed the preservation of their controlled permeability properties. At an acidic pH (pH = 2), the YF_3/PAH capsules released dextran molecules into the surrounding solution, whilst at alkaline pH (pH = 8) the dextran molecules remain captured inside.

Using inorganic nanoparticles as shell constituents increases the mechanical stiffness of the polyelectrolyte capsule, allowing it to withstand both drying and mechanical deformation. The synthesis of YF_3 nanoparticles inside polyelectrolyte shells results in a pronounced reinforcement of the capsule mechanics [59]. For relative deformations below 2%, both initial and YF_3-modified capsules undergo a linear force response (Fig. 3.11). The stiffness of the capsules is found to be 0.3 ± 0.01 N m^{-1} before the precipitation of YF_3 nanoparticles, and 2.3 ± 0.2 N m^{-1} after precipitation. Whilst untreated capsules showed an onset of buckling instability at deformations on the order of 10%, the YF_3-reinforced capsules did not buckle even for the highest deformations applied, of ca. 20%. Thus, composite polyelectrolyte/inorganic capsules are not only stiffer; they are also able to withstand much higher critical stresses. These differences are responsible for the fact that the reinforced capsules can survive drying without collapsing, despite their wall thickness being less than 100 nm.

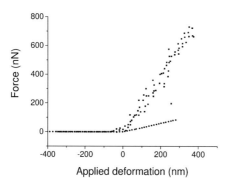

Figure 3.11 Force versus deformation curves obtained on YF_3-reinforced polyelectrolyte capsules (black squares) and on initial capsules (gray squares). (Reproduced with permission from [59].)

3.5
Conclusions and Outlook

PE multilayer capsules present a rather unique class of polymeric vesicles, with a variety of envisaged applications due to their versatile properties. Moreover, they present many opportunities for variations of these properties and the tailoring of different functions. Polyelectrolyte multilayer shells [4] provide a wide spectrum of permeability coefficients for different species, ranging from ions to macromolecules, depending to layer number, overall charge, pH, salt, or solvent. These capsules could be used as carrier systems to deliver substances to a certain location and to release it in programmed manner. In this chapter we have described some examples of pH-switchable permeability for the controlled encapsulation of polymers in defined amounts in capsules composed of weak polyelectrolytes. Such stimuli-responsive properties open avenues for their exploitation as feedback-responsive materials, where release occurs in response to factors relating to the surrounding media. We believe that the ability to control the permeability of multilayered capsules will create a stable position in the market of drug carriers, biosensing, microreactors and catalysts, construction materials, among others.

Selective permeability allows chemical reactions to be performed exclusively in the capsule interior. The *in-situ* modification of polyelectrolyte capsules by conducting syntheses inside inorganic nanoparticles creates a new class of multifunctional capsules that combines the properties of inorganic nanomaterials. These multifunctional, composite capsules may find applications for the protection, delivery, and storage of biochemical compounds that are unstable in solution or under UV/visible irradiation, where the use of capsules composed solely of polymeric components cannot be envisaged. Clearly, much further research is required in this area, most notably in understanding the mechanism of the chemical reactions that occur in the confined microsized geometric and diffusional limitations of polyelectrolyte multilayers.

Despite this lack of fundamental understanding, however, the use of these capsules – and in particular their multifunctional properties – is a rapidly developing field that not only has clear potential but has also gained much interest in industry.

Acknowledgments

The authors thank their colleagues at the MPI, for their contribution to the research and their fruitful collaboration: Dr. A. Skirtach, Dr. A. Antipov, T. Mauser, M. Prevot, D. Halozan. They also thank Prof. H. Möhwald for supporting these studies, and for helpful discussions. Funding for these investigations was provided by the Sofja Kovalevskaja Program of the Alexander von Humboldt Foundation, the German Ministry of Education and Research, and the EU-project STREP001428 "Nanocapsules for targeted delivery of chemicals". D.S. acknowledges the Marie-Curie program of 6th FP for an individual Incoming Marie-Curie Fellowship.

Abbreviations

BSA	bovine serum albumin
HA	hyaluronic acid
LbL	layer-by-layer
MF	melamine-formaldehyde
PAA	poly(acrylic acid)
PAAM	poly(acrylamide)
PAH	poly(allylamine hydrochloride)
PDADMAC	poly(diallyl dimethyl ammonium chloride)
PE	polyelectrolyte
PEM	polyelectrolyte multilayers
PEO	poly(ethylene oxide)
PLL	poly(lysine)
PMA	poly(methacrylic acid)
PNIPAM	poly(N-isopropylacrylamide)
PS	polystyrene
PSS	poly(sodium styrene sulfonate)
PVPON	poly(vinylpyrrolidone)
THF	tetrahydrofuran

References

1. G. Decher, J.-D. Hong, J. Schmitt, *Thin Solid Films* **1992**, *210-211*, 831–835.
2. G. Decher, *Science* **1997**, *277*, 1232–1237.
3. E. Donath, G.B. Sukhorukov, F. Caruso, S.A. Davis, H. Möhwald, *Angew. Chem. Int. Ed.* **1998**, *37*, 2201–2205.
4. A.A. Antipov, G.B. Sukhorukov, *Adv. Coll. Int. Sci.* **2004**, *111*, 49–61.
5. A. Fery, F. Dubreuil, H. Möhwald, *New J. Phys.* **2004**, *6*, Art. No. 18.
6. O.I. Vinogradova, *J. Phys.: Cond. Matter* **2004**, *16*, R1105–R1134.
7. H. Möhwald, E. Donath, G.B. Sukhorukov, in: G. Decher, J.B. Schlenoff (Eds.), *Multilayer Thin Films*. Ames: Wiley-VCH, Weinheim, **2003**, pp. 363–392.
8. C.S. Peyratout, L. Dähne, *Angew. Chem. Int. Ed.* **2004**, *43*, 3762–3783.
9. B. Jeong, A. Gutowska, *Trends Biotechnol.* **2002**, *20*, 305–311.
10. J. Hiller, M.F. Rubner, *Macromolecules* **2003**, *36*, 4078–4083.
11. S.E. Burke, C.J. Barrett, *Langmuir* **2003**, *19*, 3297–3303.
12. L. Zhai, F.C. Cebeci, R.E. Cohen, M.F. Rubner, *Nano Letters* **2004**, *4*, 1349–1353.
13. E. Kharlampieva, S.A. Sukhishvili, *Langmuir* **2003**, *19*, 1235–1243.
14. Z. Sui, J.B. Schlenoff, *Langmuir* **2004**, *20*, 6026–6031.
15. S.Y. Yang, M.F. Rubner, *J. Am. Chem. Soc.* **2002**, *124*, 2100–2101.
16. S.A. Sukhishvili, S. Granick, *J. Am. Chem. Soc.* **2000**, *122*, 9550–9551.
17. D.M. De Longchamp, P.T. Hammond, *Langmuir* **2004**, *20*, 5403–5411.
18. J.F. Quinn, F. Caruso, *Langmuir* **2004**, *20*, 20–22.
19. L. Richert, P. Lavalle, E. Paysan, X.Z. Shu, G.D. Prestwich, J.-F. Stolz, P. Schaaf, J.-C. Voegel, C. Picart, *Langmuir* **2004**, *20*, 448–458.
20. S.E. Burke, C.J. Barrett, *Biomacromolecules* **2003**, *4*, 1773–1783.
21. S.E. Burke, C.J. Barrett, *Macromolecules* **2004**, *37*, 5375–5384.
22. K.G. Yager, C.J. Barrett, *Cur. Op. Solid State Mater. Sc.* **2001**, *5*, 487–494.

23. I. Luzinov, S. Minko, V. Tsukruk, *Prog. Pol. Sci.* **2004**, *29*, 635–698.

24. G.B. Sukhorukov, E. Donath, S. Moya, A.S. Susha, A. Voigt, J. Hartmann, H. Möhwald, *J. Microencaps.* **2000**, *17*, 177–185.

25. G. Ibarz, L. Dähne, E. Donath, H. Möhwald, *Adv. Mater.* **2001**, *13*, 1324–1327.

26. C. Gao, S. Moya, H. Lichtenfeld, A. Casoli, H. Fiedler, E. Donath, H. Möhwald, *Macromol. Mater. Engin.* **2001**, *286*, 355–361.

27. C.Y. Gao, H. Möhwald, J.C. Shen, *ChemPhysChem* **2004**, *5*, 116–120.

28. G.B. Sukhorukov, A.A. Antipov, A. Voigt, E. Donath, H. Möhwald, *Macromol. Rapid Commun.* **2001**, *22*, 44–46.

29. A.A. Antipov, G.B. Sukhorukov, S. Leporatti, I.L. Radtchenko, E. Donath, H. Möhwald, *Coll. Surf. A* **2002**, *198-200*, 535–541.

30. C. Déjugnat, G.B. Sukhorukov, *Langmuir* **2004**, *20*, 7265–7269.

31. C. Déjugnat, D. Halozan, G.B. Sukhorukov, *Macromol. Rapid Commun.* **2005**, *26*, 961–967.

32. T. Mauser, C. Déjugnat, G.B. Sukhorukov, *Macromol. Rapid Commun.* **2004**, *25*, 1781–1785.

33. Y. Lvov, A.A. Antipov, A. Mamedov, H. Möhwald, G.B. Sukhorukov, *Nano Letters* **2001**, *1*, 125–128.

34. K. Glinel, G.B. Sukhorukov, H. Möhwald, V. Khrenov, K. Tauer, *Macromol. Chem. Phys.* **2003**, *204*, 1784–1790.

35. L. Dähne, S. Leporatti, E. Donath, H. Möhwald, *J. Am. Chem. Soc.* **2001**, *123*, 5431–5436.

36. B.-S. Kim, O.I. Vinogradova, *J. Phys. Chem. B* **2004**, *108*, 8161–8165.

37. M. Prevot, C. Déjugnat, G.B. Sukhorukov, H. Möhwald, *Abst. Pap. Am. Chem. Soc.* 226: U490-U490 206-PMSE Part 2, SEP **2003**.

38. T.C. Wang, M.F. Rubner, R.E. Cohen, *Langmuir* **2002**, *18*, 3370–3375.

39. M. Fang, P. Grant, M. McShane, G.B. Sukhorukov, V. Golub, Y. Lvov, *Langmuir* **2002**, *18*, 6338–6344.

40. T.C. Wang, B. Chen, M.F. Rubner, R.E. Cohen, *Langmuir* **2001**, *17*, 6610–6615.

41. Y. Boontongkong, R.E. Cohen, M.F. Rubner, *Chem. Mater.* **2000**, *12*, 1628–1633.

42. T.C. Wang, M.F. Rubner, R.E. Cohen, *Chem. Mater.* **2003**, *15*, 299–304.

43. S. Dante, Z.Z. Hou, S. Risbud, P. Stroeve, *Langmuir* **1999**, *15*, 2176–2182.

44. S. Joly, R. Kane, M.F. Rubner, *Langmuir* **2000**, *16*, 1354–1359.

45. I.L. Radtchenko, Ph.D. thesis, Potsdam University, Potsdam, Germany, **2003**.

46. D.E. Katsoulis, *Chem. Rev.* **1998**, *98*, 359–388.

47. D.G. Shchukin, W. Dong, G.B. Sukhorukov, *Macromol. Rapid Commun.* **2003**, *24*, 462–466.

48. I.L Radtchenko, M. Giersig, G.B. Sukhorukov, *Langmuir* **2002**, *18*, 8204–8208.

49. D.G. Shchukin, I.L. Radtchenko, G.B. Sukhorukov, *Mater. Lett.* **2003**, *57*, 1743–1747.

50. D.G. Shchukin, I.L. Radtchenko, G.B. Sukhorukov, *J. Phys. Chem. B* **2003**, *107*, 952–957.

51. D.Y. Wang, F. Caruso, *Chem. Mater.* **2002**, *14*, 1909–1913.

52. D.G. Shchukin, G.B. Sukhorukov, *Langmuir* **2003**, *19*, 4427–4431.

53. D.G. Shchukin, G.B. Sukhorukov, H. Möhwald, *Chem. Mater.* **2003**, *15*, 3947–3950.

54. D.G. Shchukin, E. Ustinovich, D.V. Sviridov, Y.M. Lvov, G.B. Sukhorukov, *Photochem. Photobiol. Sci.* **2003**, *2*, 975–977.

55. D.G. Shchukin, E. Ustinovich, H. Möhwald, G.B. Sukhorukov D.V. Sviridov, *Adv. Mater.* **2005**, *17*, 468–472.

56. A.G. Skirtach, A.A. Antipov, D.G. Shchukin, G.B. Sukhorukov, *Langmuir* **2004**, *20*, 6988–6992.

57. Z. Lu, M.D. Prouty, Z. Guo, V.O. Golub, C. Kumar, Y.M. Lvov, *Langmuir* **2005**, *21*, 2042–2050.

58. D.G. Shchukin, G.B. Sukhorukov, H. Möhwald, *Angew. Chem. Int. Ed.* **2003**, *42*, 4472–4475.

59. F. Dubreuil, D.G. Shchukin, G.B. Sukhorukov, A. Fery, *Macromol. Rapid Commun.* **2004**, *25*, 1078–1081.

4
Polymer Encapsulation of Inorganic Particles

Elodie Bourgeat-Lami and Etienne Duguet

4.1
Introduction

Hybrid organic/inorganic nanoparticles with length scales on the order of a few to several tens of nanometers have attracted considerable attention in recent years as the intimate combination of polymer and inorganic components at the nanoscale offers the promise of original properties. They are one of the fruits of the remarkable progress which has been achieved in both polymer and inorganic chemistry. Nanoparticles of noble metals, metal oxides and semiconductors such as gold, CdS or TiO_2, can be prepared in large quantities using thermal decomposition, hydrolysis, reduction and other soft-chemistry processes in solution [1]. Not only can these particles be obtained with a controlled size, but they can also be assembled into 2-D or 3-D arrays. Progress in polymer science has also permitted the design and controlled synthesis of well-defined macromolecular architectures, such as new stabilizing surfactants which are able to self-assemble and to be used for structuring mineral materials [2–5]. Typical examples of organic and inorganic colloidal systems are provided in figure 4.1.

Several routes have been developed for the synthesis of hybrid nanoparticles, and these can be classified into two main strategies according to the nature of the precursor particles. On the one hand, polymer latex particles can be decorated by mineral components through layer-by-layer self assembly [6], sol-gel nanocoating [7] or metal precipitation [8]. On the other hand, the surface of inorganic precursor nanoparticles can be modified by an adsorbed or grafted macromolecular corona or coated by a dense polymer shell. In the latter case, the encapsulation techniques are essentially derived from heterophase polymerization processes where the inorganic particles play the role of fillers or seeds. These processes allow the polymer matter to be confined at the nanoscale or mesoscale, and therefore they are suitable for nanoparticle encapsulation. In this chapter we will use the term "polymer encapsulation" to denote a polymerization performed at the surface of inorganic particles, though the core-shell morphology is far from being the exclusive morphology achieved in these systems, as will be illustrated below.

Functional Coatings. Edited by Swapan Kumar Ghosh
Copyright © 2006 WILEY-VCH Verlag GmbH & Co. KGaA, Weinheim
ISBN 3-527-31296-X

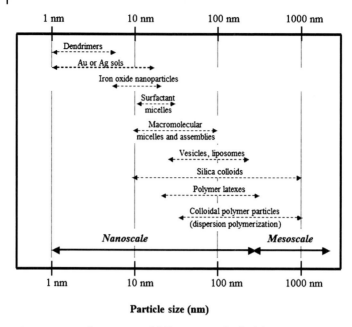

Figure 4.1 Particle size range of different types of colloidal systems.

Within the last twenty years, there has been increasing interest in the synthesis of these organic/inorganic nanoparticles, and several review articles and text books now cover this broad area of research [9–17]. In order to avoid overlapping with these previous reviews, the aim of this chapter is to describe the main strategies developed in the recent literature to synthesize organic/inorganic hybrid particles through heterophase polymerization using preformed mineral particles as seeds. The chapter is organized as follows.

In a first section, we recall the fundamental aspects associated with the formation of polymer particles through heterophase polymerization including emulsion, dispersion, and suspension polymerization processes. Since miniemulsion polymerization is reviewed in Chapter 2, only the basic principles are given here. In a second part, we provide a brief overview of the main characteristics of inorganic particles involved in encapsulation reactions. Most mineral fillers and pigments are strongly hydrophilic, and must be modified before organic polymerization in order to overcome the problem of phase separation and to promote polymer formation on their surface. The third part of the chapter describes the most important strategies, including surfactant and polymer adsorption, *in-situ* grafting, and heterocoagulation. It is not the intention in this section to provide a thorough description of the fundamental chemical and physical aspects involved in organic modification of mineral particles; rather, the intention is to provide the basic knowledge and concepts required to make our point. Numerous references, text books and recommended background reading are provided in this section, and those who wish to know more detail should refer to these articles. Typical examples of encapsulated pigments and fillers are provided in a subsequent section. We successively describe in this section the encapsulation of titanium dioxide pigments for the paint industry, the elaboration of colloidal nanocomposites based on silica or magnetic particles, and the incorporation of clay minerals into polymer latexes (so-called polymer-layered silicate nanocomposites). Some other examples (e.g., encapsulation of calcium carbonate, metallic particles and quantum dots) are also mentioned briefly at the end of the section. A special emphasis is made in the following section on the control of particle morphology, and in particular on the common abnormalities with respect to the expected ideal core-shell morphology. Finally, some properties of polymer-encapsulated minerals are briefly outlined in a last section, with emphasis on their thermal and mechanical behaviors.

4.2
Background Knowledge of Heterophase Polymerization Processes

4.2.1
General Classification

Heterogeneous processes are of great importance for the free-radical chain-growth polymerization of $CH_2=CRR'$ monomers on an industrial scale. In comparison with bulk polymerization, they allow us to overcome the rapid viscosity increase of the reaction medium with conversion, as well as its consequences, such as the difficult heat removal and the autoacceleration phenomenon. Because in many cases the continuous phase is water, they are also more environmentally friendly techniques than solution polymerization methods, where the use of organic solvents remains hazardous and expensive. Ultimately, heterophase polymerization techniques are the original routes to polymer particles ranging from a few tens of nanometers to a few millimeters in diameter.

Heterophase polymerization systems can be defined as two-phase systems in which the resulting polymer and/or starting monomer are in the form of a fine dispersion in an immiscible liquid medium defined as the "polymerization medium", "continuous phase", or "outer phase". Even if oil-in-water (o/w) systems are greatly preferred on an industrial scale, water-in-oil (w/o) systems may also be envisaged for specific purposes. Heterogeneous polymerization processes can be classified as suspension, dispersion, precipitation, emulsion, or miniemulsion techniques according to interdependent criteria which are the initial state of the polymerization mixture, the kinetics of polymerization, the mechanism of particle formation and the size and shape of the final polymer particles (Fig. 4.2) [18].

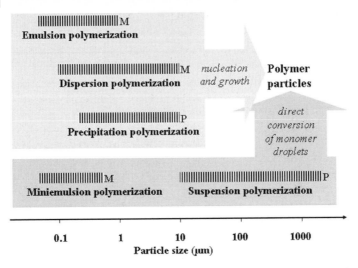

Figure 4.2 Compared features and particle size ranges of heterophase polymerization systems. Symbols M and P refer to particle size distribution: M = monodisperse particles; P = polydisperse particles. Redrawn and completed from [18a].

4.2.2
Oil-in-Water Suspension Polymerization

This may be roughly described as a bulk polymerization in which the reaction mixture is suspended as droplets in the aqueous continuous phase. Therefore, the initiator, monomer, and polymer must all be insoluble in water. The suspension mixture is prepared by the addition of a solution of initiator in monomer to the preheated aqueous suspension medium. Droplets of the organic phase are formed and maintained in suspension by the use of: (i) vigorous agitation throughout the reaction; and (ii) hydrophilic macromolecular stabilizers dissolved in water (e.g., low molar-mass polymers such as poly(vinyl alcohol), polyvinylpyrrolidone, and hydroxymethylcellulose). Each droplet acts as a small bulk polymerization reactor for which the normal kinetics apply. Polymer is produced in the form of beads, the average diameter of which is close to that of the initial monomer droplets (0.01 to 2 mm), even if inadvertent droplet breaking and coalescence widen the bead size distribution. Polymer beads are easily isolated by filtration, provided that they are rigid and not tacky. Therefore, the suspension polymerization process is unsuitable for the preparation of polymers that have low glass transition temperatures. This method is widely used for styrene, methyl methacrylate, and vinyl chloride monomers, for instance.

4.2.3
Oil-in-Water Emulsion Polymerization

Emulsion polymerization is another heterogeneous process of great industrial importance, and allows the elaboration of aqueous colloidally stable dispersions of polymer particles, known as latexes [19]. In "conventional" emulsion polymerization, the polymer particles are formed by starting from an insoluble (or scarcely soluble) monomer emulsified by the aid of a surfactant above its critical micelle concentration (CMC). The monomer is originally distributed between coarse emulsion droplets, surfactant micelles and the water phase, where a small proportion of monomer (depending on its solubility) is molecularly dissolved. Unlike suspension polymerization, the initiator is soluble in water, and this leads to a strongly different particle formation mechanism. Polymerization thus begins in the aqueous phase by the formation of free radicals through the initiator thermolysis and the addition of the first monomer units. These oligomeric radical species (oligoradicals) are rapidly captured by the monomer-swollen micelles, where propagation is supported by absorption of monomer diffusing from the monomer droplets through the aqueous phase to maintain equilibrium. Therefore, stabilized nuclei are produced leading to primary particles, and these gradually grow until the monomer is completely consumed. The size of these particles is determined by the number of primary latex particles formed and the time during which they grow. The polymer particles generally have final diameters in the range of 0.05 to 1 μm – that is, considerably smaller than for suspension polymerization. One of the important features of emulsion polymerization is also the ability to control particle morphology, for example, the formation of core-shell particles and other equilibrium morphologies by successive additions of different monomers.

In emulsifier-free (hereafter referred to as "soap-free") polymerizations, the polymerization is carried out in the same way as described above, except that no surfactant is used. Nucleation occurs by oligoradical precipitation into unstable nuclei which collide to form larger particles. Polymerization takes place mainly within these monomer-swollen particles, and the particles grow in similar manner to conventional emulsion polymerization.

Polymers prepared by emulsion polymerization are used either directly in the latex form or after isolation by coagulation or spray-drying of the latex.

4.2.4
Precipitation and Dispersion Polymerizations

In precipitation polymerization, the reaction mixture is initially homogeneous, as in solution polymerization, but it is a precipitant for the polymer. Thus, the initially formed macromolecules collapse and coagulate to create particle nuclei, which gradually flocculate into irregularly shaped and polydisperse particles. Such a process concerns for instance the synthesis of polytetrafluoroethylene in water or polyacrylonitrile in bulk.

In the case of dispersion polymerization, the polymerization medium is not a precipitant but a poor solvent for the resulting polymer [20]. Thus, the macromolecules swell rather than precipitate, and the polymerization proceeds largely within these individual particles, leading to more monodisperse products. For ensuring their stability, macromolecular stabilizers must be used, as in suspension polymerization. Finally, another characteristic of dispersion polymerization reactions is the diameter of the polymer particles (in the range 0.1 to 10 μm); this is generally much larger than in emulsion polymerization, although small polymer particles can also be obtained in the presence of reactive stabilizers [21] or block copolymers [22].

4.2.5
Oil-in-Water Miniemulsion Polymerization

Miniemulsion polymerization may be roughly described as a suspension polymerization leading to polymer particles in the range of submicronic sizes. Indeed, particles are obtained by direct conversion of small monomer droplets without serious exchange kinetics being involved [23]. Nevertheless, due to the small droplets size, the initiator can be either oil- or water-soluble. In a first step, miniemulsion droplets of 30 to 500 nm are formed by shearing (high-pressure homogenizer or ultrasound) a system containing the dispersed phase, the continuous phase, a surfactant, and an hydrophobe playing the role of an osmotic pressure agent for preventing the interdroplet mass transfer phenomenon (known as Ostwald ripening). Therefore, polymer particles are obtained by direct conversion of monomer droplets and their final size can be controlled by altering the shearing conditions. The advantages of miniemulsion polymerization are mainly associated to its versatility and applicability to non-radical polymerizations and to the encapsulation of resins, liquids and preformed particles. The potential of the miniemulsion polymerization technique to create organic/inorganic hybrids is described fully in Chapter 2, and will not be reconsidered here.

4.2.6
Applications of Polymer Latexes

The elaboration of particles of colloidal dimensions is desirable in many industrial domains. Some obvious cases where the colloidal state is required are, for instance, paints, inks and lacquers [24]. The reasons for using colloidal suspensions are varied. The most important benefits are a very large specific surface area, a great versatility in terms of particle size and surface properties, the ability to introduce functional groups and to selectively design the particles' interior, and – finally – the possibility of controlling the rheological properties of the suspension.

The largest industrial uses of polymer latexes are latex paints and adhesives, but other major applications include paper coatings and textiles. Latex paints mostly involve acrylic polymers, while natural and synthetic styrene-butadiene rubbers (SBR) are the main binders of latex adhesives. The most common types of polymer latexes used in coating applications are listed in Table 4.1.

Table 4.1 Principal industrial applications of polymer latexes.

Applications	Type of polymer used
Paints	Vinyl acetate/dibutyl maleate copolymers, pure acrylics, styrene/acrylic copolymers, vinyl acetate/butyl acrylate copolymers, styrene/acrylate copolymers
Adhesives	Natural rubber, styrene-butadiene rubber (SBR), vinyl acetate, acrylics, chloroprene and copolymers
Paper coating	Styrene, butadiene, acrylonitrile, acrylic esters, unsaturated carboxylic acid
Carpet backing	Carboxylated SBR
Cements and concretes	Polyvinyl acetate, SBR, polyacrylics

In all of these applications the particles are soft and must be capable of forming a film at temperatures close to room temperature. In addition, they must display favorable interactions with the surface of the substrate on which they are applied. In this respect, a review of recent publications and patents showed that structured latex particles still represent an active area of research, the principal objective being to improve the performances of the coating formulation. Indeed, as will be illustrated below in the case of organic/inorganic colloids, structured latexes represent a possible means to overcome the usual compromises between a good film quality and optimal properties.

Besides the above-mentioned large-scale industrial applications that constitute the greatest part of latex production, it is worth mentioning that colloidal particles have also found major developments in the life sciences (drug encapsulation, drug delivery, colloidal supports for biodiagnostics, etc.), in microfluidics (lab-on-a-chip devices for fluid management) and in nanotechnologies (colloidal templates for designing patterned surfaces in soft lithography techniques, macroporous materials issued from 3D colloidal crystals, etc.) [25,26].

Whatever the domain of application, formulations based on polymer latexes contain a significant amount of inorganic pigments, fillers and extenders of various nature, depending on the properties that must be achieved. Extenders are mostly used for economical reasons, but they are also known to improve the mechanical properties and durability of the resulting materials. A brief overview of the principal types of mineral particles used in coating formulations together with their main characteristics is provided in the following section.

4.3
Inorganic Particles

As previously mentioned, inorganic pigments and fillers are finely divided solid particles which are introduced into polymers to provide the best compromise between a low cost and a high level of properties. Mineral fillers are widely present in almost all of the above-mentioned industrial domains, including the paint, adhesive and textile industries. Among the advantages of inorganic fillers are properties such as stiffness, mechanical strength, chemical inertness, thermal and abrasion resistances, and optical properties (transparency, opacity). Most common mineral fillers are carbonates, sulfates, aluminosilicates (clays, zeolites) and metal oxides (SiO_2, Al_2O_3, TiO_2...). There exist many different grades of mineral fillers according to whether they are synthetic or natural, and depending on their chemical composition or on the method which has been used for their synthesis. For example, fumed silica, which is obtained by high-temperature vapor hydrolysis of silicium tetrachloride, is in many instances very different from precipitated grades, which are prepared from the reaction of sodium silicate with hydrochloric acid solutions. Besides, it is known that the high reactivity of oxide surfaces evolves from the high density of hydroxyl groups which can further react to give strong bonds (through olation, oxolation, esterification) or weak bonds (hydrogen bonds) with the surrounding medium [27]. The concentration of hydroxyl groups is extremely dependent on parameters such as thermal treatments, and may change significantly from one synthetic procedure to another. The reactivity of inorganic surfaces is also very dependent on conditions such as pH and ionic strength. The pH of aqueous suspensions of metal oxides is a very important characteristic since it controls the degree of ionization of the hydroxyl groups which, in turn, depends on the acido-basicity of the oxide surface. The acid–base character of oxide nanoparticles can be estimated by the isoelectric point (IEP) of the surface, which corresponds to the pH value of the aqueous suspension at which the surface charge is zero. A low IEP value indicates an acidic surface, while a high IEP indicates a basic surface. The surface charge of the oxide is changing consequently from positive (at pH lower than IEP) to negative at higher pH (Fig. 4.3).

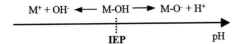

Figure 4.3 Ionization reactions of mineral oxide surfaces as a function of pH.

There exist many different methods to estimate the electrical charge of solid oxides and hydroxides, but electrophoresis is by far the method most frequently used [28]. An extensive list of IEPs of metal oxides was proposed by Parks some forty years ago [29]. Grades of inorganic fillers also differ in particle size and shape and in properties such as surface tension, thermal resistance, etc. Finally, it should be noted that many mineral powder systems – and especially pigments – are produced

Table 4.2 Common pigments and fillers involved in encapsulation reactions. Isolectric pH values, particles dimensions and fields of application.

Inorganic material	Chemical formula	Particles dimension [µm]	Isoelectric point	Field of application
Metal oxides				
Titanium dioxide	TiO_2 (anatase)	0.2–1	6.2	Paints
	TiO_2 (rutile)		4.7	
Alumina	Al_2O_3	0.01–0.6	9.1	Biomedical
Zinc oxide	ZnO	/	9	Electronics
Iron oxide	Fe_2O_3 (maghemite)	/	7–8	Biomedical
Silica	SiO_2	0.02–0.5	2.2	Elastomers, paints
Aluminosilicates				
Kaolin	$Al_2Si_2O_5(OH)_4$	/	4.8	
Montmorillonite		Platelets	/	Engineered plastics
Laponite		Disc-shaped	/	Plastics
Metals				
Silver	Ag	/	/	Optics
Gold	Au	/	/	Optics
Aluminum	Al	/	/	Decorative coatings
Insoluble salts				
Calcium carbonate	$CaCO_3$	0.4–2	9.6	Paper coating
Barium sulfate	$BaSO_4$	0.5–3		Coating
Sulfides and sulfites				
Calcium sulfite	$CaSO_3$	/	/	Composites
Cadmium sulfide	CdS	1–5	/	Optics
Zinc sulfide	ZnS	/	/	Optics

as agglomerated particles with a broad size distribution. Typical examples of inorganic fillers which have been specifically involved in encapsulation reactions are listed in Table 4.2, together with their fields of application and their most important characteristics. The reader should refer to relevant text books on pigments and fillers for a more complete description [30–33].

4.4
Polymer Encapsulation of Inorganic Particles

4.4.1
Synthetic Strategies

It is generally admitted that when mixing a mineral substance with a polymer, the incompatibility between both materials most often leads to phase segregation. This situation is analogous to polymer blends and, in most cases, the two components phase separate into discrete domains. In order to improve adhesion and promote the formation of finely divided inorganic domains into polymer matrices, it is common to introduce adhesion promoters in the product formulation. These molecules (also called "compatibilizers" or "coupling agents") are capable of reacting with both the inorganic filler and the polymer, thus creating chemical bonds between the two components. They also promote surface wetting of the filler and create better homogeneity of the blend. This "compatibilization" concept, which is of major importance in many technological applications (e.g., latex paints and composite materials) also holds for the elaboration of polymer-encapsulated minerals through heterophase polymerization. Indeed, the preparation of organic/inorganic particles in dispersed media usually requires some efforts to establish favorable interactions between the core and shell materials (Fig. 4.4). By analogy with the use of polymeric dispersants in slurry formulations, the role of compatibilizers in encapsulation technologies is to promote wetting of the mineral surface by the polymer, a prerequisite conditions for successful encapsulation.

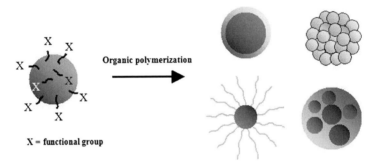

Figure 4.4 General synthetic strategy used for the elaboration of polymer-encapsulated inorganic particles through heterophase polymerization.

Compatibilizers used in encapsulation reactions can be either chemically grafted onto the mineral surface or physically adsorbed through iono-covalent or H-bonding interactions. While silane and titanate coupling agents fall into the first category, surfactant molecules or macromolecules belong to the second category. In addition, they can be either reactive or nonreactive. When they are nonreactive, their role is to modify the mineral/water interfacial properties and promote the subsequent adsorption of monomer and/or oligomeric entities during a polymerization reaction. Polymerization then takes place at the vicinity of the inorganic surface according to the so-called admicellar polymerization mechanism. In the case when they are reactive, they interact with the components of the polymerization and form *in situ* a covalent bond which process also promotes affinity of the growing polymer chains for the mineral surface.

4.4.1.1 **Macromolecule and Surfactant Adsorption:**
The Admicellization/Adsolubilization Concept

The adsorption of macromolecules and/or surfactant molecules onto inorganic surfaces is an essential requirement in most encapsulation processes. If the polymer layer is sufficiently thick, the adsorption can be regarded as a special case of polymer encapsulation by preformed polymer chains. However, one major objective of these surface modifications is to control the interfacial properties and to promote the subsequent adsorption of monomers and/or oligomers, for example during an emulsion-like polymerization reaction.

The principle of adsorption of surfactant molecules onto mineral surfaces is now well established [34]. Nonionic, anionic or cationic emulsifiers each exhibit strong interactions with the surface of inorganic particles, depending on the suspension pH [35]. For instance, it is well known that block copolymers or nonionic surfactants containing poly(ethylene oxide) units adsorb onto metal oxide surfaces through interactions between the basic ether oxygens in the polymer backbone and acidic or H-bonding sites on the surface of the inorganic particles. The adsorption of octyl and nonyl phenol polyoxyethylenic surfactants with different oxyethylenic chain lengths has been thoroughly investigated by several authors on many different mineral powders (ground quartz, silica, kaolin and alumina) [36–39]. It is commonly admitted that, at low concentrations, the emulsifier adsorbs by its oxyethylenic chains onto a few sites of the mineral surface [40–42]. Further adsorption then occurs by an aggregation process of the surfactant molecules. At high emulsifier concentrations, the adsorbed surfactant aggregates form micelles on the inorganic surface, as evidenced by several techniques [43,44]. As expected, the adsorbed amount is strongly dependent on pH since the latter controls the surface density of hydroxyl groups. However, the pH is a predominant parameter especially when adsorption takes place through electrostatic interactions between the charged head of the surfactant and the surface charge of the mineral. This is indeed the case of ionic emulsifiers such as sodium dodecyl sulfate (SDS), alkylbenzene sulfonates or long-chain organic quaternary ammonium ions for instance, which are known to adsorb strongly onto hydrophilic inorganic surfaces of opposite charge, forming adsorbed bilayers of surfactant aggregates [45]. The adsorption of ionic emulsifiers is accompanied by the precipitation of the inorganic colloid, since the surface charge of the mineral is neutralized upon addition of increasing concentrations of the surfactant. Restabilization occurs at higher concentrations of the soap due to the formation of a surfactant bilayer (Fig. 4.5).

As depicted in Figure 4.5, the surfactant bilayer created at the solid/aqueous solution interface provides hydrophobic loci for the solubilization of monomer or radicals; these can polymerize further on the modified inorganic surface and effectively coat the solid particles according to an emulsion-like polymerization reaction. Three steps are involved in the coating reaction. In the first step, the emulsifier adsorbs onto the mineral surface, forming micelle-like aggregates. In a second step, the monomer is solubilized in the adsorbed micelles. Finally, the polymerization takes place as a conventional emulsion polymerization in the monomer-swelled admicelles (Fig. 4.6).

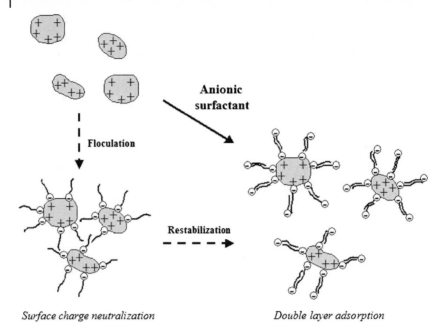

Figure 4.5 Principle of pigment flocculation and redispersion in water upon the adsorption of a surfactant of opposite charge.

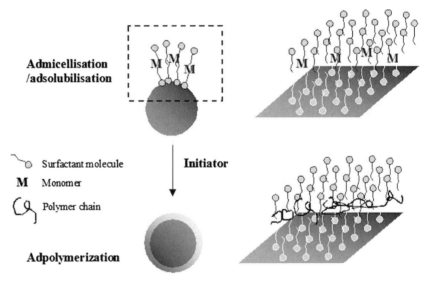

Figure 4.6 Schematic illustration of pigment encapsulation through an emulsion-like polymerization reaction. The process involves: 1) formation of surfactant bilayers (admicellization); 2) solubilization of monomer (adsolubilization); and 3) free radical polymerization (admicellar polymerization).

Although the first patents on encapsulation reactions through admicellar polymerization date back to 1964 [46], Wu, Harwell and O'Rear [47] first explored this method in the open literature in the mid-1980s for the formation of thin films of polystyrene on the surface of a porous aluminum oxide powder. During the same period, Hasegawa and Meguro also described the coating reaction of fine inorganic powders including iron oxide and titanium dioxide pigments through emulsion polymerization of styrene adsolubilized into adsorbed SDS bilayers [48,49]. The admicellization/adpolymerization concept was next extended to a large variety of monomers (isoprene, butadiene, methyl methacrylate, vinyl acetate), surfactants and inorganic substrates, including precipitated silica [50]. Today, admicellar polymerization is one of the principal methods of polymer encapsulation with potential industrial developments. Indeed, this method does not require cost-expensive reagents, and involves conventional emulsion polymerization processes with no specific needs to adapt already existing industrial plants. However, as will be illustrated below in the case of titanium dioxide pigments, one critical aspect of encapsulation reaction through admicellar polymerization is that of pigment agglomeration. In this regard, surfactant molecules play a determinant role as they must not only interact with the mineral surface, but must also be able to stabilize the polymer-encapsulated inorganic particles. The technology must therefore be adapted to every particular case in order to face this challenging aspect, and it seems unlikely that it can be generalized to a large variety of pigments on an industrial scale without much effort.

The adsorption of polymers onto mineral surfaces has also been extensively studied. The adsorbed amounts depend on various parameters. First, the nature of the polymer (random or block copolymer) will determine the energy of interactions between the segments of the macromolecules and the inorganic surface. In particular, the molecular weight of the polymer and the solvency of the medium have strong influences on the quality of adsorption. The nature of the mineral surface is also a critical parameter. Of course, the adsorption of polymers depends not only on the chemical composition of the inorganic particles but also on the ionic strength and, principally, on the pH of the surrounding solution. The suspension pH controls the surface charge of the mineral and determines consequently the nature of the interaction (electrostatic or hydrogen bonding). Adding polymers to a dispersed system may also have drastic consequences on the rheology and stability of the colloidal dispersion [51]. The adsorbed polymer can prevent the inorganic particles from aggregation by steric stabilization, but it can also induce bridging flocculation. The adsorbed amounts and the thickness of the adsorbed layer are also important characteristics of polymer adsorption. The adsorption of poly(N-vinyl pyrrolidone) (PVP) from water onto pyrogenic silica was studied by Cohen Stuart et al. [52], who found the adsorbed amount to be influenced not only by the molecular weight of the polymer sample but also by its polydispersity. These authors showed that PVP exhibited rather high adsorption affinity from water, and that hydrogen bonding was the driving force for the adsorption [53]. As demonstrated by Parnas et al. [54], the adsorption of PVP onto silica is also strongly dependent on the surface chemistry. The incorporation of vinyl, ethyl and isobutyl silylating agents significantly increased the amount of adsorbed PVP. Despite the interest in such systems, very few reports were made on the use of

PVP in encapsulation technologies, although PVP is an important stabilizer of dispersion polymerization reactions. Dispersion polymerization of styrene was performed in the presence of silica particles using PVP as a steric stabilizer [55], but no evidence was found that PVP played an active role in the encapsulation process (although one would expect the silica surface properties to be seriously affected by the presence of PVP). The adsorption of other nonionic water-soluble polymers such as poly(ethylene oxide) (PEO) and hydroxypropyl cellulose (HPC) has also been investigated by several authors, and in some cases were found to play a determinant role in encapsulation reactions. For example, Furusawa et al. described the adsorption of HPC on silica and showed that the saturated amount of adsorption was dependent on temperature [56]. The adsorption value (ca. 1.5 mg m^{-2}) observed at temperatures higher than the lower critical solution temperature (LCST) of the polymer was 1.5-fold as large as the value obtained at room temperature. The authors reported that the dense layer of HPC adsorbed at the LCST allowed synthesis of the core-shell silica/polystyrene composite latexes. When the emulsion polymerization reaction was performed in the presence of an anionic emulsifier at concentrations higher than its CMC, a completely different morphology was obtained. The composite particles exhibited a raspberry shape (Fig. 4.7), which morphology was attributed to the heterogeneous coagulation of the polymer particles formed in the continuous phase on the hydrophobized silica surface.

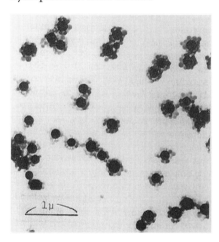

Figure 4.7 Electron micrograph of silica-polystyrene composite particles showing large silica particles surrounded by small heterocoagulated polystyrene particles. The silica beads have been made hydrophobic through the adsorption of hydroxypropyl cellulose. Reprinted from [56]; © 1986, with permission from Elsevier.

4.4.1.2 Introduction of Reactive Functional Groups on Mineral Surfaces

In recent years, much attention has been paid to the synthesis of polymers covalently attached to inorganic solids to form organophilic coatings for applications in field such as chromatography and catalysis [57–59]. Modification of the filler surface also provides interesting properties in particulate-filled composites, promoting for instance the adhesion of the inorganic material with the polymer matrix and lowering the critical surface tension of the filler particles. Graft polymerization, which involves the growth of polymer chains from the inorganic surface by a chain propagation reaction mechanism, can be accomplished provided that surface

Table 4.3 Some commercial silane and organotitanate coupling agents.

Coupling agent	Chemical structure	Acronym
Vinyl trimethoxysilane	$CH_2=CHSi(OCH_3)_3$	VTMS
3-Trimethoxysilyl propyl methacrylate	$CH_2=C(CH_3)COOCH_2CH_2CH_2Si(OCH_3)_3$	MPS
3-Trimethoxysilyl propane thiol	$HSCH_2CH_2CH_2Si(OCH_3)_3$	MPTS
Amino propyl trimethoxysilane	$H_2NCH_2CH_2CH_2Si(OCH_3)_3$	APS
Glycidoxy propyl trimethoxysilane	$CH_2(O)CHCH_2O(CH_2)_3Si(OCH_3)_3$	GPMS
Diisopropyl methacryl isostearoyl titanate	$((H_3C)_2HCO)_2Ti \big\langle {}^{OCOCH_2C_{16}H_{33}}_{OCOC(CH_3)=CH_2}$	KR7
Trimethacryl isopropyl titanate	$(CH_2=C(CH_3)COO)_3TiOCH(CH_3)_2$	/
Diisopropyl diisostearoyl titanate	$((H_3C)_2HCO)_2Ti(OCOCH_2C_{16}H_{33})_2$	KR TTS

active sites are available for reaction with a monomer. In the case of oxides, which contain a certain amount of surface hydroxyl groups, chemical modification of the surface can be achieved by using coupling agents [60–62]. The surface hydroxyl groups M–OH are reacted with organically substituted molecules of the type RM'X$_3$, where M' designates a metal or a semi-metal such as Ti, Si, Zr, or Al, X an hydrolysable group (halogen atom, amine, alkoxy), and R a nonhydrolyzable organic group (vinyl, allyl, methacryl, azo, peroxide, thiol groups, etc.). Thus, M–O–M' oxo bonds are created and, on the basis of the R groups extending outward from the surface, the particle can participate in the polymerization reaction. The most common coupling agents are chlorosilanes, alkoxysilanes and organotitanates. A list of some commercial coupling agents is provided in Table 4.3.

Apart from the use of silane or titanate molecules, inorganic particles can also be functionalized with charged molecules that bind strongly to their surface. Again, these molecules are preferentially those bearing a functionality reactive in the polymerization process – for example, an initiating group, a monomer, or a chain transfer agent. Electrostatic by nature, the interaction between these ionic compounds and the mineral surface is mostly controlled by pH. For instance, 2,2'-azo(bis)isobutyramidine dihydrochloride (AIBA; Table 4.4) was shown to adsorb irreversibly onto silica in the range of pH 4 to 10 [63]. The silica sol was precipitated by addition of the cationic initiator, and a nonionic emulsifier was introduced in the suspension to stabilize the system. This methodology produces inorganic nanoparticles with initiator-anchored functionalities and was also applied with success to layered silicate materials, as will be discussed later in Section 4.4.2.4.

Reactive groups can be also be incorporated onto inorganic surfaces by means of macromonomers or monomeric surfactants (also called "surfmers") that interact favorably with the inorganic surface through electrostatic or hydrogen-bonding attractions. For example, hydrophilic PVP-based macromonomers having styrene end-groups, proved to be efficient compatibilizers for the coating reaction of large colloidal silica particles by polystyrene during polymerization reactions performed in a mixture of ethanol and tetrahydrofuran (THF) [64]. The encapsulation proba-

Table 4.4 Main monomers and initiators used during the synthesis of organic/inorganic composite particles through emulsion polymerization.

Nomenclature		Chemical structure
1	2,2'-azo(bis) isobutyramidine dihydrochloride (AIBA) [27,42–45]	
2	Pyrrole [46]	
3	Aniline [47]	
4	4-vinyl pyridine [36–41]	
5	N-[(ω-methacryloyl)-ethyl] trimethyl ammonium chloride	

bly took place via the copolymerization of styrene with the terminal group of the macromonomer adsorbed onto the inorganic surface. PEO-based macromonomers were reported to behave in a similar way, as will be detailed further in the case of silica (see Section 4.4.2.2). In a series of articles, Nagai and coworkers also described the aqueous polymerization of surface-active monomers to coat inorganic silica particles. Not only did the reactive surfactant allow the incorporation of polymerizable groups on the oxide surface, but it also promoted monomer (ad) swelling, as illustrated schematically in Figure 4.8.

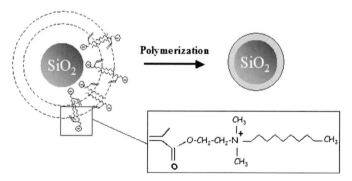

Figure 4.8 Principle of silica encapsulation through emulsion polymerization using surface-active monomers.

Table 4.5 Main surfmers and macromonomers involved in the preparation of organic/inorganic composite particles through emulsion polymerization.

Nomenclature	Chemical structure
6	*N*-dimethyl-*N*-[(ω-methacryloyl)-ethyl] alkyl ammonium chloride [32]	
7	*N*-(decadecyl styrene) trimethyl ammonium chloride [33]	
8	*N*-[(ω-methacryloyl)-decadecyl] trimethyl ammonium chloride [33]	
9	Polyethylene oxide monomethylether mono methacrylate [34]	$CH_3O\text{-}(CH_2\text{-}CH_2O)_n\text{-}CO(CH_3)=CH_2$
10	ST-PVP: 3, poly(*N*-vinyl pyrrolidone) Styrene [35]	

A full description of these systems will not be attempted here, as they are more fully discussed in the following section. Tables 4.4 and 4.5 provide a nonexhaustive list of monomers, initiators, surfmers or macromonomers which have been involved in encapsulation reactions.

4.4.1.3 Heterocoagulation

The term "heterocoagulation" generally refers to the aggregation of dissimilar particles. It is important in many modern technologies and processes such as surface coating, the manufacture of core/shell composites, ceramics processing, and in mineral flotation. The heterocoagulation process also provides an easy access to composite particles as reported elsewhere [65,66]. The general concept involves mixing together dispersions of oppositely charged particles. Typically, small organic latexes are deposited on the surface of large inorganic particles that constitute the core of the heterocoagulate, or vice-versa (Fig. 4.9).

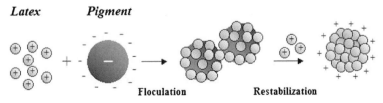

Latex   *Pigment*

Floculation   Restabilization

Figure 4.9 Schematic diagram illustrating the heterocoagulation of small, positively charged latex particles onto larger, negatively charged pigments.

For example, Kato et al. described the interaction of preformed amphoteric latex particles with the surface of titanium dioxide pigments [67]. The polymer latexes were synthesized in the presence of a zwitterionic emulsifier, N, N'-dimethyl n-lauryl betaine at pH 7.0 and showed an isoelectric point in the range of pH 7–8. Strong interactions were observed between pH 3 and 8 where the latexes were positively charged while TiO_2 particles were negatively charged. As evidenced by turbidity measurements, the mixed heterocoagulated suspensions were destabilized upon addition of an increased number of latex particles due to neutralization of the surface charge of the pigment, but restabilization occurred with further addition of the latexes. Similarly, the adsorption of cationic polystyrene latex particles onto spherical rutile titanium dioxide particles has been investigated by Vincent and colleagues [65]. The ionic strength of the suspension was shown to significantly influence the adsorption behavior. More latex particles were heterocoagulated on the TiO_2 surface when the electrolyte concentration was increased due to diminution of the electrostatic repulsion between neighboring adsorbed particles. Core-shell particles can also be obtained from heterocoagulated structures by annealing the particles at temperatures above the glass transition temperature of the surface polymer, as demonstrated by Pham and Kumacheva using titanyl-coated silica particles as the core [68]. It was shown that the total particle concentration and the number ratio N_S/N_L of small to large particles were determinant parameters to avoid particle aggregation and to produce well-defined monodisperse heterocoagulates. The N_S/N_L ratio must also exceed a critical value in order to ensure a high coverage of the seed surface and near-complete polymer spreading upon annealing, leading consequently to reasonably smooth shells.

4.4.2
Examples of Polymer-Encapsulated Pigments and Fillers

Ceramic materials are the starting materials for a large variety of applications, and play a critical role for instance in the elaboration of paper coatings, as pigments in the paint industry, or as binders for the construction industry. In the paint industry, high-quality coatings with a high gloss and color strength are generally required. In gloss and semi-gloss paints, the pigment is predominantly titanium dioxide, whilst in matt paints significant quantities of extenders such as calcium carbonate, China clay and silica are included in the paint formulation. Two basic requirements of paints are to be decorative and protective. In both applications, it is essential that the surface is uniformly and completely covered by the coating formulation. Opacity is therefore an essential requirement of latex paints, and this is most readily achieved using a pigment of high refractive index. In this respect TiO_2 is the most widely used, although ZnO or S_bO_2 can also be employed. Whatever the paint composition, high-quality coatings are obtained for optimal pigment dispersion, which is to say for optimum particle size and stability. In order to improve and to stabilize a paint dispersion, it is common to use polymeric dispersants [69]. However, this is generally not sufficient and many pigments require a surface treatment to maximize their efficiency. Indeed, one problem of paint formulation is not only to promote pigment dispersion in the latex blend but also to maintain a minimum distance between individual pigment particles in the dried film. In practice, this never occurs. However, if the polymer could be introduced onto the surface of the pigment particles so that the pigment became the core and the polymer the shell, optimal disposition of the inorganic particles within the polymeric film could be achieved. This would result in optimal light scattering and hence better opacity and good film properties [70].

4.4.2.1 Polymer Encapsulation of Titanium Dioxide Pigments

Owing to the major technological importance of TiO_2 pigments in the paint industry, we should not be surprised to observe that most published works in this field are in the patent literature [46,71]. Among the various reported strategies, emulsion polymerization is by far the most frequently used approach. In a typical procedure, the pigment particles are dispersed into water with the help of surfactant. High-shear mixing (e.g., by applying ultrasound) may also be carried out to help dissociate the pigment agglomerates [72]. A monomer or a mixture of monomers is then introduced into the suspension medium, and a water-soluble free radical initiator is subsequently added to start polymerization. In most examples, the coating takes place through admicellization/adpolymerization and involves the adsorption of surfactant molecules in a bilayer fashion, as detailed in the previous section. As mentioned above, this process is of course highly dependent upon the geometry of the bilayer structure and on the surfactant packing density, which in turn are functions of the soap concentration. Too- low an emulsifier concentration may lead to incomplete pigment coverage, while too-high a concentration may result in the formation of free polymer particles that do not participate in the coating. The nature of the surfactant also plays an important role in the coating mechanism. For example, Solc [71c,f] and Hasegawa et al. [73] claimed the use of water-soluble anionic surfactants, while Martin [71d,h] recommended the use of nonionic oxyethylenic amphiphiles. As shown by Hoy and Smith, it can also be advantageous to use a combination of surfactants. For example, these authors demonstrated that more uniform coverages and better coating efficiencies could be achieved by using an amphipathic polymer in combination with a companion surfactant (Fig. 4.10) [71g,74].

When successful, the coating provides an efficient means of controlling pigment interparticle spacing and enables better paint performances to be achieved, such as higher hiding power, tinting strength, gloss, scrub resistance, or stability compared to conventional paint formulations.

The fundamental aspects of polymer encapsulation of TiO_2 pigments were widely studied during the 1990s by the group of German and van Herk [75–81], with the successive research theses of Caris [82] and Janssen [83]. With the intention of achieving high coating efficiencies, Caris first used diisopropyl methacryl isostearoyl titanate molecules (Table 4.3) to covalently attach polymer chains onto the surface of the TiO_2 pigments, according to the coupling strategy described in the previous section [75,76]. As mentioned earlier, the role of the titanate is to promote anchoring of the growing polymer chains, and hence favor surface polymer-

veloped during the past twenty years to coat the individual silica particles with a protecting polymer layer. Since silica has a polar surface, and hence, is perfectly wetted by water, strategies need again to be developed in order to promote polymer formation on its surface. This section describes the main techniques and briefly reports on the recent advances in this area.

Silica encapsulation has been widely studied both in the open literature and in the patent literature. The first studies were reported by Hergeth et al., who described the elaboration of composite particles made from quartz powders and polyvinyl acetate through seeded emulsion polymerization [85]. These authors showed that the number of seed particles must exceed a minimal value to prevent formation of new particles and, thus promote seed particles' growth. The polymerization was proved to take place in the vicinity of the surface according to the admicellar polymerization mechanism described previously, and the so-produced "interfacial" polymer was shown to display physical properties different than those of the bulk polymer.

During the 1990s, Espiard and coworkers described the use of a polymerizable silane coupling agent, 3-trimethoxysilyl propyl methacrylate (MPS; Table 4.3), to covalently attach the polymer shell on the seed surface [86–88]. The silane molecule allowed the grafting of a significant amount of polymer since the early stages of polymerization. The grafted polymer chains formed tight loops on the silica surface into which the free polymer chains were entangled. Composite films containing up to 40% by weight of silica were produced from these latexes. These films were fully transparent up to high silica contents, and showed remarkable mechanical properties similar to those of vulcanized elastomers reinforced with solid particles [89,90] (see Section 4.6.3). This strategy was next largely applied in various papers [91–93] and patents [94] to produce aqueous dispersions of silica/polymer composite particles with a core-shell morphology.

Always in an attempt to compatibilize the core and shell materials, Yoshinaga and co-workers described the synthesis of a series of oligomeric silane molecules and their use in encapsulation reactions [95]. However, as the polymerizations were performed in the absence of surfactant, the resulting composite particles were not colloidally stable.

Since silica colloids are acidic (the IEP is ca. pH 2), basic molecules have been reported to adsorb strongly onto their surface via electrostatic attraction and acid–base chemistry (see Section 4.4.1.2). Such an approach can be advantageously used in polymerization reactions in order to provide a controlled and specific interaction of the growing polymer chains with the silica surface through a cooperative assembly-growth process. Typical examples of this general concept involve the use of cationic initiators, cationic monomers or cationic monomeric surfactant molecules (typical examples are provided in Tables 4.4 and 4.5). These molecules can be regarded as coupling agents with a double function. Part of the molecule is attracted onto silica, while the other part participates in the polymerization process by means of initiation or copolymerization reactions.

Nagai and co-workers [96] described, for example, the aqueous polymerization of a series of surface-active cationic monomers having different alkyl chain length

(C$_n$Br, structure 6, Table 4.5) immobilized on the surface of silica gel [96b]. Using a similar procedure, Yoshinaga [97] described the spontaneous formation of small polymer plots on the surface of silica particles by self-polymerization of a series of adsorbed cationic surface-active monomers in THF solutions.

In a series of articles, Armes et al. [98] synthesized polymer/silica nanocomposite particles through soapless emulsion copolymerization of styrene and acrylic monomers using 4-vinyl pyridine (4-VP) as a basic comonomer (structure 4, Table 4.4). The presence of the 4-VP comonomer ensured strong interaction of the vinyl polymers with the acidic silica surface, resulting in the formation of nanocomposite colloids with a "currant-bun" morphology characterized by silica beads assembled into colloidal aggregates cemented together by the polymer synthesized during the emulsion polymerization reaction. The small silica particles emerging from the composite surface were shown to participate to the colloidal stability of the particles. Similar conclusions have been addressed by Tiarks et al. concerning the miniemulsion polymerization of styrene using silica nanoparticles as pickering stabilizer [99].

Based on this same general idea, colloidal dispersions of nanocomposite particles made from silica cores and polymeric overlayers have been successfully prepared using appropriate cationic radical initiators, as described in a recent Japanese patent [100]. Recently, Luna-Xavier et al. also demonstrated the successful formation of nanosize silica/PMMA composite colloids using AIBA as cationic initiator and a nonionic polyoxyethylenic surfactant (NP$_{30}$) [63,91,101]. Composite particles made from silica beads surrounded by small heterocoagulated PMMA latexes or a thin polymer layer were produced, depending on the size of the silica beads (Fig. 4.11).

The role of the suspension pH and the influence of the monomer, silica and initiator concentrations on the assembly process have been investigated in depth, and analyzed in a quantitative manner. Electrostatic attraction between the polymer end groups and the negatively charged silica surface proved to be the driving force of

a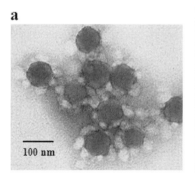

b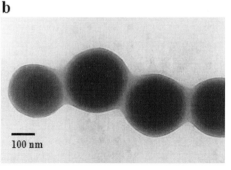

Figure 4.11 Transmission electron micrographs of the silica/PMMA nanocomposite particles obtained through emulsion polymerization in alkaline solutions using AIBA as a cationic initiator. a) Dp SiO$_2$ = 68 nm; b) Dp SiO$_2$ = 230 nm. Reprinted from [101a], © 2001, with permission from Elsevier.

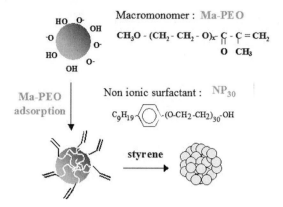

Figure 4.12 Schematic representation of the macromonomer-mediated assembly process of polymer latexes onto colloidal silica nanoparticles.

polymer encapsulation at high pH, whereas polymerization in the adsorbed surfactant bilayers appeared to be the predominant mechanism at lower pH.

Of relevance to this section, Reculusa et al. also recently demonstrated that the addition of a small amount (only 1.5 µmole/m^2) of a monomethylether mono methylmethacrylate poly(ethylene oxide) macromonomer (structure 9, Table 4.5) allowed the direct self-assembly of nanometric polystyrene latex particles on the surface of submicronic silica particles through an *in-situ* nucleation and growth process (Fig. 4.12) [102]. The two sets of particles were assembled in a raspberry-like morphology via the formation of hydrogen bonds at the interface of the inorganic and organic colloids. The size and shape of the assembly can be easily controlled by varying the sizes and stoichiometries of the colloidal components (Fig. 4.13).

Finally, it is worth noting also the recent studies of Percy and Armes on the surfactant-free synthesis of colloidal silica/PMMA nanocomposites in the absence of auxiliary comonomers and/or initiator using a commercial dispersion of silica beads in isopropyl alcohol [103]. The silica particles presumably contained long alkyl chains on their surface, and the assembly process was suspected to be driven in this case by simple hydrophobic interactions.

4.4.2.3 Polymer Encapsulation of Magnetic Particles

Magnetic particles embedded in polymer latexes are used in numerous applications such as conducting materials, catalyst carriers, inks for magnetic printers, and high-density recording media [104]. In the biomedical field, magnetic latexes are essentially dedicated to *in-vitro* diagnostics, for example cellular labeling, cell sorting, biomolecule purification, and detoxification of biological fluids [105]. The role of the magnetic component is to ensure that the particles is moving in a gradient magnetic field, and this is of great interest for robotized equipment. Although their magnetic susceptibility is much lower, magnetite Fe_3O_4 or maghemite γ-Fe_2O_3 particles are generally preferred to metallic iron, cobalt or nick-

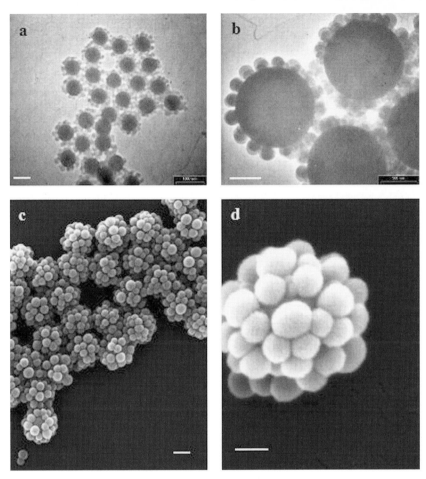

Figure 4.13 TEM (a,b) and SEM (c,d) images illustrating the macromonomer-mediated self-assembly process of colloidal polystyrene particles onto submicronic silica spheres through emulsion polymerization. (a,c) Dp SiO_2 = 1 μm (scale bar: 1 μm); (b,d) Dp SiO_2 = 500 nm (scale bar: 200 nm). Note the homogeneous distribution of the polymer particles on the silica surface. Reproduced with permission from *Chem. Mater.* **2002**, *14*, 2354–2359. © 2002 American Chemical Society.

el particles for chemical stability and biocompatibility reasons. In order to increase magnetic activity, many of these nanoparticles must be gathered together into one larger polymer particle. The polymer shell is also used not only to protect the inorganic component and to endow biocompatibility, but also to induce reactive chemical functions capable of immobilizing biological species.

Prior to their encapsulation, the magnetic nanoparticles must be thoroughly dispersed, which is a much more difficult task than with nonmagnetic particles. Indeed, the prevention of agglomeration is due to balancing not only the gravitational force and van der Waals interactions, but also the magnetic attractive forces [106]. Stable magnetic dispersions in liquids are often called "ferrofluids", because when they are concentrated enough and placed in a gradient magnetic field, the particles and the carrier fluid move together in the direction of the field. From a magnetic property point of view, these nanoparticles are superparamagnetic, meaning that they respond to a magnetic field but lose their magnetization when the field is removed. Such a behavior is strongly size-dependent and generally observed with magnetic particles with a diameter of only 10–20 nm.

In the case of aqueous ionic ferrofluids, the surface of the particles is charged and therefore the particles repel each other, if their proximity falls below a certain distance. This electrostatic stabilization is pH-sensitive and, in a pH ranging from 6 to 10, the particles flocculate (the IEP of iron oxides corresponds to pH 7). The most common synthetic route is the coprecipitation of hydrated divalent and trivalent iron salts in the presence of a strong base [107]. The IEP can be shifted to pH 2 using either citrate ligand, or silica coating; thus, above pH 4 the coated iron oxide particles are peptizable [108]. Conversely, the IEP can be shifted to pH higher than 10.5 by aminopropylsilane treatment, ensuring colloidal stability at pH lower than 8 [109]. Magnetic nanoparticles may also be dispersed on the basis of their steric (entropic) stabilization by adsorbed fatty acid layer(s) in both aqueous and organic media [104]; such dispersions are often termed "surfacted ferrofluids".

Because the iron oxide nanoparticles may be precipitated under mild conditions, pioneering studies dealt with the *in-situ* precipitation of magnetite or maghemite within the pores of preformed monosized polystyrene particles and the subsequent capping of the pores with a polymer layer to seal in the magnetic nanoparticles [110]. Commercial products are so-prepared with particle diameters of a few micrometers, iron oxide loadings up to 30 wt.%, and surface-reactive groups such as amine, hydroxyl, carboxylic, thiol and aldehyde for self-made derivatization (Dynabeads®; Dynal Biotech, Oslo). Another strategy is based on the heterocoagulation route for the synthesis of nanocomposite particles made up of inorganic cores arranged on a polymer particulate substrate as a magnetic surface layer and possibly encapsulated in a third polymer layer [111] (Fig. 4.14). This third layer is generally obtained by emulsion polymerization using the heterocoagulates as seeds. When *N*-isopropylacrylamide was used as monomer, original thermally sensitive magnetic latexes were obtained [112].

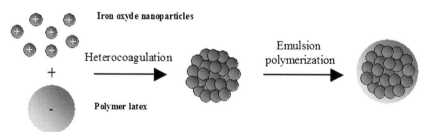

Figure 4.14 Schematic diagram showing the process of encapsulation of Fe_2O_3 particles with a trilayer structure.

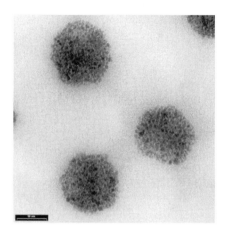

Figure 4.15 Transmission electron micrograph of magnetic poly
(hydroxyethylmethacrylate-*co*-methacrylic acid) particles. Reprinted
from [116], © 2001, with permission from Elsevier.

Emulsion polymerization techniques have also been used for the direct encapsulation of magnetic particles [113,114]. A double layer of surfactant was generally
used (sodium oleate combined with sodium dodecyl benzene sulfonate). The
method yielded up to 20 wt.% of encapsulated magnetite into polystyrene polydisperse latex particles. Sometimes, the formation of large amounts of coagulum
could not be avoided [113a].

Water-in-oil emulsion routes were also investigated by the use of hydrophilic
monomers. The emulsion-seeded copolymerization of methacrylic acid, hydroxyethyl methacrylate and a cross-linker resulted in a stable hydrophilic polymeric
shell containing up to 3 wt.% of magnetic particles [115]. Higher magnetic contents
(18 wt.%) were obtained through the use of double-hydrophilic poly(ethylene oxide)-bloc-poly(methacrylic acid) copolymers for synthesizing and stabilizing the
iron oxide nanoparticles [116]. After drying, the coated particles spontaneously
repeptized into a mixture of monomers which was emulsified into decane with the
aid of a small amount of emulsifier along with ultrasonication. Subsequent polymerization generated magnetic latex particles with regular shapes, homogeneous
compositions, and a relatively low size polydispersity (Fig. 4.15).

4.4.2.4 Polymer-Layered Silicates (PLS) Nanocomposites

Composite materials based on layered silicates have been studied for a long time.
This is due to the fact that these natural minerals are easily available and can significantly improve the properties of the host polymer. Most natural silicates have a
sheet-like structure and consist of silica tetrahedral bonded to alumina or magnesia
tetrahedral in a number of ways. Layered silicates commonly involved in nanocomposite synthesis belong to the structural family of the 2:1 phyllosilicates. These are
usually from the clay group and more specifically from the smectite group. Among
them, Montmorillonite (MMT) is by far the clay most frequently used, although

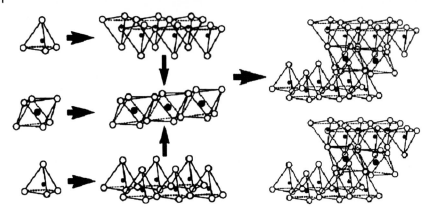

Figure 4.16 Schematic representation of smectite clay minerals formation and structure.

some recent studies have been conducted also on synthetic clays such as Laponite. Their lamellar structure consists of two-dimensional layers with a central sheet of $M_{2-3}(O)_6$ octahedra (M being either a divalent or a trivalent cation), sandwiched between two external sheets of $Si(O,OH)_4$ tetrahedra (Fig. 4.16).

The layer thickness is around 1 nm, and the lateral dimensions of these layers may vary from a few tens of nanometers to several microns, depending on the particular silicate. Individual particles consist of a stack of a given number of face-to-face associated unit layers. Isomorphic substitution of Si(IV) by Al(III) in the tetrahedral sheet, or of Al(III) by Mg(II) in the octahedral sheet, generates negative charges. These negative charges are counterbalanced by hydrated alkali or alkaline earth cations (Na^+, Ca^{2+}) located on the exterior surfaces of the unit layer packages, as well as in the interlayer. These cation layers are exchangeable and weakly bonded. Smectite clays are characterized by a moderate cation exchange capacity (CEC) generally expressed in milliequivalents per 100 grams (mEq 100 g^{-1}), with typical CECs being in the range of 60 to 120 mEq 100 g^{-1}. In addition to interlayer charges, undercoordinated metal ions (Mg^{2+}, Si^{4+}, Fe^{3+} or Al^{3+}), located on the broken edges of the crystals, can react with water molecules to form hydroxyl groups in order to complete their coordination sphere. The contribution of these edge sites to the CEC is approximately 20%, and depends on the size and shape of the clay particles. The smaller the particles' size, the more important the edges contribution. Another specificity of smectite clays is their capacity to adsorb water and other polar molecules between the sheets, thus producing a significant expansion of the interlayer spacing. Only smectite clays have this particular property of increasing the interlamellar space. Other 2:1 clay minerals such as mica and vermiculite do not expand due to their excessively high layer charge which results in strong irreversible electronic interactions between the sheets. For comprehensive and detailed information on clay colloidal chemistry and surface properties, the reader is referred to excellent text books in the domain [117].

Interest among the polymer science community for polymer-layered silicate (PLS) nanocomposites began during the mid-1990s with pioneering studies from the group of Toyota on the incorporation of Montmorillonite into nylon-6 [118]. These authors showed that the incorporation of a tiny amount of clay within the polymeric matrix allowed a large array of properties – for example, mechanical resistance, barrier properties and fire retardancy – to be significantly improved. However, due to the large surface area of the layered silicate particles and the strong interparticulate interactions which characterize most of these systems, the incorporation of clays into polymers is not straightforward and requires the development of adapted processing routes in order to yield controllable and well-defined nanophase morphologies. Three main routes are currently reported: exfoliation/adsorption; *in-situ* intercalative polymerization; and melt intercalation [119]. For similar reasons to those mentioned previously for mineral oxides, all three strategies usually require pretreatment of the clay minerals in order to improve their compatibility with the polymer matrix and to achieve a good dispersion. An organophilic clay can be produced from a naturally occurring hydrophilic clay by treating the clay mineral with silane coupling agents, or by exchanging interlayer cations with organic cations such as alkylammonium ions. During the past decade, a variety of organic cations have been incorporated into the galleries of clay minerals, including cationic surfactants, and various primary, tertiary or quaternary alkyl ammonium ions containing functional reactive groups (allyl, styryl, carboxyl, etc.). Unfortunately, it is beyond the scope of this chapter to detail all possible solutions. In general, ammonium ions render the silicate surface hydrophobic, which then makes possible the intercalation of a variety of polymers. Compared to the great deal of information available on the ion-exchange process, it should be noted that much less attention has been paid to the silanization of clay particles. The reaction occurs in this case on the broken edges of the crystal in a similar way as for the grafting of silane reagents on the surface hydroxyls of mineral oxides [120–122].

It is generally admitted that many parameters may influence the final morphology of the composite material, such as the type of clay, the nature of the polymer, and the type of polymerization performed for elaboration of the nanocomposite material. Three main structures are usually reported: segregated; intercalated; or exfoliated (Fig. 4.17). The exfoliated morphology consists of individual silicate lay-

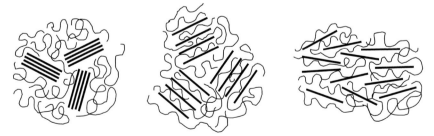

Figure 4.17 Different morphologies encountered in polymer-layered silicate nanocomposites. (Left) Phase segregated; (center) intercalated; (right) exfoliated nanocomposites.

ers dispersed in the polymer matrix as a result of extensive polymer penetration and delamination of the silicate crystallites, whereas a finite expansion of the clay layers produces intercalated nanocomposites. In general, the greatest property enhancements are observed for exfoliated nanocomposites which could be regarded as the "ideal" morphology, although in practice many systems fall short of the idealized nanostructure.

Although numerous studies have been devoted to *in-situ* intercalative polymerization in solution or in bulk [123], only a limited number of contributions have dealt with the synthesis of clay/polymer nanocomposites through emulsion polymerization [124–131].

The first papers were published by Lee et al., who reported the successful elaboration of intercalated nanocomposites based on MMT and PMMA [132], polystyrene [133] or copolymers of styrene and acrylonitrile [134] through conventional emulsion polymerization. Confinement of the polymer chains in the interlayer gallery space was evidenced by differential scanning calorimetry (DSC) and thermal gravimetric analysis (TGA) measurements, and was suspected to originate from ion-dipole interactions between the organic polymers and the MMT surface. Unfortunately, as the composite particles were precipitated, no information was provided on their morphology. However, with the clay being used as supplied, it is very unlikely that special interactions were taking place between the exfoliated clay layers and the growing latex particles in the diluted suspension medium. It can be anticipated rather that the polymer particles were physically entrapped between the clay layers consequent to flocculation and drying of the composite suspension, as shown schematically in Figure 4.18. For steric and energetic considerations, the polymer latex particles could no longer move from the interlayer space, and this re-

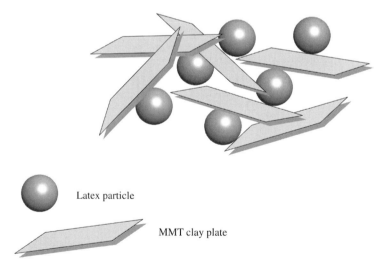

Latex particle

MMT clay plate

Figure 4.18 Suspected morphology of polymer/MMT composite materials produced through conventional emulsion polymerization without any pretreatment of the clay particles.

sulted in the polymer chains being confined in the vicinity of the clay surface, as demonstrated by the authors.

Studies involving the use of organically modified clay particles in heterophase polymerization are rather scarce. Indeed, we are aware of only two reports that combine the emulsion or suspension polymerization approaches and ion-exchange reaction. In one of these reports, AIBA is immobilized in the clay interlayer region to yield exfoliation of MMT in the PMMA matrix through suspension polymerization [135]. In another relevant study, it was demonstrated that exfoliated structures could be obtained by post-addition of an aqueous dispersion of layered silicates (either MMT or laponite) into a polymethyl methacrylate latex suspension produced in the presence of suitable cationic compounds (cationic initiator, monomer or surfactant) [136]. Since the latex particles were cationic and the clay platelets anionic, strong electrostatic forces were developed at the polymer/clay interface.

Following this line, Negrete-Herrera et al. recently synthesized PLS nanocomposites based on a synthetic clay: Laponite and a film-forming copolymer of styrene and butyl acrylate (polystyrene-*co*-butylacrylate) through emulsion polymerization. The main advantage of using Laponite instead of Montmorillonite is the dimension of the crystals (e.g., 1 nm thick and 40 nm long), which is of the same order of magnitude as the diameters of the polymer latexes. In the first paper of their series, these authors demonstrated that nanocomposite colloids based on Laponite could be readily obtained by grafting of a polymerizable silane on the clay surface [137]. The emulsion polymerization reaction was accomplished in conventional manner using potassium persulfate as initiator and sodium dodecyl sulfate as surfactant. Stable composite latexes with diameters in the range of 50 to 150 nm were successfully produced, provided that the original clay suspension was stable enough. The clay plates were found to be located at the external surface of the polymer latex particles (Fig. 4.19a). These authors demonstrated in a subsequent study that a similar morphology was achieved when the clay was previously ion-exchanged with a cationic initiator (AIBA, structure 1, Table 4.4) or a cationic monomer (MADQUAT, structure 5, Table 4.4) as illustrated in transmission electron microscopy (TEM) images (Fig. 4.19b and c) [138].

The resulting latex suspensions were cast into film materials by evaporating the water. Analysis of an ultrathin cross-section of the film by cryo-TEM indicated successful composite latex particle coalescence and film formation, the clay plates forming a three-dimensional honey-like structure as a consequence of their original localization at the external surface of the polymer particles (Fig. 4.20).

It seems obvious, with regard to the above film nanostructure, that such PLS materials could find applications in the coating industry as flame retardants without damaging the mechanical performances or the optical transparency of the material [139]. Although flame-retardant paints have been known for many years, this is still an active domain of research in this area, and the incorporation of clays into these materials may open new perspectives.

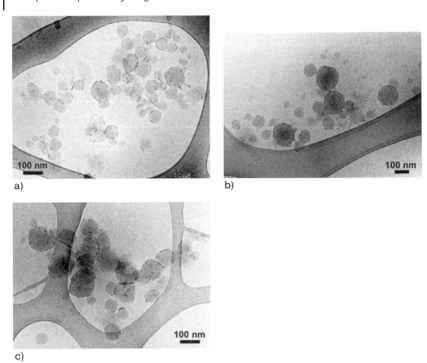

a)

b)

c)

Figure 4.19 Cryo-TEM images of poly(styrene-*co*-butyl acrylate)/laponite nanocomposite particles prepared through emulsion polymerization using: (a) γ-MPS; (b) AIBA; and (c) MADQUAT as reactive compatibilizers.

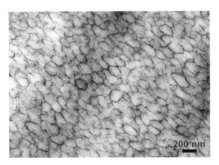

Figure 4.20 Cryo-TEM image of a thin section of the film material produced by coalescence of laponite/poly(styrene-*co*-butyl acrylate) composite latexes, showing the internal microstructure of the film.

4.4.2.5 Miscellaneous

As mentioned previously, polymer-encapsulated pigments are of major implication in the coating industry and, consequently, a variety of monomers have been used to produce the shell, including vinyl acetate as well as combinations of vinyl acetate, isobutyl acrylate, styrene, *n*-butyl acrylate, methyl methacrylate, acrylonitrile, and isobutyl vinyl ether. In principle, a huge number of inorganic solids – including metal powders such as iron, steel, titanium, cobalt, nickel, gold, platinum or copper, metal oxides (CuO, ZnO, Al_2O_3, CeO_2), and various fillers, such as talc, mica, barytes, calcium carbonate, china clay, and dolomite – can be covered with polymers according to the general synthetic procedures described above for TiO_2, silica, iron oxide or clays. However, in practice, the syntheses need to be adapted to each individual situation. For example, metallic surfaces do not display the same reactivity as metal oxides or carbonates and sulfates, and specific strategies must be developed. Similarly, it is clear that the surface properties of metal oxides may suffer significant changes from one oxide to another depending on the pH of the suspension and the ionic strength, as well as on the acidic or basic character of the surface. Some selected examples are reported in the following paragraphs.

Anisotropic lamellar pigments have been used in anticorrosive coatings for many years. The most commonly used pigment for this purpose is ferric mica [140], but fine metallic zinc, "bronze" alloy or aluminum powders having flake morphologies have also been reported for decorative paint applications [30]. Aluminum flakes with large diameters, when thoroughly dispersed, provide a high metallic sheen to a film, paint or ink, one which has a polished, mirror-like finish. Dispersions of metal powders in aqueous suspension require the metal to be passivated (oxidized) to prevent reaction in the liquid phase. For example, Batzilla et al. described the encapsulation reaction of commercial aluminum pigments through an emulsion-like polymerization process [141]. These authors reported the use of phosphorus-containing protecting agents to provide a good dispersion of the Al-pigment and control its reactivity. They also recommended the use of monomers with strong adhesion to the metal surface (e.g., carboxylic acid derivatives). In a related study, silver nanoparticles have been recovered with a thin polymer layer via emulsion polymerization of styrene and MMA monomers in the presence of oleic acid [142]. The fatty acid derivative was shown to be essential in order to achieve pigment encapsulation.

Another class of extenders used widely in coating applications is calcium carbonate ($CaCO_3$). Yu et al. recently reported the preparation of $CaCO_3$/polystyrene composite particles using a polymerizable silane derivative previously attached onto the mineral surface [143]. The pretreated $CaCO_3$ particles were shown to act as comonomers during the emulsion polymerization process in similar manner as discussed previously for the case of silica.

Semiconductor particles can also be used advantageously in coating applications to provide specific optical response to the material. As an example, Kumacheva et al. recently described the synthesis of monodisperse nanocomposite particles with inorganic CdS nanocrystals sandwiched between a PMMA core and a P(MMA-co-BA) outer copolymer shell layer. The particles are prepared by emulsion polymerization in three steps (Fig. 4.21) [144]. In a first step, polymer latexes are used as host matrices for CdS nanocrystals formation [145,146]. To do so, monodisperse poly(methyl methacrylate-co-methacrylic acid) (PMMA-PMAA) latex particles were ion-exchanged with a $Cd(ClO_4)_2$ solution. The Cd^{2+} ions thus introduced into the electrical double layer were further reduced into CdS nanoclusters by addition of a Na_2S solution. The CdS-loaded nanocomposite particles were subsequently recov-

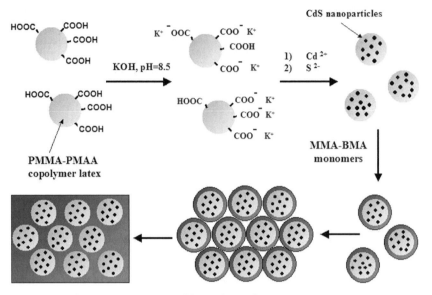

Figure 4.21 Schematic representation of the synthesis of PMMA-*co*-PMAA/CdS/PMMA-*co*-BuA multilayered hybrid particles with a periodic structure. Redrawn and adapted from [146].

ered by a film-forming polymer shell by reacting MMA and butyl acrylate monomers. The resulting colloidal nanocomposites were finally assembled in 3D periodic arrays consisting of rigid PMMA-PMAA/CdS core particles regularly distributed within the soft polymer matrix.

Periodic structures of polyacrylic/silver colloids have been elaborated in a similar way using Ag$^+$ ions as precursors. The method clearly opens a new avenue for producing optically responsive materials with a controlled periodicity. Nanocomposite materials with several other functions can be elaborated by this technique using different types of nanoparticles and organic polymers, as illustrated in a recent study of the synthesis of multilayered gold-silica-polystyrene core shell particles through seeded emulsion polymerization [147]. Here, silica-coated gold colloids were encapsulated by polystyrene using MPS as silane coupling agent according to the procedure described in Section 4.4.2.2 for silica. These particles were subsequently transformed into hollow spheres by chemical etching of the silica core in an acidic medium.

Following a related procedure, Kamata et al. reported the elaboration of multishell composite particles and hollow polybenzyl methacrylate (PBzMA) beads containing movable gold cores [148]. The core was made from silica-coated gold colloids, while the shell was produced by atom transfer radical polymerization (ATRP) using a procedure similar to that described in Section 4.4.4. The silica shell sandwiched between the gold core and the polymer outer layer was selectively dissolved using aqueous HF to generate the hollow particles, the morphology of which is illustrated in Figure 4.22.

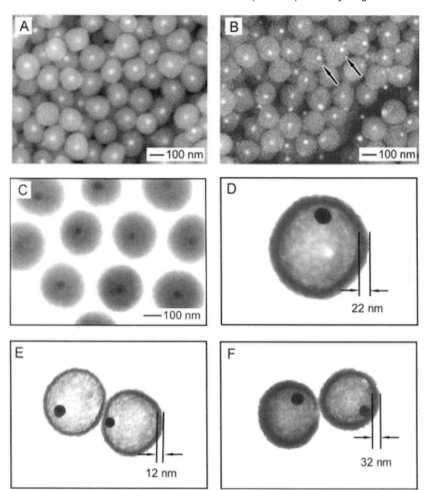

Figure 4.22 (A,B) Backscattering SEM and (C,D) TEM images of Au-SiO$_2$-PBzMA particles before (A,C) and after (B,D) HF etching. The polymerization time was 4 h, and the polymer shell was ~22 nm thick. (E,F) TEM images of Au-Air-PBzMA synthesized using different polymerization times: (E) 3 h, (F) 6 h. The polymer shells were ~2 nm and ~32 nm thick, respectively. Reprinted with permission from [148]. © 2003, American Chemical Society.

4.4.3
Other Heterophase Polymer Encapsulation Techniques

In addition to the emulsion polymerization technique, miniemulsion, dispersion and suspension polymerizations also provide versatile routes to polymer encapsulation of inorganic particles.

4.4.3.1 Miniemulsion Polymerization

In contrast to emulsion polymerization, which involves the formation of large monomer droplets dispersed in the aqueous phase, in a miniemulsion process the monomer phase is stabilized in the form of small droplets with diameters in the range of typically 30 to 500 nm, which perform the role of nanoreactors. This concept of nanoreactor is particularly attractive for the elaboration of a variety of nanoparticulate materials that could not be achieved by conventional emulsion polymerization, including polycondensates, metals, ceramics, hybrid polymers, and nanocomposites as recently reviewed by Landfester [23b]. Following this nanoreactor concept, the miniemulsion technique has been applied successfully to the encapsulation of a variety of inorganic particles including titanium dioxide pigments [149], carbon black [150], calcium carbonate [150a], magnetic nanoparticles [151], and organic pigments [152,153]. The principle of pigment encapsulation through miniemulsion polymerization is discussed in detail in Chapter 2, and will not be treated further here.

4.4.3.2 Dispersion Polymerization

The use of dispersion polymerization to produce composite particles was first reported by Bourgeat-Lami and Lang, who synthesized silica/polystyrene colloids in a mixture of ethanol and water using MPS as coupling agent and PVP as steric stabilizer [55,154]. In a preliminary article, these authors first demonstrated that the use of a coupling agent was a prerequisite condition for encapsulation to succeed [55]. In fact, no polymer was produced around the silica particles when polymerization was carried out in the absence of MPS. In a following study, the diameter of the silica beads was varied over a large range (from 13 to 630 nm), and the authors described the influence of the number of silica beads on the shape and composition of the composite particles. The average number of silica beads per polymer particle was statistically determined from the TEM images, and found to vary from one to several thousand when increasing the number of silica beads in the suspension medium. While the particles containing only one silica bead had an irregular contour, those loaded with a larger amount of silica presented a spherical shape (Fig. 4.23a). Nanocomposite particles with some exotic morphologies were also obtained under specific experimental conditions.

With regard to the mechanism of nanocomposite particle formation, it was argued that small segregated polymer domains were formed on the silica surface from the early stage of polymerization by copolymerization of styrene with the methacrylate group of the coupling agent. These polymer nuclei continued to grow until they formed a continuous contour by coalescence with neighboring growing latex particles, the shape of which was determined by the initial number of growing polymer domains. Regular core-shell morphologies were thus obtained using poly(styrene)-*b*-poly(ethylene oxide) block copolymers as stabilizers. Indeed, in this case, the number of polymer particles synthesized was much larger than the number of silica beads, and the polymer nuclei could recover the inorganic surface from the start of polymerization. Subsequent growth allowed the formation of a thin and smooth polymer layer surrounding the silica particles. Silica-free latexes were also

a)

b)

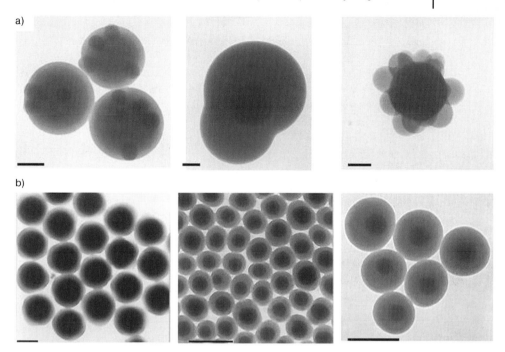

Figure 4.23 TEM images of polystyrene-coated MPS-grafted silica beads obtained through dispersion polymerization using (a) PVP and (b) PS-*b*-PEO block copolymers as steric stabilizers. Scale bar: 250 nm. Reprinted from [16b] with permission.

formed under these conditions, but these could be easily separated from the composite particles by centrifugation (Fig. 4.23b).

Following a similar route, Sondi et al. also described the formation of a protective poly(*tert*-butyl acrylate) layer on the surface of MPS-functionalized silica nanoparticles [155]. The amount of bound polymer was found to depend on the MPS-grafting density, which in turn was a function of the initial MPS concentration. The silica particles, the surface of which was efficiently recovered by both grafted and un-grafted polymers, showed an improved resistance to chemical etching. These studies highlighted the potential interest of encapsulated mineral oxide particles in photo-resistant technologies.

As described above for emulsion polymerization, macromonomers can also be used advantageously to promote polymer encapsulation of inorganic particles. PVP-based macromonomers with styrene end-groups were shown to be efficient compatibilizers during the coating reaction of large colloidal silica particles by dispersion polymerization of styrene into a mixture of ethanol and THF [64]. The encapsulation was shown to take place via the copolymerization of styrene, with the terminal group of the macromonomer adsorbed onto the inorganic surface.

Finally, it is worth mentioning that dispersion polymerization has also been used by Matijevic et al. [156,157] for the coating of a series of metal oxide particles

(α-Fe_2O_3, CeO_2, CuO, TiO_2, SiO_2) with polyaniline and polypyrrole, two conducting polymers in a mixture of ethanol and water using poly(vinyl alcohol) as steric stabilizer. In the serendipitous studies of Armes et al., silica/conducting polymer colloidal nanocomposites were also produced by the oxidative polymerization of either aniline or pyrrole via dispersion polymerization in aqueous media on spherical or "stringy" silica nanoparticles [158]. Electrically conducting organic polymers have attracted the attention of scientists for several decades due to their unique physical and chemical properties [159]. The elaboration, properties and applications of conductive coatings are reviewed in Chapter 6, and will not be discussed further here.

4.4.3.3 Suspension Polymerization

In the past, few reports have dealt with the encapsulation of inorganic particles through a suspension polymerization technique. Bakhshaee et al. reported a study with carbon blacks [160], while Vincent et al. described the encapsulation of silica particles in PMMA beads [161]. A more comprehensive study was later reported by Duguet and co-workers regarding the encapsulation of alumina particles (diameter 0.5 µm) in PMMA microspheres [162]. Following their preliminary surface treatment with 3-(trimethoxysilyl)propylmethacrylate, the alumina particles were readily dispersed in the initiator/MMA solution. This first dispersion was then used to form droplets suspended in water according to the double dispersion principle, by using poly(vinyl alcohol) as a suspension stabilizer. When polymerization was complete, composite alumina/PMMA beads were recovered, with thermogravimetric analysis indicating an alumina content of up to 25 wt.%. As generally observed in typical MMA suspension polymerization, the greater the quantity of PVA, the finer the beads. More surprisingly, the alumina to MMA weight ratio also appeared to be critical to achieve the desired particle-size distribution in the final product: the greater the amount of alumina, the larger the composite beads.

A nonconventional suspension polymerization process was also investigated by utilizing a water-agarose gel as a suspending phase. Reactor agitation was only at an early stage in the process to fix the droplet size, and stopped when the temper-

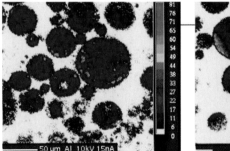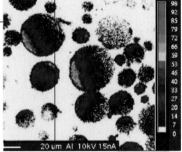
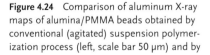

Figure 4.24 Comparison of aluminum X-ray maps of alumina/PMMA beads obtained by conventional (agitated) suspension polymerization process (left, scale bar 50 µm) and by the suspension polymerization process in the water-agarose gelled system (right, scale bar 20 µm). Reproduced from [162a] with permission.

ature reached the gelling point of the agarose solution (42 °C). Polymerization was performed at 70 °C and, when complete, the gelled phase was reverted to an aqueous solution simply by stirring and warming to 90 °C. Under such conditions, apparently similar composite alumina/PMMA beads were obtained, though the aluminum X-ray map obtained by electron probe microanalysis showed that the lack of agitation and a higher density of alumina favored the concentration of inorganic particles into the lower hemisphere of the beads, resembling a loaded dice (Fig. 4.24).

4.4.4
Surface-Initiated Polymerizations

Apart from the formation of dense coatings, core-shell organic-inorganic particles can also be elaborated by templating inorganic colloids with polymer brushes using conventional or living polymerization techniques. The main point that differentiates these systems from those described previously is that the suspension medium is a good solvent for the polymer being formed, and this results in the growth of a hairy solvated polymer layer around each individual particle. Upon solvent evaporation, collapse of the polymer chains provides an effective, protective, dense coating similar to that achieved during the formation of core-shell colloid through heterophase polymerization.

In contrast to encapsulation reactions achieved by heterophase polymerization, which mainly involves free radical polymerization, the grafting of polymers to inorganic surfaces (i.e., metal oxides, clay minerals, metals and semiconductor nanoparticles) can be performed in a variety of ways through anionic [163], cationic [57], or catalytic polymerizations. For example, hybrid nanoparticles with a block copolymer shell structure have been synthesized by ring-opening polymerization (ROP) of norbornenyl groups immobilized onto gold colloids [164]. ROP of ε-caprolactone has also been recently conducted from the surface of silica gel and cadmium sulfide nanoparticles [165]. Amine or alcohol groups have been attached to the inorganic surface using silane coupling agents in order to provide a covalent anchoring of the macromolecules into hydrolytically stable polymer brushes. Poly(ethylene oxide) nanoparticles were grown in a similar manner from the surface of silica particles using glycidoxypropyl trimethoxysilane as co-initiator and coupling agent [166]. Between 1980 and 1996, Tsubokawa described the grafting of various polymers from the surface of silica nanoparticles through cationic and free radical polymerizations using covalently grafted peroxyester and diazo functions [167]. In azo-initiated reactions, ungrafted polymer chains were also produced, since the initiator was attached to the surface by only one end. In order to achieve a better control of the grafting density, Prucker and Rühe described the designed synthesis of an asymmetric diazo silane coupling agent [168]. The azo compound contained a cleavable ester group in order to facilitate degrafting of the polymer chains for analytical purposes, and was monofunctional to control self-assembly of the chlorosilane azo initiator on the silica surface. Polymerization was performed in toluene at 60 °C, and again afforded grafted and free polymer chains. The grafting density was found to depend upon the silane concentration and polymerization time. Despite the confinement of the growing active centers on the silica surface, high molecular-weight polymers were formed without the occurrence of excessive branching or crosslinking reactions.

In recent years, several groups have reported the synthesis of polymer-grafted nanoparticles with controlled molecular weights and molecular weight distributions, using controlled radical polymerization (CRP) techniques. One key advantage of CRP in comparison to conventional free radical processes is the possibility of synthesizing well-defined polymers, which can be grown with the desired thickness and composition. The technique allows the formation of densely grafted and hairy outer polymer layers on nanoparticulate inorganic surfaces. Owing to the narrow molecular weight polydispersity of the polymer chains, the grafted particles can self-organize into 2D arrays with controlled interparticle distances function of the degree of advancement of the reaction.

CRP is usually divided into three categories: (i) ATRP; (ii) reversible radical addition fragmentation chain transfer (RAFT); and (iii) nitroxide-mediated polymerization (NMP). All three techniques permit the polymer molecular weight, the polydispersity, and the polymer architecture to be accurately controlled, and have been used to build up highly dense polymer brushes from inorganic particles. A list of macro-initiators which have been developed recently for this purpose is provided in Table 4.6.

The ATRP technique was described extensively by von Werne and Patten in the early 2000s using silica nanoparticles as inorganic colloid [169]. A general synthetic strategy basically involves the covalent attachment of various halide-functionalized ATRP macroinitiators (see Table 4.6) on the inorganic surface, and the subsequent controlled/living growth reaction of the polymer chains from the anchored initiator molecules [170,171]. Not only silica but also aluminum oxide particles [172], magnetic colloids [173,174], gold [175] and photoluminescent cadmium sulfide nanoparticles [176] have been used as macro initiators. When cast from solution, the resulting nanocomposite film materials exhibit hexagonal ordering of the inorganic cores and properties arising from the inorganic component (Fig. 4.25).

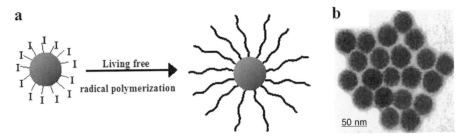

a

Living free radical polymerization

b

50 nm

Figure 4.25 (a) Reaction scheme for the synthesis of polymer-grafted inorganic particles via controlled radical polymerization using chemically anchored CRP macroinitiators. (b) TEM illustration of the CdS/SiO$_2$/PMMA hybrid nanoparticles produced by atom transfer radical polymerization initiated from the CdS/SiO$_2$ nanoparticles surface. Adapted with permission from [176b]. © 2001, American Chemical Society.

Table 4.6 Chemical structures of the macro-initiators involved in the CRP polymerization of a variety of monomers from nanoparticulate inorganic surfaces.

Conventional free radical polymerization	Mineral/monomer
(structure: O–Si with cyclohexene ring bearing COOR (C=O) and COOH groups)	Silica/methyl methacrylate, styrene, acrylonitrile
—Si—OOR	Silica/methyl methacrylate
—Si—(CH₂)₃ O–C(=O)–(CH₂)₃–C(CH₃)(CN)–N=N–C(CH₃)(CN)–CH₃	Silica gel/styrene
—R—NH–C(=O)–(CH₂)₂–C(CH₃)(CN)–N=N–C(CH₃)(CN)–(CH₂)₃–C(=O)–OH	Silica gel/styrene
—Si(OEt)₂–(CH₂)₃–O–CH₂–C(H)(OH)–CH₂–O–C(=O)–(CH₂)₂–C(CH₃)(CN)–N=N–C(CH₃)(CN)–(CH₂)₂–COOH	Silica gel/styrene

Nitroxide-mediated polymerization	Mineral/monomer
(structure: O–Si–(chain)–O–CH₂–phenyl with nitroxide alkoxyamine group)	Colloidal silica/styrene, maleic anhydride
—Si(OEt)₂–(CH₂)₁₀–C(=O)–O–CH₂–CH(phenyl)–O–N(ᵗBu)–CH–P(=O)(OEt)₂ (ᵗBu)	Silica gel/styrene

Atom-transfer radical polymerization	Mineral/monomer
—O–Si–(CH₂)₃–O–C(=O)–CH(CH₃)–Br , CuBr : dNbipy	Silica/styrene

Table 4.6 Continue

Atom-transfer radical polymerization	Mineral/monomer
$\equiv$Si$-$O$-$Si$-$(CH$_2$)$_3$$-O-$C($=$O)$-$C(CH$_3$)(CH$_3$)$-$Br , CuBr : dNbipy	Silica/styrene, SStNa, DEA, NaVBA, DMA*
$\equiv$Si$-$O$-$Si$-$(CH$_2$)$_3$$-O-$C($=$O)$-$C(CH$_3$)(CH$_3$)$-$Br , CuBr : dNbipy	CdS/SiO$_2$/MMA**
$\equiv$Si$-$O$-$Si$-$(CH$_2$)$_2$$-C_6H_4$$-$CHCl$-CH_3$, CuCl : dNbipy	Silica/styrene
$\equiv$Si$-$O$-$Si$-$(CH$_2$)$_{11}$$-O-$C($=$O)$-$CHCl$-C_6H_5$, CuCl : dNbipy	Silica/styrene
HO$-$C($=$O)$-$CH$_2$CH$_2$$-$Cl , CuCl : dNbipy	MnFe$_2$O$_4$/styrene
$\equiv$Si$-$O$-$C($=$O)$-$CH(CH$_3$)$-$Br , CuBr/PMDETA	Alumine/MMA**
Br(CH$_3$)C$-$... H$_3$C$-$N$^+$(CH$_3$)(CH$_3$)$-$(CH$_2$)$_{11}$$-O-$C($=$O)$-$C(CH$_3$)$_2$$-$Br , CuBr : HMTETA	Montmorillonite/MMA**
HS$-$(CH$_2$)$_{11}$$-O-$C($=$O)$-$C(CH$_3$)$_2$$-$Br , CuBr : Me$_6$tren	Gold/butyl acrylate Gold/MMA**

Reversible addition fragmentation chain transfer (RAFT)	Mineral/monomer
$\equiv$Si$-$(CH$_2$)$_{11}$O$-$C($=$O)$-$(CH$_2$)$_2$$-$C(CH$_3$)(CN)$-N=N-$C(CH$_3$)(CN)$-CH_3$ + Ph$-$C(CH$_3$)$_2$$-S-$C($=$S)$-$Ph	Silica gel/styrene

* SStNa: sodium styrene sulfonate; DEA: 2-(diethyl amino ethyl) methacrylate; NaVBA: sodium 4, vinyl benzoate; DMA: 2-(dimethyl amino ethyl) methacrylate.
** MMA: methyl methacrylate.

For example, when silica-coated photoconductive CdS nanoparticles were used as inorganic particles, the resultant materials were shown to retain the photoluminescent properties of the core. However, no mention was made about the colloidal stability of these systems.

The ATRP technique has also been recently reported to be effective in aqueous media using hydrophilic water-soluble acrylic monomers [177]. Finally, it should be mentioned that hollow polymeric microspheres have been produced through ATRP by templating silica microspheres with PBzMA, and subsequently removing the core by chemical etching [178]. These studies illustrate the potential of graft polymerizations in the production of nanostructured particles.

It has been shown recently that the so-called stable free radical polymerization (SFRP) can also be used to initiate the polymerization of vinyl monomers from inorganic surfaces [179]. Although a library of nitroxide and nitroxide-based alkoxyamine compounds has been recently reported in the literature for the living free radical polymerization of a variety of monomers, extrapolation of the NMP technique to the grafting of inorganic surfaces requires the development of adequate surface-active initiators and has been much less explored. Reactive unimolecular alkoxyamine initiators carrying trichlorosilyl [180,181] or triethoxysilyl [182] end-groups for further attachment onto mineral substrates have been synthesized, and used with success for example in the growth reaction of polymer chains with controlled molecular weights and well-defined architectures from the surface of silica particles. One of the prime advantage of these unimolecular systems is the possibility to control accurately the structure and concentration of the initiating species. However, a major drawback is the multi-step reaction required for synthesis of the functional alkoxyamine. Thus, Parvole et al. [183] and Kasseh et al. [184] reported a bimolecular system based on the strategy of Rühe and colleagues [168], in which the NMP process is initiated from an azo or a peroxide initiator attached to fumed silica. Whilst successful, this approach still involves a two-step chemical reaction to synthesize the functional azoic or peroxidic initiator. Therefore, Bartholome and colleagues recently described a versatile one-step synthetic strategy based on the simultaneous reaction of a polymerizable silane, a source of radical and N-*tert*-butyl-N-[1-diethylphosphono-(2,2-dimethylpropyl)] nitroxide used as spin trap (Fig. 4.26) [185].

As for ATRP, the SFRP technique can also be used advantageously for the designed construction of nanoparticles and nanomaterials with new shapes and structures. Following this line, shell-crosslinked polymeric capsules have been elaborated in a multistep procedure by templating colloidal silica with polymeric compounds and crosslinking the polymer shell [180]. Micrometric silica beads were first modified by grafting on their surface a chlorosilane alkoxylamine initiator (see Table 4.6). Copolymers were then grown from the surface-attached initiator using an appropriate amount of sacrificial "free" alkoxylamine. The copolymer chains were designed to carry maleic anhydride functional groups for further crosslinking reactions. A diamine crosslinker was added in a third step to effect interchain coupling via the formation of a bisimide. The inorganic silica template was finally removed in a last step by chemical etching. In an alternative strategy, styrene

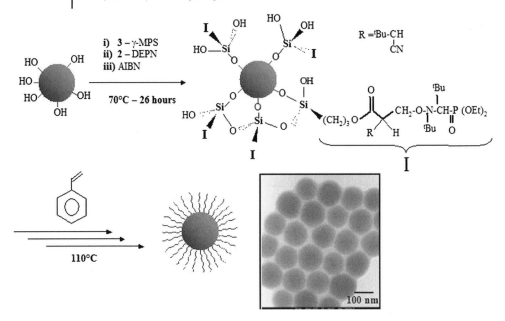

Figure 4.26 Reaction scheme for one-step covalent bonding of a DEPN-based alkoxyamine initiator onto silica particles and subsequent grafting of polystyrene from the functionalized silica surface.

monomer was copolymerized with 4-vinylbenzocyclobutene, and the resulting nanocomposite core/shell particles were heated at 200 °C for thermal crosslinking.

4.5
Organic/Inorganic Particle Morphology

Both thermodynamic and kinetic factors are known to affect the morphology of conventional polymer latex particles. Moreover, emulsion polymerization processes performed in the presence of polymeric latex particles as seeds lead to original biphasic morphologies as the consequence of phase separation phenomena [186]. As a single example described in this section, particles of polystyrene partially crosslinked with divinylbenzene were used as microsized seeds [186g]. A mixture of monomers (styrene and divinylbenzene, with a weight composition different from that of the seeds) was allowed to swell the precursors overnight. A standard emulsion polymerization was then carried out during 24 h at various temperatures, and particles with dumbbell-like or snowman-like morphologies were observed. On the whole, the phase separation process was favored by increasing either the monomer to polymer swelling ratio, the seed crosslink density, the seed particle size, the temperature, or the crosslinker concentration in the mixture of swelling monomers (all other parameters being constant). By varying the nature of the seeds and the monomer mixture, more exotic structures were obtained such as ovoid-like and red blood corpuscle-like [186a], golf ball-like [186b], raspberry-like, and octopus-like [186d]. Theoretical explanations taking into account geometric as well as thermodynamic considerations have been proposed to predict the final particles' morphology [186e,f].

Although the influence of encapsulation reactions on particles' morphology has never been investigated in depth, different types of morphologies have been reported in the literature for encapsulated minerals. Each of these may be described as a variation on the ideal core-shell morphology, where a single inorganic core is coated with a polymer shell of constant thickness (Fig. 4.27). The number of possible morphologies is lower than for polymeric seeds, since inorganic particles are neither able to be swelled by monomers, nor are deformable.

When the surface of the inorganic particles is efficiently modified to promote interactions with monomers and/or oligoradicals (as described earlier), there is every chance that the ideal core-shell morphology is obtained. Nevertheless, morphologies frequently encountered in encapsulation reactions are also internal occluded domains of several inorganic particles, or of agglomerated pigments into the composite colloid (see Fig. 4.27). The occurrence of the formation of such morphologies has been described earlier, and appeared either to be inherent to the dispersion and suspension-derived encapsulation techniques, or to be the consequence of aggregation phenomena in emulsion polymerization routes (see below).

"Inverted" raspberry-like morphologies (the mineral particles being located at the surface of the latex spheres) have also been discussed in Sections 4.4.2.2 and 4.4.2.4 about colloidal silica and layered silicates, respectively. These are mainly a consequence of the surfactant-like behavior of the inorganic particles in specific situations. This was clearly illustrated in a recent report by Landfester, who showed that silica or clays can be used as pickering stabilizers of miniemulsion polymerizations, resulting therefore in the formation of "armored" latexes, the surface of which was recovered by the small inorganic particles [99,131].

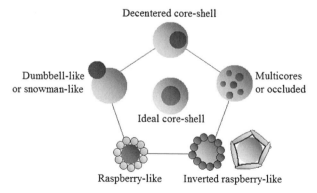

Figure 4.27 Morphologies of organic/inorganic latex particles obtained through an heterophase polymerization technique (cross-sections).

Raspberry-like morphologies are generally obtained by heterocoagulation processes, and may be converted in core-shell particles by annealing [68]. However, raspberry-like morphologies may also be obtained spontaneously when the inorganic particles are large enough and when their surface is only moderately modified for promoting interactions. Under such conditions, capture of the oligomeric radical species by the inorganic surface is possible, but the latex particles grow independently, since the wetting is not efficient enough. Such behavior was observed when silica seeds were previously allowed to react with a monomethylether mono methyl-methacrylate poly(ethylene oxide) macromonomer (1.5 µmol m^{-2}) prior to styrene emulsion polymerization (as described in Section 4.4.2.2; see Fig. 4.13) [102]. It was also reported that the average number of polystyrene nodules per silica particle could be tuned by varying the diameter and the concentration of the silica seeds – that is, by adjusting the particle number ratio N_{latex}/N_{silica}. When this ratio is equal to 1, original dumbbell-like or snowman-like morphologies were obtained [187]. Multipod-like morphologies with five, six, or eight polystyrene nodules per silica particle (Fig. 4.28)

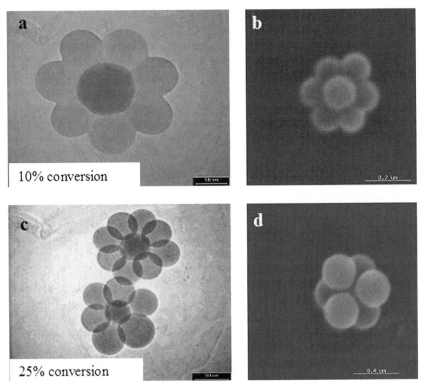

Figure 4.28 Daisy-shaped and multipod-like silica/polystyrene particles produced by emulsion polymerization. The silica particles have been surface-modified by a moderate amount of the MPS silane coupling agent (namely 0.1 molec nm^{-2}). Reproduced with permission from *NanoLetters* **2004**, *4*, 1677–1682. ©Copyright 2004, American Chemical Society.

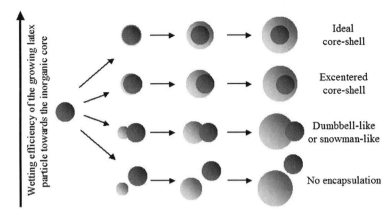

Figure 4.29 Schematic morphology evolution of organic/inorganic latex particles produced by seeded emulsion polymerization as a function of their respective affinities (cross-sections).

were reported in the case of silica seeds previously treated with small amounts of methacryloxymethyltriethoxysilane (in the range 0.1 to 1 molecule nm^{-2}) [188]. This simple method will most likely open the way for the fabrication of large amounts of original assemblies.

Excentered core-shell morphologies may be considered as asymmetrical structures in which the inorganic phase is situated at the border of the composite particles. In fact, they are very probably the result of an intermediate situation between the ideal core-shell morphology and the snowman-like one, as depicted schematically in Figure 4.29.

Finally, it would be too simplistic to consider that the nature and the adsorbed amount of promoting molecules are the only two parameters that allow the control of hybrid particles' morphology. Because interfacial tensions are heavily involved, the nature of the emulsion stabilizer may also significantly influence the final morphology. For example, by simply changing the type of surfactant used in the polymerization recipe, the particle morphology can change from excentered core-shell to core-shell, and vice versa. Clearly, systematic studies will need to be performed in order to rationalize the role of the numerous parameters responsible to obtain a given morphology. Moreover, predictive models – similar to those developed for composite latex particles containing two or more polymers – should be elaborated in order to better understand and describe the morphologies of organic/inorganic colloids in the different situations mentioned above.

Although such morphologic predictions are not yet available, much effort has been made in recent years to characterize the morphology of organic/inorganic latexes, using a variety of analytical techniques. Direct observation techniques such as scanning and transmission electron microscopy are by far the most frequently used as they allow direct imaging of the particles' shape and composition. Samples are prepared as for conventional latexes by depositing one drop of the latex suspension onto a carbon-coated copper grid and drying in air. Since the mineral par-

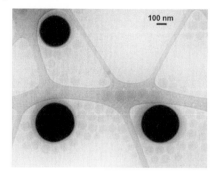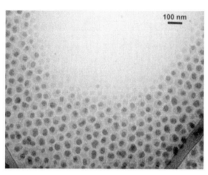

Figure 4.30 Cryo-TEM images of silica/poly(butyl acrylate) composite latexes produced in emulsion polymerization using MPS-grafted silica particles as the seed (unpublished results). The silica particle diameter is 530 nm (left) and 50 nm (right), respectively.

ticles appear darker than the polymer, they can be easily localized on the electron micrographs. When required, staining with heavy atom compounds such as uranyl acetate or phosphotungstic acid is performed in order to produce higher contrasts and to identify the different phases. Other complementary techniques such as atomic force microscopy [55] or dark field electron microscopy [189] have also been reported, though these are much less popular. Cold-stage (cryo) imaging is another useful SEM or TEM procedure for soft polymers (T_g below the ambient temperature). In this procedure, the particle suspension is frozen in liquid nitrogen, introduced into the microscope sample chamber at low temperature, and particles are observed as frozen in ice. Typical cryo-TEM images of clay/polymer and silica/polymer composite particles are shown in Figures 4.19 and 4.30, respectively.

The surface of composite particles can also be characterized using indirect spectroscopy techniques such as X-ray photoelectron spectroscopy (XPS) and aqueous electrophoresis (see reports of Armes and colleagues [98,103,190a]). These authors used XPS to analyze the surface composition of a series of vinyl polymer-silica colloidal nanocomposites produced in the presence of an auxiliary 4-VP monomer that helped cement together the ultrafine silica particles and the polymer into composite nanoparticles. When 4-VP was involved as the monomer, XPS analysis showed the surface composition of the particles to be similar to the bulk composition. This suggested a "currant-bun" morphology, with the silica particles being distributed uniformly throughout the composite latex spheres. In contrast, when styrene was used as the comonomer, the resulting poly(styrene-co-vinyl pyridine)/silica nanocomposite particles were distinctly silica-rich, suggesting a "core-shell" morphology (with the silica particles forming the shell).

A deeper insight into the nanomorphology of the composite particles was provided by electron spectroscopic imaging (ESI) [190b]. The technique allows the establishment of elemental map compositions, and involves the use of conventional TEM and spectrometry. The method can be described briefly as follows. When an electron beam passes through the sample, interaction with electrons of different elements results in characteristic energy losses. Electrons with a well-defined energy are selected by a prism-mirror system. If only elastic electrons are selected, a bright field image is obtained. However, when monochromatic inelastic electrons are selected, electron spectroscopic images are formed that show the local concentration of a given element (bright region). As shown in Figure 4.31, ESI analysis enabled the "currant-bun" morphology characteristics of the poly(4-VP)/silica nanocomposite particles to be confirmed, and also allowed identification of the "inverted" core-shell morphology of the polystyrene/silica nanocomposite particles produced in the absence of auxiliary monomer, with the silica sol forming again a well-defined monolayer surrounding the nanocomposite cores (Fig. 4.31b) [103].

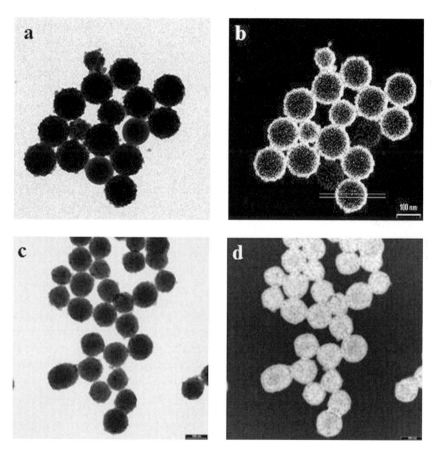

Figure 4.31 Typical bright-field (a,c) and energy-filtered (25 eV; b,c) transmission electron micrographs obtained for the polymer/silica nanocomposite particles (scale bar 100 nm). The top images show a "core-shell"-type morphology for polystyrene/silica nanocomposite particles, while the bottom images depict the "currant bun" morphology that is characteristic of poly(4-vinylpyridine)/silica nanocomposite particles. Reprinted with permission from [190]. © 2005, American Chemical Society.

Another important aspect related to the control of composite particles' morphology in encapsulation reactions is the occurrence of severe coagulation during polymerization (see above). Latex instability is most often reliable for pigment destabilization, and can be monitored indirectly by sudden changes in polymerization kinetics. For example, the temporary fall in polymerization rate noted by Caris [78] during PMMA encapsulation of titanium dioxide pigments was attributed to pigment agglomeration during the course of the coating reaction. Hasegawa [191] also reported that $CaCO_3$ powders have a strong tendency to aggregate over a certain range of SDS surfactant concentrations. The decrease in polymerization rate with increasing SDS concentration was found to follow the sedimentation curves of the inorganic pigment, supporting the fact that aggregated rather than individual pigments have been encapsulated. The lower the overall particle number, the slower the polymerization rate. Such instability appeared to be even more pronounced in the case of titanium dioxide pigments, for which it was reported that coagulation can take place under certain experimental conditions. Guyot et al. have also demonstrated [189], using dark-field and bright-field microscopy, that hydrophobic MPS-grafted silica nanoparticles, incorporated into poly(ethyl acrylate) latexes by *in-situ* emulsion polymerization, formed agglomerates at the surface of the latex particles. The aggregation was additionally proven by dynamic light scattering, with the argument being made that addition of the surfactant by a semi-continuous feed process was responsible for the composite particles' agglomeration.

4.6
Properties and Applications

As detailed in the previous sections, the coating of inorganic particles with organic shells is of interest in many applications. The inorganic materials may display, for example, a number of benefits such as heat resistance and optical, magnetic, conductive, or mechanical properties. In turn, the polymer coating may protect the inorganic core from interactions with the environment or improve the processability of the composite due to the more flexible character of the polymer. Most of these properties depend not only on the chemical composition but also on the dispersion state of the inorganic particles into the composite material, and can vary significantly with the size, shape, and concentration of the mineral.

4.6.1
Improved Film Properties in Latex Paints

As mentioned above (see Section 4.4.2.1), the coating of individual pigment particles with polymers enables circumvention of the inherent problem of pigment agglomeration in water-based paints. An optimum spacing of individual particles can be reached, and this results in a more efficient hiding of the encapsulated pigments. Uniform distribution of pigments in latex paints may also improve film appearance and performance, with higher gloss, less tack and better adhesion (due to less internal stress development) being observed during film formation. Another advantage gained by coating pigments with a continuous organic polymer layer is that they are protected from interactions with other pigments or additives, and from environmental aggression such as UV irradiation and pH variations. Better storage stability, color stability and durability are also obtained after coating [46]. In a similar vein, it has been claimed recently that the properties of silver halide photographic materials could be improved by coating the photographic film with an aqueous dispersion of core-shell type composite particles, with colloidal silica as the core and vinylic polymers as the shell. In this respect, higher pressure durability, anti-adhesive properties in presence of water and higher tensile strengths were obtained, while the glossy character, transparency and brittleness of the film were not affected by the presence of the inorganic particles.

4.6.2
Optical, Thermal, and Conductive Properties of Organic/Inorganic Colloids

Remarkable light-sensitive effects were observed by Haga et al. [192,193] for encapsulated zinc oxide and cadmium sulfide particles in comparison to conventional dispersion blend composites. It was reported that the high photoconductivity of the composite materials was due to enhanced interfacial interactions between the mineral and the polymer matrix. Haga also studied the thermal behavior of encapsulated titanium dioxide pigments, and found that the TiO_2/PMMA or TiO_2/PS composites had two thermal relaxation regions as opposed to a unique relaxation peak observed on the DSC curves for the polymer matrix. In addition, the encapsulated materials showed better thermal stability than the pure polymer. Polymer–clay hybrids also showed substantially enhanced thermal stability [130]. The improved thermal properties of the nanocomposites have been attributed in many cases to the strong fixation and reduced molecular motion of the polymer chains intercalated in the clay galleries. The conductivity of metal oxide–polypyrrole nanocomposites was also significantly enhanced in comparison to bare polypyrrole [194]. Here again, the increase in conductivity was attributed to improved interactions between the conducting polymer particles as a consequence of the nanocomposite structure.

4.6.3
Mechanical Properties of Filled Thermoplastics

In the field of polymers, one practical goal is to provide a simple adjustment of mechanical properties (for example, in elastomers a strong internal mechanical reinforcement is required) by incorporating high surface/volume interactive inorganic fillers. However, when conventional filler blending occurs under high-viscosity conditions it is difficult to control the morphological structure of the resultant material and the degree of particle dispersion. Heterogeneous distributions of the agglomerated filler particles and poor polymer–filler interactions often result from mechanical blending, giving independent phases of the organic polymer and inorganic particles at the micron level. The mechanical properties of polymer blends can be substantially improved by modifying the surface of the mineral powder using encapsulation techniques. It appears that interactions at the near interface between the inorganic core and the polymer promote adhesion, and this will play a very important role in the reinforcement mechanism. It has also been shown [e.g., 87,89,90] that polymer grafting on the surface of functionalized silica particles during an emulsion polymerization reaction results in remarkable mechanical properties of the latexes, similar to those obtained with vulcanized elastomers reinforced with solid particles (Fig. 4.32).

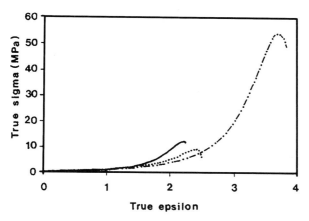

Figure 4.32 Stress–strain curves of poly(ethyl acrylate-*co*-hydroxyethyl methacrylate) films containing 6 wt.% MPS-functionalized silica (_ .. _ ..), nonfunctionalized silica (....), and no silica (-) Reproduced from [89b], with permission.

Although less spectacular, an increase in tensile strength has been reported for PMMA/CaSO$_3$ composites made by *in-situ* emulsion polymerization compared to composites made from mechanical blending [195].

4.7
Conclusions and Future Challenges

Although the synthesis of polymer colloids has been known for almost a century, the elaboration of organic/inorganic particles is only in its infancy. Some fifty years ago, early methods focused primarily on pigment encapsulation for the paint industry, and were mainly motivated by the problem of pigment dispersion. Since then, research interest in polymer-encapsulated minerals and the so-called colloidal nanocomposites has "exploded" such that the subject now relates to a wide variety panel of inorganic particles including mineral oxides, aluminosilicates, metal colloids, and semiconductors.

A variety of synthetic approaches have been developed in order to promote polymer formation on the inorganic surface and to build up the targeted nanocomposite morphology. Most of these methods are based on surface modification techniques, the main purpose being to incorporate suitable functional groups that can further participate in the polymerization reaction. These methods have been shown to be effective and to permit good control over the particles' morphology. Although remarkable advances have been made during recent years, much remains to be done in relation not only to the fundamental aspects associated with the elaboration of this new class of materials but also to the development of new concepts and compatibilization techniques that will broaden the range of organic/inorganic composite particles and assemblies that can be elaborated through heterophase polymerization.

In addition to core-shell particles, as has been reported in a number of recent reviews, organic/inorganic colloids will be able to exhibit a much larger variety of morphologies with potential advantages over the conventional core-shell structure. Among such interest is the elaboration of dissymmetrical organic/inorganic colloids (e.g., dumbbell-like and snowman-like particles). Likewise, colloids with a multilayer shell structure (e.g., mineral particles sandwiched between a polymer core and a polymer shell) might be of interest for specific applications and broaden the range of nanostructures that could be achieved with heterophase polymerization.

Although at present it seems possible to synthesize nanocomposite colloids with new morphological characteristics and outstanding properties, it will in the near future be necessary to face the demands of the coating industry in terms of inorganic solid fraction and/or overall solid content. The majority of commercial polymer latexes have a solid content of at least 40%; indeed, some have a polymer content of almost 70%. In addition, it may be necessary to introduce a significant proportion of fillers into these latexes, according to the application required. Both of these aspects represent clear challenges, and future efforts will undoubtedly lead to the development of robust and reliable strategies that can be extrapolated to large-scale production.

Abbreviations

4-VP	4-vinyl pyridine
AIBA	2,2'-azo(bis)isobutyramidine dihydrochloride
ATRP	atom transfer radical polymerization
CEC	cation exchange capacity
CMC	critical micelle concentration
CRP	controlled radical polymerization
DSC	differential scanning calorimetry
ESI	electron spectroscopic imaging
HPC	hydroxypropyl cellulose
IEP	isoelectric point

LCST   lower critical solution temperature
MMA   methyl methacrylate
MMT   Montmorillonite
MPS   3-trimethoxysilyl propyl methacrylate
NMP   nitroxide-mediated polymerization
PBzMA   polybenzyl methacrylate
PEO   poly(ethylene oxide)
PLS   polymer-layered silicate
PMMA-PMAA   poly(methyl methacrylate-*co*-methacrylic acid)
PVP   poly(*N*-vinyl pyrrolidone)
RAFT   reversible addition fragmentation chain transfer
ROP   ring-opening polymerization
SBR   styrene-butadiene rubber
SDS   sodium dodecyl sulfate
SEM   scanning electron microscopy
SFRP   stable free radical polymerization
TEM   transmission electron microscopy
TGA   thermal gravimetric analysis
THF   tetrahydrofuran
XPS   X-ray photoelectron spectroscopy

References

1. G. Schmid, *Nanoparticles*, Wiley VCH, Weinheim, **2003**.
2. F. Caruso, *Colloids and colloid assemblies: synthesis, modification, organization and utilization of colloid particles*, Wiley VCH, Weinheim, **2004**.
3. L.M. Liz-Marzan, P.V. Kamat, *Nanoscale Materials*, Kluwer Academic Publishers, Springer Verlag, Boston, **2003**.
4. J. Zhang, Z.-L.Wang, J. Liu, S. Chen, G-Y. Liu, *Self-Assembled Nanostructures*, Nanostructure Science and Technology Series, MA Kluwer Academic Publishers, Boston, **2002**.
5. P. Knauth, J. Schoonman, *Nanostructured Materials: Selected Synthesis Methods, Properties and Applications*, Electronic Materials Science and Technology Series, MA Kluwer Academic Publishers, Boston, **2002**.
6. (a) F. Caruso, H. Lichtenfeld, M. Giersig, H. Möhwald, Electrostatic self-assembly of silica nanoparticle-polyelectrolyte multilayers on polystyrene latex spheres, *J. Am. Chem. Soc.* **1998**, *120*, 8523–8524; (b) A.S. Susha, F. Caruso, A.L. Rogach, G.B.

Sukhorukov, A. Kornowski, H. Möhwald, M. Giersig, A. Eychmüller, H. Weller H, Formation of luminescent spherical core-shell particles by the consecutive adsorption of polyelectrolyte and CdTe(S) nanocrystals on latex colloids, *Colloid Surface A* **2000**, *163*, 39–44; (c) F. Caruso, Hollow capsule processing through colloidal templating and self-assembly, *Chem. Eur. J.* **2000**, *6*, 413–419.
7. (a) R.A. Caruso, M. Antonietti, Sol-gel nanocoatings: An approach to the preparation of structured materials, *Chem. Mater.* **2001**, *13*, 3272–3282; (b) F. Caruso, Nanoengineering of particle surfaces, *Adv. Mater.* **2001**, *13*, 11–22.
8. (a) H. Tamai, S. Hamamoto, F. Nishiyama, H. Yasuda. Ultrafine metal particles immobilized on styrene/acrylic acid copolymer particles, *J. Colloid Interf. Sci.* **1995**, *171*, 250–253; (b) C.-W. Chen, M.-Q. Chen, T. Serizawa, M. Akashi, In situ synthesis and the catalytic properties of platinum colloids on polystyrene microspheres with surface-grafted poly(N-isopropylacrylamide), *Chem.*

Commun. **1998**, 831–832; (c) A. Dok-outchaev, J. Thomas James, S.C. Koene, S. Pathak, G.K. Surya Prakash, M.E. Thompson, Colloidal metal deposition onto functionalized polystyrene microspheres, *Chem. Mater.* **1999**, *11*, 2389–2399; (d) P.H. Wang, C.-Y. Pan, Ultrafine palladium particles immobilized on polymer microspheres, *Colloid Polym. Sci.* **2001**, *279*, 171–177.

9. C.H.M. Hofman-Caris, Polymers at the surface of oxide nanoparticles, *New J. Chem.* **1994**, *18*, 1087–1096.

10. J.A. Fendler, F.C. Meldrum, The colloid chemical approach to nanostructured materials, *Adv. Mater.* **1995**, *7*, 607–632

11. (a) A.M. van Herk, Encapsulation of inorganic particles, in: *Polymeric Dispersions: Principles and Applications*, J.M. Asua (Ed.), NATO ASI Series E: Applied Sciences 335, Kluwer Academic, Dordrecht, **1997**, pp. 435–450; (b) A.M. van Herk, A.L. German, Microencapsulated pigments and fillers, in: *Microspheres Microcapsules & Liposomes, Vol. 1, Preparation & Chemical Applications*, R. Arshady (Ed.), Citus Books, London, **1999**, Chap. 17, pp. 457–486.

12. A.D. Pomogalio, Hybrid polymer/inorganic nanocomposites, *Russ. Chem. Rev.* **2000**, *69*, 53–80.

13. G. Kickelbick, Concepts for the incorporation of inorganic building blocks into organic polymers on a nanoscale, *Prog. Polym. Sci.* **2002**, *28*, 83–114.

14. V. Castelvetro, C. De Vita, Nanostructured hybrid materials from aqueous polymer dispersions, *Adv. Colloid Interface Sci.* **2004**, *108-109*, 167–185.

15. G. Kickelbick, U. Schubert, Organic functionalization of metal oxide nanoparticles, in: *Synthesis, Functionalization and Surface Treatment of Nanoparticles*, M.I. Baraton (Ed.), American Scientific Publishers, Los Angeles, **2003**, Chap. 6, pp. 1–12.

16. (a) E. Bourgeat-Lami, Organic/inorganic nanocomposites by multiphase polymerization, in: *Dendrimers, Assemblies and Nanocomposites*, R. Arshady, A. Guyot (Eds.), MML Series 5, Citus Books, London, **2002**, Chap. 5, pp. 149–194; (b) E. Bourgeat-Lami, Organic-inorganic nanostructured colloids. *J. Nanosci. Nanotechnol.* **2002**, *2*, 1–24; (c) E. Bourgeat-Lami, Or-

ganic/inorganic nanocomposite colloids, in: *Encyclopedia of Nanoscience and Nanotechnology*, H.S. Nalwa (Ed.), American Scientific Publishers, Los Angeles, **2004**, Vol. 8, pp. 305–332.

17. J.E. Mark, C.Y.-C. Lee, P.A. Bianconi, *Hybrid Organic-Inorganic Composites*, ACS Symposium Series 585, Washington DC, **1995**.

18. (a) R. Arshady, Manufacturing methodology of microspheres, in: *Preparation and Chemical Applications*, R. Arshady (Ed.), MML Series 1, Citus Book, London, **1999**, Chap. 4, pp. 85–124; (b) K. Tauer, Heterophase Polymerization, in: *Encyclopedia of Polymer Science and Technology*, H.F. Mark (Ed.), Wiley & Sons, Inc., 3rd edition, **2004**; (c) M. Antonietti, K. Tauer, 90 years of polymer latexes and heterophase polymerization: more vital than ever, *Macromol. Chem. Phys.* **2003**, *204*, 207–219; (d) K. Tauer, Latex particles, in: *Colloids and Colloid Assemblies*, F. Caruso (Ed.), Wiley VCH, Weinheim, **2004**, pp. 1–51.

19. (a) I. Piirma, *Emulsion Polymerization*, Academic Press, New York, **1982**; (b) G.W. Poehlein, R.H. Ottewill, J.W. Goodwin, *Science and Technology of Polymer Colloids*, Vol. II, Martinus Nijhoff, Boston MA, **1983**; (c) D.H. Everett, *Basic principles of colloid science*, Royal Society of Chemistry, London, **1988**; (d) W.B. Russel, D.A. Saville, W.R. Schowalter, *Colloidal Dispersions*, Cambridge University Press, Cambridge, **1989**; (e) R.J. Hunter, *Introduction to Modern Colloid Science*, Oxford University Press, Oxford, **1993**; (f) J.M. Asua, Emulsion polymerization: from fundamental mechanisms to process developments, *J. Polym. Sci. Part A: Polym. Chem.* **2004**, *42*, 1025–1041.

20. K.E.J. Barret, *Dispersion Polymerisation in Organic Media*, Wiley, London, **1975**.

21. (a) S. Kawaguchi, M.A. Winnik, K. Ito, Dispersion copolymerization of n-butyl methacrylate with poly(ethylene oxide) macromonomers in methanol-water. Comparison of experiment with theory, *Macromolecules* **1995**, *28*, 1159–1166; (b) J. Liu, L.M. Gan, C.H. Chew, C.H. Quek, H. Gong, L.H. Gan, The particle size of latexes from dispersion polymerization of styrene using poly(ethylene oxide) macromonomers as a polymerizable stabi-

lizer, *J. Polym. Sci., Part A: Polym. Chem.* **1997**, *35*, 3575–3583; (c) E. Bourgeat-Lami, A. Guyot, Thiol-ended poly(ethylene oxide) as reactive stabilizer for dispersion polymerization of styrene, *Colloid Polym. Sci.* **1997**, *275*, 716–729.

22. (a) G. Riess, C. Labbe, Block copolymers in emulsion and dispersion polymerization, *Macromol. Rapid Commun.* **2004**, *25*, 401–435; (b) H. Imai, S. Kawaguchi, K. Ito, Syntheses of poly(ethylene oxide-b-styrene oxide) macromonomers and their application to emulsion and dispersion copolymerization with styrene, *Polym. J. (Tokyo)* **2003**, *35*, 528–534.

23. (a) J.M. Asua, Miniemulsion polymerization, *Prog. Polym. Sci.* **2002**, 27, 1283–1346; (b) K. Landfester, Polyreactions in miniemulsion, *Macromol. Rapid Commun.* **2001**, *22*, 896–936; (c) K. Landfester, The generation of nanoparticles in miniemulsions, *Adv. Mater.* **2001**, *13*, 765–768.

24. (a) K.O. Calvert, in: *Polymer Lattices and their Applications*, Applied Science Publishers, London, **1982**; (b) D. Urban, K. Takamura, in: *Polymer Dispersions and their Industrial Applications*, Wiley VCH, Weinheim, **2002**.

25. Y. Xia, B. Gates, Y. Yin, Y. Lu, Monodispersed colloidal spheres: old materials with new applications, *Adv. Mater.* **2000**, *12*, 693–713.

26. O.D. Velev, A.M. Lenhoff, Colloidal crystals as templates for porous materials, *Curr. Opin. Colloid Interf. Sci.* **2000**, *5*, 56–63.

27. R.K. Iler, *The chemistry of silica*, John Wiley & Sons, New York, **1979**.

28. R.J. Hunter, *Zeta potential in Colloid Science*, Academic Press, London, **1981**.

29. G.A. Parks, The isoelectric points of solid oxides, solid hydroxides, and aqueous hydroxo complex systems, *Chem. Rev.* **1965**, 65, 177–198.

30. R.F. Conley, Function and Character of Particulate Surfaces, in: *Practical Dispersion. A guide to understanding and formulating slurries*, R.F. Conley (Ed.), VCH Publishers, New York, **1996**, Chap. 10, pp. 267–306.

31. G.D. Parfitt, K.S.W. Singh, *Characterization of Powder Surfaces*, Academic Press, London, **1976**.

32. G. Buxbaum, *Industrial Inorganic Pigments*, John Wiley & Sons, Weinheim, **1998**.

33. A.G. Abel, Pigments for paints, in: *Paints and Surface Coatings. Theory and Practice* (2nd edition), R. Lambourne, T.A. Strivens (Eds.), Ellis Horwood, Chichester, **1987**, Chap. 3.

34. For a review in this field, see: F. Tiberg, J. Brinck, L. Grant, Adsorption and surface-induced self-assembly of surfactants at the solid-aqueous interface, *Curr. Opin. Colloid Interf. Sci.* **2000**, *4*, 411–419.

35. K. Meguro, The interaction between pigment and surfactant, in: *Presence and Future in Science and Technology*, Moeller, Ungeheur and Ulmer (Eds.), 19th FATIPEC Congress, Ludwigsburg, **1988**, Vol. I, pp. 49–63.

36. K. Esumi, Interactions between surfactants and particles: dispersion, surface modification and adsolubilization, *J. Colloid Interf. Sci.* **2001**, *241*, 1–17.

37. R. Denoyel, J. Rouquerol, Thermodynamic (including microcalorimetry) study of the adsorption of nonionic and anionic surfactants onto silica, kaolin and alumina, *J. Colloid Interf. Sci.* **1991**, *143*, 555–572.

38. P. Somasundaran, E.D. Snell, Q. Xu, Adsorption behaviour of alkylarylethoxylated alcohols on silica, *J. Colloid Interf. Sci.* **1991**, *144*, 165–173.

39. D.N. Furlong, J.R. Aston, Adsorption of polyoxyethylated nonyl phenols at silica/aqueous solution interfaces, *Colloids Surf.* **1982**, *4*, 121–129.

40. R. Denoyel, F. Rouquerol, J. Rouquerol, *Adsorption from Solution*, C. Rochester (Ed.), Academic Press, London, **1982**.

41. M.W. Rutlanda, T.J. Senden, Adsorption of the poly(oxyethylene) nonionic surfactant $C_{12}E_5$ to silica: a study using atomic force microscopy, *Langmuir* **1993**, 412–418.

42. J.S. Clunie, B.T. Ingram, *Adsorption from Solution at the Solid/Liquid Interface*, G.D. Parfitt, C.H. Rochester (Eds.), Academic Press, San Diego, **1983**.

43. P. Levitz, H. Van Damme, D. Keravis, Fluorescence decay study of the adsorption of non ionic surfactants at the solid-liquid interface. 1. Structure of the adsorption layer on a hydrophilic solid, *J. Phys. Chem.* **1984**, *88*, 2228–2235.

44. G.G. Warr, Surfactant adsorbed layer structure at solid/liquid solution interfaces: impact and implications of AFM imaging studies, *Curr. Opin. Colloid Interf. Sci.*, **2000**, *5*, 88–94.

45. J.H. Scamehorn, R.S. Schechter, W.J. Wade, Adsorption of surfactants on mineral oxide surfaces from aqueous solutions. I. Isomerically pure anionic surfactants, *J. Colloid Interf. Sci.* **1982**, *85*, 463–478.

46. (a) P. Newmann, Coated pigments, US Patent 3,133,893 (**1964**); (b) Polymer-coated pigments, French Patent 1,504,833 (**1967**).

47. (a) J. Wu, J.H. Harwell, E.A. O'Rear, Two-dimensional reaction solvents: surfactant bilayers in the formation of ultrathin films, *Langmuir* **1987**, *3*, 531–537; (b) J.H. O'Haver, J.H. Harwell, L.R. Evans, W.H. Waddell, Polar copolymer-surface modified precipitated silica, *J. Appl. Polym. Sci.* **1996**, *59*, 1427–1435.

48. M. Hasegawa, K. Arai, S. Saito, Effect of surfactant adsorbed on encapsulation of fine inorganic powder with soapless emulsion polymerization, *J. Polym. Sci., Part A: Polym. Chem.*, **1987**, *25*, 3231–3239.

49. K. Meguro, T. Yabe, S. Ishioka, K. Kato, K. Esumi, Polymerization of styrene adsolubilized in surfactant adsorbed bilayer on pigments, *Bull. Chem. Soc. Jpn.* **1986**, 3019–3021.

50. (a) J.H. O'Haver, J.H. Harwell, E.A. O'Rear, L.J. Snodgrass, W.H. Waddell, In situ formation of polystyrene in adsorbed surfactant bilayers on precipitated silica, *Langmuir* **1994**, *10*, 2588–2593; (b) W.H. Waddell, J.H. O'Haver, L.R. Evans, J.H. Harwell, Organic Polymer Surface Modified Precipitated Silica, *J. Appl. Polym. Sci.* **1995**, *55*, 1627–1641; (c) J.H. O'Haver, J.H. Harwell, L.R. Evans, W.H. Waddell, Polar copolymer-surface modified precipitated silica, *J. Appl. Polym. Sci.* **1996**, *59*, 1427–1435.

51. Th.F. Tadros, *The effects of polymers on dispersion properties*, Academic Press, London, **1982**.

52. M.A. Cohen Stuart, G.J. Fleer, B.H. Bijsterbosch, The adsorption of poly(vinyl pyrrolidone) onto silica. I. Adsorbed amount, *J. Colloid Interf. Sci.* **1982**, *90*, 310–320.

53. M.A. Cohen Stuart, G.J. Fleer, B.H. Bijsterbosch, Adsorption of poly(vinyl pyrrolidone) on silica. II. The fraction of bound segments, measured by a variety of techniques, *J. Colloid Interf. Sci.* **1982**, *90*, 321–334.

54. R.S. Parnas, M. Chaimberg, V. Taepaisitphongse, Y. Cohen, The adsorption of polyvinylpyrrolidone and poly(ethylene oxide) onto chemically modified silica, *J. Colloid Interf. Sci.* **1989**, *129*, 441–450.

55. E. Bourgeat-Lami, J. Lang, Encapsulation of inorganic particles by dispersion polymerization in polar media. 1. Encapsulation of inorganic particles by dispersion polymerization in polar media, *J. Colloid Interf. Sci.* **1998**, *197*, 293–308.

56. K. Furusawa, Y. Kimura, T. Tagawa, Synthesis of composite polystyrene lattices with silica particles in the core, *J. Colloid Interf. Sci.* **1986**, *109*, 69–76.

57. (a) N. Fery, R. Hoene, K. Hamann, Graft reactions on silicon dioxide surfaces, *Angew. Chem., Int. Ed. Engl.* **1972**, *11*, 337; (b) R. Kroker, M. Schneider, K. Hamann, Polymer reactions on powder surfaces, *Prog. Org. Coat.* **1972**, *1*, 23–43; (c) R. Laible, K. Hamann, Formation of chemically bound polymer layers on oxide surfaces and their role in colloidal stability, *Adv. Colloid Interf. Sci.* **1980**, *13*, 65–99.

58. M. Jiang, S. Wang, X. Jin, A new approach to graft polymerization on oxide surfaces, *J. Mater. Sci. Lett.* **1990**, *9*, 1239–1240.

59. (a) N. Tsubokawa, H. Ishida, K. Hashimoto, Effect of initiating groups introduced onto ultrafine silica on the molecular weight polystyrene grafted onto the surface, *Polym. Bull.* **1993**, *31*, 457–464; (b) N. Tsubokawa, Functionalization of carbon black by surface grafting of polymers, *Prog. Polym. Sci.* **1992**, *17*, 417–470; (c) D.L. Huber, G. Carlson, K. Gonsalves, The formation of polymer monolayers: from adsorption to surface initiated polymerizations, in: *Interfacial Aspects of Multicomponent Polymer Materials*, Proceedings of the American Chemical Society Symposium, pp. 107–122, D.J. Lohse, T.P. Russell, L.H. Sperling (Eds.), Plenum Press, New York, **1997**; (d) K. Yoshinaga, Surface modification of inorganic particles, in: *Fine Particles*, Surfactant Science Series 92, T. Sugimoto (Ed.), Marcel Dekker, New York,

2000, pp. 626–646; (e) Y. Cohen, W. Yoshida, V. Nguyen, N. Bei, J.-D. Jou, Surface modification of inorganic oxide surfaces by graft polymerization, in: *Oxide Surfaces*, Surfactant Science Series 103, J.A. Wingrave (Ed.), Marcel Dekker, New York, **2001**, pp. 321–353.

60. P. Plueddemann, *Silane Coupling Agents*, Second Edition, Plenum Press, New York **1991**.

61. D.E. Leyden, *Silanes, Surfaces and Interfaces*, Gordon & Breach, New York, **1985**.

62. P. Van Der Voort, E.F. Vansant, Silylation of the silica surface: a review, *J. Liq. Chromatogr. Relat. Technol.* **1996**, *19*, 2723–2752.

63. J.L. Luna-Xavier, E. Bourgeat-Lami, A. Guyot, The role of initiation in the synthesis of silica/poly(methyl methacrylate) (PMMA) nanocomposite latex particles through emulsion polymerization, *Colloid Polym. Sci.* **2001**, *279*, 947–958.

64. K. Yoshinaga, T. Yokoyama, T. Kito, Polystyrene coating of monodispersed colloidal metal oxides by grafting to hydrophilic macromer adsorbed on the surface, *Polym. Adv. Technol.* **1993**, *4*, 38–42.

65. N.J. Marston, B. Vincent, N.G. Wright, The synthesis of spherical rutile titanium dioxide particles and their interaction with polystyrene latex particles of opposite charge, *Prog. Colloid Polym. Sci.* **1998**, *109*, 278–282.

66. (a) E. Bleier, E. Matijevic, Heterocoagulation. I. Interaction of monodispersed chromium hydroxide with polyvinyl chloride latex, *J. Colloid Interf. Sci.* **1976**, *55*, 510–523; (b) H. Sasaki, E. Matijevic, E. Barouch, Heterocoagulation. VI. Interactions of monodispersed hydrous aluminum oxide sol with polystyrene latex, *J. Colloid Interf. Sci.* **1980**, *76*, 319–329; (c) K. Csoban, E. Pefferkorn, Perikinetic aggregation induced by chromium hydrolytic polymer and sol, *J. Colloid Interf. Sci.* **1998**, *205*, 516–527.

67. K. Kato, M. Kobayashi, K. Esumi, K. Meguro, Interaction of pigments with polystyrene latex prepared using a zwitterionic emulsifier, *Colloids Surf.* **1987**, *23*, 159–170.

68. H.H. Pham, E. Kumacheva, Core-shell particles: building blocks for advanced

polymer materials, *Macromol. Symp.* **2003**, *192*, 191–205.

69. (a) R.F. Conley, *Practical Dispersion – A Guide to Understanding and Formulating Slurries*, VCH Publishers, New York, **1996**; (b) J.D. Schofield, Extending the boundaries of dispersant technology, *Prog. Org. Coat.* **2002**, *45*, 249–257.

70. W.-H. Hou, T.B. Lloyd, F.M. Fowkes, Pigmented Polymer Particles with Controlled Morphologies, in: *Polymer Latexes: preparation, characterization and applications*, E.S. Daniels, T. Sudol and M. El-Aasser (Eds.), ACS Symposium Series 492, Washington DC, **1992**, Chap. 25, pp. 405–421.

71. (a) P.R. Sperry, R.J. Wiersema, K. Nyi, Dispersing paint pigments, US Patent 4,102,843 **(1978)**; (b) J.D. Schofield, Dispersible inorganic pigment, US Patent 4,349,389 **(1982)**; (c) J. Solc, Colloidal size hydrophobic polymers particulate having discrete particles of an inorganic material dispersed therein, US Patent 4,421,660 **(1983)**; (d) R.W. Martin, Polymer-encapsulated dispersed solids and methods, Eur. Patent Appl. 0104498 **(1984)**; (e) R.W. Martin, Method of encapsulating finely divided solid particles, US Patent 4,608,401 **(1986)**; (f) J. Solc, Method for preparing colloidal size particulate, US Patent 4,680,200 **(1987)**; (g) K.L. Hoy, C.W. Glancy, J.M.O. Lewis, Micro-composite systems and processes for making same, Eur. Patent Appl. 0392065 **(1990)**; (h) R.W. Martin, Encapsulating finely divided solid particles in stable suspensions, US Patent 4,771,086 **(1992)**; (i) U. Hees, M. Kluge, F.W. Raulfs, H. Schoepke, K. Siemensmeyer, R. Van Gelder, J. Weiser, H. Heissler, S. Adams, G. Renz, P.A. Simpson, Method for treating particulate pigments, WO 2004113454 **(2004)**.

72. (a) R.L. Templeton-Knight, A process for encapsulating inorganic pigment with polymeric materials, *J. Oil Color Chem. Assoc.* **1990**, *73*, 459–564; (b) R.L. Templeton-Knight, Encapsulation of inorganic particles by emulsion polymerization, *Chem. Ind.* **1990**, 512–515; (c) J.P. Lorimer, T.J. Mason, D. Kershaw, I. Livsey, R.L. Templeton-Knight, Effect of ultrasound on the encapsulation of titanium dioxide pigment, *Colloid Polym. Sci.* **1991**, *269*, 392–397;

73. (a) M. Hasegawa, K. Arai, S. Saito, Uniform microencapsulation of fine inorganic powders with soapless emulsion polymerization, *J. Polym. Sci., Part A: Polym. Chem.* **1987**, *25*, 3117–3125; (b) M. Hasegawa, K. Arai, S. Saito, Selective adsorption of polymer on freshly ground solid surfaces in soapless emulsion polymerization, *J. Appl. Polym. Sci.* **1987**, *33*, 411–418.

74. K.L. Hoy, O.W. Smith, A process for the polymerization of monomers on the surface of hiding pigments dispersed in water, *ACS Polym. Mater. Sci. Eng. Preprint* **1991**, *65*, 78–79.

75. C.H.M. Caris, L.P.M. van Elven, A.M. van Herk, A. German, *Polymerization at the Surface of Inorganic Submicron Particles*, 19th FATIPEC Conference Book **1988**, *3*, 341–354.

76. C.H.M. Caris, L.P.M. van Elven, A.M. van Herk, A. German, Polymerization of MMA at the surface of inorganic submicron particles, *Br. Polym. J.* **1989**, *21*, 133–140

77. C.H.M. Caris, A.M. van Herk, A.L. German, Polymerization at the surface of TiO_2 pigments in emulsion-like system, 20th FATIPEC Conference Proceedings **1990**, 325–329.

78. C.H.M. Caris, R.P.M. Kuijpers, A.M. van Herk, A.L. German, Kinetics of (co)polymerization at the surface of inorganic submicron particles by emulsion-like systems, *Macromol. Chem. Macromol. Symp.* **1990**, *35/36*, 535–548.

79. E.A.W.G. Janssen, A.M. van Herk, A.L. German, Encapsulation of inorganic filler particles by emulsion polymerization. *ACS Div. Polym. Chem. Polym. Preprint* **1993**, *34*, 532–533.

80. R.Q.F. Janssen, A.M. van Herk, A.L. German, Stability of polymer encapsulated TiO_2 in aqueous emulsion systems, *J. Oil Color Chem. Assoc.* **1993**, *11*, 455–461.

81. R.Q.F. Janssen, A.M. van Herk, A.L. German, Polymer encapsulation of TiO_2 in aqueous emulsion systems, *22nd FATIPEC Conference Proceedings* **1994**, *1*, 104–118.

82. C.H.M. Caris, *Polymer encapsulation of inorganic submicron particles in aqueous dispersion*, PhD Thesis, Eindhoven University of Technology, The Netherlands **(1990)**.

83. R.Q.F. Janssen, *Polymer encapsulation of titanium dioxide. Efficiency, stability and compatibility*, PhD Thesis, Eindhoven, University of Technology, The Netherlands **(1995)**.

84. Y. Haga, T. Watanabe, R. Yosomiya, Encapsulating polymerization of titanium dioxide, *Angew. Makromol. Chem.* **1991**, *189*, 23–34.

85. (a) W.D. Hergeth, M. Peller, P. Hauptmann, Polymerizations in the presence of seeds. II. Monitoring the emulsion polymerization in the presence of fillers by means of ultrasound, *Acta Polymerica* **1986**, *37*, 468–469; (b) W.D. Hergeth, P. Starre, K. Schmutzler, S. Wartewig, Polymerizations in the presence of seeds: III. Emulsion Polymerization of vinyl acetate in the presence of quartz powder, *Polymer* **1988**, *29*, 1323–1328; (c) W.D. Hergeth, U.J. Steinau, H.J. Bittrich, G. Simon, K. Schmutzler, Polymerization in the presence of seeds: Part IV: Emulsion polymers containing inorganic filler particles, *Polymer* **1989**, *30*, 254–258; (d) W.D. Hergeth, U.J. Steinau, H.J. Bittrich, K. Schmutzler, S. Wartewig, Submicron particles with thin polymer shells, *Progr. Colloid Polym. Sci.* **1991**, *85*, 82–90.

86. (a) P. Espiard, *Encapsulation de silices colloidales par polymérisation en émulsion: latex filmogènes*, PhD Thesis, Université Lyon I, France **(1992)**; (b) P. Espiard, A. Revillon, A. Guyot, J.E. Mark, Nucleation of Emulsion Polymerization in the Presence of Small Silica Particles in *Polymer Latexes: preparation, characterization and applications*, Chap. 24, pp. 387–403, E.S. Daniels, T. Sudol, M. El-Aasser (Eds.), ACS Symposium Series 492, Washington DC, **1992**.

87. J.C. Daniel, P. Espiard, A. Guyot, Nouveaux pigments à base de silices et de polymères, compositions filmogènes les contenant, films obtenus à partir desdites compositions et procédé de préparation, Eur. Pat. 505230, Rhône-Poulenc **(1992)**.

88. (a) E. Bourgeat-Lami, P. Espiard, A. Guyot, Poly(ethyl acrylate) latexes encapsulating nanoparticles of silica. 1. Functionalization and dispersion of silica, *Polymer* **1995**, *36*, 4385–4389; (b) P. Espiard, A. Guyot, Poly(ethyl acrylate) latexes encapsulating nanoparticles of silica. 2. Grafting

Process onto silica, *Polymer* **1995**, *36*, 4391–4395.

89. (a) P. Espiard, A. Guyot, J. Perez, G. Vigier, L. David, Poly(ethyl acrylate) latexes encapsulating nanoparticles of silica. 3. Morphology and mechanical properties of reinforced films, *Polymer* **1995**, *36*, 4397–4403; (b) E. Bourgeat-Lami, P. Espiard, A. Guyot, C. Gauthier, L. David, G. Vigier, Emulsion polymerization in the presence of colloidal silica, *Angew. Makromol. Chem.* **1996**, *242*, 105–122.

90. E. Bourgeat-Lami, P. Espiard, A. Guyot, S. Briat, C. Gauthier, G. Vigier, J. Perez, Composite polymer colloid nucleated by functionalized silica in *Hybrid Organic-Inorganic Composites,* Chap. 10, pp. 112–124, J.E. Mark, C.Y.-C. Lee, P.A. Bianconi (Eds.), ACS Symposium Series 585, Washington DC, **1995**.

91. E. Bourgeat-Lami, J.-L. Luna-Xavier, A. Guyot, Hybrid silica/polymer colloids, in: *Organic/Inorganic Hybrid Materials*, Vol. 628, CC3.5, R.M. Laine, C. Sanchez, C. Brinker, E. Gianellis (Eds.), Mater. Res. Soc., Symp. Proc., **2000**.

92. K. Zhang, H. Chen, X. Chen, Z. Chen, Z. Cui, B. Yang, Monodisperse silica-polymer core-shell microspheres via surface grafting and emulsion polymerization. *Macromol. Mater. Eng.* **2003**, *288*, 380–385.

93. (a) Z. Zeng, J. Yu, Z.X. Guo, Preparation of epoxy-functionalized polystyrene/silica core-shell composite nanoparticles, *J. Polym. Sci. Part A. Polym. Chem.* **2004**, *42*, 2253–2262; (b) Z. Zeng, J. Yu, Z.X. Guo, Preparation of carboxyl-functionalized polystyrene/silica core-shell composite nanoparticles, *Macromol. Chem. Phys.* **2004**, *205*, 2197–2204.

94. K. Shiratsuchi, H. Hokazono, Aqueous dispersion of core-shell type composite particles with colloidal silica as the cores and with organic polymers as the shells and production method thereof, US Patent 5,856,379 (**1999**).

95. (a) K. Yoshinaga, R. Horie, F. Saigoh, T. Kito, N. Enomoto, H. Nishida, M. Komatsu, Modification of mono-dispersed silica colloid particles with polymer silane coupling agents, *Polym. Adv. Technol.* **1992**, *3*, 91–93; (b) K. Yoshinaga, K. Sueishi, H. Karakawa, Preparation of monodispersed polymer-modified colloidal silica

particles of less than 50 nm diameter, *Polym. Adv. Technol.* **1995**, *7*, 53–56; (c) K. Yoshinaga, M. Iwasaki, M. Teramoto, H. Karakawa, Control of polymer layer thickness in coating of monodispersed colloidal silica, *Polym. Polym. Comp.* **1996**, *4*, 163–172; (d) K. Yoshinaga, K. Kondo, A. Kondo, Capabilities of polymer-modified monodisperse colloidal silica particles as biomedical carrier, *Colloid Polym. Sci.* **1997**, *275*, 220–226.

96. (a) K. Nagai, Y. Ohishi, K. Ishiyama, N. Kuramoto, Polymerization of surface-active monomers. III. Polymer encapsulation of silica gel particles by aqueous polymerization of quaternary salt of dimethylaminoethyl methacrylate with lauryl bromide, *J. Appl. Polym. Sci.* **1989**, *38*, 2183–2189; (b) K. Nagai, Polymerization of surface-active monomers and applications, *Macromol. Symp.* **1994**, *84*, 29–36.

97. K. Yoshinaga, F. Nakashima, T. Nishi, Polymer modification of colloidal particles by spontaneous polymerization of surface-active monomers, *Colloid Polym. Sci.* **1999**, *277*, 136–144.

98. (a) C. Barthet, A.J. Hickey, D.B. Cairns, S.P. Armes, Synthesis of novel polymer-silica colloidal nanocomposites via free-radical polymerization of vinyl monomers, *Adv. Mater.* **1999**, *11*, 408–410; (b) M.J. Percy, C. Barthet, J.C. Lobb, M.A. Khan, S.F. Lascelles, M. Vamvakaki, S.P. Armes, Synthesis and characterization of vinyl polymer-silica colloidal nanocomposites, *Langmuir* **2000**, *16*, 6913–6920; (c) J.I. Amalvy, M.J. Percy, S.P. Armes, Synthesis and characterization of novel film forming polymer/silica colloidal nanocomposites, *Langmuir* **2001**, *17*, 4770–4778.

99. F. Tiarks, K. Lanfester, M. Antonietti, Silica nanoparticles as surfactants and fillers for latexes made by miniemulsion polymerization, *Langmuir* **2001**, *17*, 5775–5779.

100. T. Sakurai, H. Murata, T. Mizutani, Y. Kimura, M. Miyamoto, High-solids aqueous composite silica-polymer dispersions with good storage stability, their manufacture and use in coatings, Japanese Patent 11-209622 (**1999**).

101. (a) J.L. Luna-Xavier, A. Guyot, E. Bourgeat-Lami, Synthesis and characteri-

zation of silica/poly(methyl methacrylate) nanocomposite latex particles using a cationic azo initiator, *J. Colloid Interf. Sci.* **2001**, *250*, 82–92; (b) J.L. Luna-Xavier, A. Guyot, E. Bourgeat-Lami, Preparation of nano-sized silica/poly(methyl methacrylate) composite latexes by heterocoagulation, *Polym. Int.* **2004**, *53*, 609–617.

102. S. Reculusa, C. Poncet-Legrand, S. Ravaine, C. Mingotaud, E. Duguet, E. Bourgeat-Lami, Syntheses of raspberry-like silica/polystyrene materials, *Chem. Mater.* **2002**, *14*, 2354–2359.

103. (a) M.J. Percy, S.P. Armes, Surfactant-free synthesis of colloidal poly(methyl methacrylate)/silica nanocomposites in the absence of auxiliary comonomers, *Langmuir* **2002**, *18*, 4562–4565; (b) M.J. Percy, J.I. Amalvy, D.P. Randall, S.P. Armes, S.J. Greaves, J.F. Watts, Synthesis of vinyl polymer-silica colloidal nanocomposites prepared using commercial alcoholic silica sols, *Langmuir* **2004**, *20*, 2184–2190.

104. For a review, see: (a) R. Arshady, D. Poulinquen, A. Halbreich, J. Roger, J.N. Pons, J.-C. Bacri, M. de F. Da Silva, U. Häfeli, Magnetic nanospheres and nanocomposites, in: *Dendrimers, Assemblies, Nanocomposites*, Chap. 6, pp. 283–329, R. Arshady, A. Guyot (Eds.), MML Series 5, Citus Book, London, **2002**; (b) A. Elaïssari, F. Sauzedde, F. Montagne, C. Pichot, Preparation of magnetic lattices, in: *Colloidal Polymers: Synthesis and Characterization*, Chap. 11, pp. 285–318, A. Elaïssari (Ed.), Surfactant Science Series 115, Marcel Dekker, New York, **2003**; (c) J. Lefort, Mémoire sur les oxydes ferroso-ferriques et leur combinaisons, *C. R. Acad. Sci.* **1852**, *34*, 488–491.

105. (a) J.T. Kemshead, J. Ugelstad, Magnetic separation techniques: their application to medicine, *Mol. Cell. Biochem.* **1985**, *67*, 11–18; (b) A.K. Gupta, M. Gupta, Synthesis and surface engineering of iron oxide nanoparticles for biomedical applications, *Biomaterials* **2005**, *26*, 3995–4021.

106. R.E. Rosensweig, *Ferrohydrodynamics*, Cambridge University Press, Cambridge, **1985**.

107. R. Massart, Preparation of aqueous magnetic liquids in alkaline and acidic media, *IEEE Trans. Magn.* **1981**, *17*, 1247–1248.

108. (a) R.M. Cornell, U. Schertmann, Iron oxides in the laboratory; preparation and characterization, VCH, Weinheim, **1991**; (b) J. Bacri, R. Percynski, D. Salin, V. Cabuil, R. Massart, Magnetic colloidal properties of ionic ferrofluids, *J. Magn. Magn. Mater.* **1990**, *85*, 27; (c) S. Mornet, F. Grasset, J. Portier, E. Duguet, Maghemite-silica nanoparticles for biological applications, *Eur. Cells Mater.* **2002**, 3 suppl. 2, 110.

109. S. Mornet, J. Portier, E. Duguet, A method for synthesis and functionalization of ultrasmall superparamagnetic covalent carriers based on maghemite and dextran, *J. Magn. Magn. Mater* **2005**, *93*, 127–134.

110. (a) J. Ugelstad, T. Ellingsen, A. Berge, O.B. Helgee, Magnetic polymer particles and process for the preparation thereof, PCT Patent Application WO83/03920 (**1983**); (b) J. Ugelstad, A. Berge, T. Ellingsen, R. Schmid, T.N. Nilsen, P.C. Mork, P. Stenstad, E. Hornes, O. Olsvik, Preparation and application of mono-sized polymer particles, *Prog. Polym. Sci.* **1992**, *17*, 87–161.

111. (a) K. Furusawa, K. Nagashima, C. Anzai, Synthetic process to control the total size and component distribution of multilayer magnetic composite particles, *Colloid Polym. Sci.* **1994**, *272*, 1104–1110; (b) J. Lee, M. Senna, Preparation of monodispersed polystyrene microspheres uniformly coated by magnetite via heterogeneous polymerization, *Colloid Polym. Sci.* **1995**, *273*, 76–82; (c) F. Caruso, A.S. Susha, M. Giersig, H. Möhwald, Magnetic core-shell particles: preparation of magnetite multilayers on polymer latex microspheres, *Adv. Mater.* **1999**, *11*, 950–953.

112. (a) F. Sauzedde, A. Elaïssari, C. Pichot, Hydrophilic magnetic polymer latexes. 1. Adsorption of magnetic iron oxide nanoparticles onto various cationic latexes, *Colloid Polym. Sci.* **1999**, *277*, 846–855; (b) F. Sauzedde, A. Elaïssari, C. Pichot, Hydrophilic magnetic polymer latexes. 2. Encapsulation of adsorbed

iron oxide naoparticles, *Colloid Polym. Sci.* **1999**, *277*, 1041–1050.

113. (a) N. Yanase, H. Noguchi, H. Asakura, T. Suzuta, Preparation of magnetic latex particles by emulsion polymerization of styrene in the presence of a ferrofluid, *J. Appl. Polym. Sci.* **1993**, *50*, 765–776; (b) H. Noguchi, N. Yanase, Y. Uchida, T. Suzuta, Preparation and characterization by thermal analysis of magnetic latex particles, *J. Appl. Polym. Sci.* **1993**, *48*, 1539–1547.

114. A. Kondo, H. Kamura, K. Higashitani, Development and application of thermo-sensitive magnetic immunomicros-pheres for antibody purification, *Appl. Microbiol. Biotechnol.* **1994**, *41*, 99–105.

115. (a) P.A. Dresco, V.S. Zaitsev, R. Gambi-no, B. Chu, Preparation and properties of magnetite and polymer magnetite nanoparticles, *Langmuir* **1999**, *15*, 1945–1951; (b) V.S. Zaitsev, D.S. Fil-imonov, I.A. Presnyakov, R.J. Gambino, B. Chu, Physical and chemical properties of magnetite-polymer nanoparticles and their colloidal dispersions, *J. Colloid Interface Sci.* **1999**, 212, 49–57.

116. K. Wormuth, Superparamagnetic latex via inverse emulsion polymerization, *J. Colloid Interf. Sci.* **2001**, *241*, 366–377.

117. (a) F. Wypych, K.G. Satyanarayana, *Clay Surfaces. Fundamentals and Applications*, Interface Science and Technology Series, Academic Press, London, **2004**; (b) H. van Olphen, *An introduction to Clay Colloid Chemistry*, Wiley Interscience, New York, **1977**.

118. (a) A. Usuki, Y. Kojima, M. Kawasumi, A. Okada, Y. Fukushima, T. Kuruachi, O. Kamigaito, Synthesis of nylon-6 clay hybrid, *J. Mater. Res.* **1993**, *8*, 1179–1184; (b) Y. Kojima, A. Usuki, M. Kawasumi, A. Okada, Y. Fukushima, T. Kuruachi, O. Kamigaito, Synthesis of nylon 6-clay hybrid by montmorillonite intercalated with ε-caprolactam, *J. Polym. Sci. Polym. Chem.* **1993**, *31*, 983–986; (c) Y. Kojima, A. Usuki, M. Kawasumi, A. Okada, Y. Fu-kushima, T. Kuruachi, O. Kamigaito, Mechanical properties of nylon-6 clay hy-brids, *J. Mater. Res.* **1993**, *8*, 1185–1189.

119. (a) M. Alexandre, P. Dubois. Polymer-lay-ered silicate nanocomposites: prepara-tion, properties and uses of a new class of materials, *Mater. Sci. Eng.* **2001**, *28*,

1–63; (b) E. Duguet, S. Rey, J. Mérida-Robles, Intercalation Polymerization, *Encyclopedia of Polymer Science and Technology*, H.F. Mark (Ed.), Wiley & Sons, Inc., 3rd edition, **2004**, vol. 10, pp. 250–272.

120. D.F. Schmidt, G. Qian, E.P. Giannelis, Nanocomposites from novel synthetic organo-layer silicates, *Polym. Mater. Sci. Eng.* **2000**, *82*, 215–216.

121. A. Gültek, T. Seçkin, Y. Onal, M.G. Iç-duygu, Poly(methacrylic) acid and γ-methacryloxy propyl trimethoxy silane/clay nanocomposites prepared by in situ polymerization. *Turk. J. Chem.* **2002**, *26*, 925–937.

122. N. Negrete-Herrera, J.-M. Letoffe, J.-P. Reymond, E. Bourgeat-Lami, *J. Mater. Chem.* **2005**, 15, 863–871.

123. (a) F. Dietsche, Y. Thomann, R. Tho-mann, R. Mülhaupt, *J. Appl. Polym. Sci.* **2000**, *75*, 396–405; (b) A.S. Moet, A. Ake-lah, *Mater. Lett.* **1993**, *18*, 97–102; (c) A. Akelah, A.S. Moet, *J. Mater. Sci.* **1996**, *31*, 3589–3596; (d) A. Tabtiang, S. Lum-long, R.A. Venables, *Eur. Polym. J.* **2000**, *36*, 2559–2568; (e) L.P. Meier, R.A. Shelden, W.R. Caseri, U.W. Suter, Poly-merization of styrene with initiator ioni-cally bound to high surface area mica: grafting via an unexpected mechanism, *Macromolecules* **1994**, *27*, 1637–1642.

124. C.S. Chou, E.E. Lafleur, D.P. Lorah, R.V. Slone, K.D. Neglia, Aqueous nanocom-posite dispersions: processes, composi-tions and uses thereof. US Patent 6,838,507 (**2005**).

125. X. Tong, H. Zhao, T. Tang, Z. Feng, B. Huang, Preparation and characteriza-tion of poly(ethyl acrylate)/bentonite nanocomposites by in situ emulsion polymerization, *J. Polym. Sci., Part A: Polym. Chem.* **2002**, *40*, 1706–1711.

126. Y.S. Choi, M.H. Choi, K.H. Wang, S.O. Kim, Y.K. Kim, I.J. Chung, Synthesis of exfoliated PMMA/Na-MMT nanocom-posites via soap-free emulsion polymer-ization, *Macromolecules* **2001**, *34*, 8978–8985.

127. D. Wang, J. Zhu, Q. Yao, C.A. Wilkie, A comparison of various methods for the preparation of polystyrene and poly(methyl methacrylate) clay nanocom-posites, *Chem. Mater.* **2002**, *14*, 3837–3843.

128. G. Chen, Y. Ma, Z. Qi, Preparation and morphological study of an exfoliated polystyrene/montmorillonite nanocomposite, *Scr. Mater.* **2001**, *44*, 125–128.

129. Y.K. Kim, Y.S. Choi, K.H. Wang, I.J. Chung. Synthesis of exfoliated PS/Na-MMT nanocomposites via emulsion polymerization, *Chem. Mater.* **2002**, *14*, 4990–4995.

130. S. Bandyopadhyay, E.P. Giannelis, A.J. Hsieh, Thermal and thermomechanical properties of PMMA nanocomposites, *Polym. Mater. Sci. Eng.* **2000**, *82*, 208–209.

131. B. zu Putlitz, K. Landfester, H. Fischer, M. Antonietti, The generation of armored latexes and hollow inorganic shells made of clay sheets by templating cationic miniemulsions and latexes, *Adv. Mater.* **2001**, *13*, 500–503.

132. D.C. Lee, L.W. Jang, Preparation and characterization of PMMA-clay hybrid composite by emulsion polymerization, *J. Appl. Polym. Sci.* **1996**, *61*, 1117–1122.

133. M.W. Noh, D.C. Lee, Synthesis and characterization of PS-clay nanocomposite by emulsion polymerization, *Polym. Bull.* **1999**, *42*, 619–626.

134. (a) M.W. Noh, L.W. Jang, D.C. Lee, Intercalation of styrene-acrylonitrile copolymer in layered silicate by emulsion polymerization, *J. Appl. Polym. Sci.* **1999**, *74*, 179–188; (b) M.W. Noh, D.C. Lee, Comparison of characteristics of SAN-MMT nanocomposites prepared by emulsion and solution polymerization, *J. Appl. Polym. Sci.* **1999**, *74*, 2811–2819.

135. X. Huang, W.J. Brittain, Synthesis of a PMMA-layered silicate nanocomposite by suspension polymerization, *Polymer Preprints* **2000**, *41*, 521–522.

136. X. Huang, W.J. Brittain, Synthesis and characterization of PMMA nanocomposites by suspension and emulsion polymerization, *Macromolecules* **2001**, *34*, 3255–3260.

137. N. Negrete-Herrera, J.-M. Letoffe, J.-L. Putaux, L. David, E. Bourgeat-Lami, Aqueous dispersions of silane-functionalized laponite clay platelets. A first step toward the elaboration of water-based polymer/clay nanocomposites, *Langmuir* **2004**, *20*, 1564–1571.

138. N. Negrete-Herrera, S. Persoz, J.-L. Putaux, L. David, E. Bourgeat-Lami, Synthesis of polymer latex particles decorated by organically-modified laponite clay platelets by emulsion polymerization, *J. Nanosci. Nanotechnol.* **2006**, *6*, 421–431.

139. J.W. Gilman, C.L. Jackson, A.B. Morgan, R. Harris, E. Manias, E.P. Giannelis, M. Wuthenow, D. Hilton, S.H. Phillips, Flammability properties of polymer-layered silicate nanocomposites. Polypropylene and polystyrene nanocomposites, *Chem. Mater.* **2000**, *12*, 1866–1873.

140. A. Kalendova, P. Tamchynova, V. Stengl, J. Subrt, Behaviour of surface-treated mica and other pigments with lamellar particles in anticorrosive coatings, *Macromol. Symp.* **2002**, *187*, 367–376.

141. Th. Batzilla, A. Tulke, Preparation of encapsulated aluminium pigments by emulsion polymerization and their characterization, *J. Coat. Technol.* **1998**, *70*, 77–83.

142. L. Quaroni, G. Chumanov, Preparation of polymer-coated functionalized silver nanoparticles, *J. Am. Chem. Soc.* **1999**, *121*, 10642–10643.

143. J. Yu, J. Yu, Z.-X. Guo, Y.-F. Gao, Preparation of CaCO$_3$/polystyrene inorganic/organic composite particles, *Macromol. Rapid Commun.* **2001**, *22*, 1261–1264.

144. E. Kumacheva, Synthesis of hybrid colloidal polymer-inorganic composites, World Patent WO 2004081072 (**2004**).

145. H. Yao, Y. Takada, N. Kitamura, Electrolyte effects on CdS nanocrystal formation in chelate polymer particles, *Langmuir* **1998**, *14*, 595–601.

146. J. Zhang, N. Coombs, E. Kumacheva, A new approach to hybrid nanocomposite materials with a periodic structure, *J. Am. Chem. Soc.* **2002**, *124*, 14512–14513.

147. S. Gu, J. Onishi, E. Mine, Y. Kobayashi, M. Konno, Preparation of multilayered gold-silica-polystyrene core-shell particles by seeded polymerization, *J. Colloid Interf. Sci.* **2004**, *279*, 284–287.

148. K. Kamata, Y. Lu, Y. Xia, Synthesis and characterization of monodispersed core-shell spherical colloids with movable cores, *J. Am. Chem. Soc.* **2003**, *125*, 2384–2385.

149. (a) B. Erdem, D. Sudol, V.L. Dimonie, M. El-Aasser, Encapsulation of inorganic particles via miniemulsion polymerization. I. Dispersion of titanium dioxide

particles in organic media using OLOA 370 as stabilizer, *J. Polym. Sci., Part A: Polym. Chem.* **2000**, *38*, 4419–4430; (b) B. Erdem, D. Sudol, V.L. Dimonie, M. El-Aasser, Encapsulation of inorganic particles via miniemulsion polymerization. II. Preparation and characterization of styrene miniemulsion droplets containing TiO$_2$ particles, *J. Polym. Sci., Part A: Polym. Chem.* **2000**, *38*, 4431–4440; (c) B. Erdem, D. Sudol, V.L. Dimonie, M. El-Aasser, Encapsulation of inorganic particles via miniemulsion polymerization. III. Characterization of encapsulation, *J. Polym. Sci., Part A: Polym. Chem.* **2000**, *38*, 4441–4450.

150. (a) N. Bechthold, F. Tiarks, M. Willert, K. Landfester, M. Antonietti, Miniemulsion polymerization: Applications and new materials, in: *Polymers in Dispersed Media*, pp. 549–555, (Eds.), Macromol. Symp. 151, Weinheim, Wiley VCH, **2000**; (b) F. Tiarks, K. Landfester, M. Antonietti, Encapsulation of carbon black by miniemulsion, *Macromol. Chem. Phys.* **2001**, *202*, 51–60.

151. D. Hoffmann, K. Landfester, M. Antonietti, Encapsulation of magnetite in polymer particles via the miniemulsion polymerization process, *Magnetohydrodynamics* **2001**, *37*, 221.

152. S. Lelu, C. Novat, C. Graillat, A. Guyot, E. Bourgeat-Lami, Encapsulation of an organic phthalocyanine blue pigment into polystyrene latex particles using a miniemulsion process, *Polym. Int.* **2002**, *51*, 1–7.

153. C.S. Chern, Y.C. Liou, Miniemulsion polymerization of styrene in the presence of a water-insoluble blue dye, *Polymer* **1998**, *9*, 3767–3776.

154. (a) E. Bourgeat-Lami, J. Lang, Encapsulation of inorganic particles by dispersion polymerization in polar media. 2. Effect of silica size and concentration on the morphology of silica-polystyrene composite particles, *J. Colloid Interf. Sci.* **1999**, *210*, 281–289; (b) F. Corcos, E. Bourgeat-Lami, C. Novat, J. Lang, Poly (styrene-b-ethylene oxide) copolymers as stabilizer for the synthesis of silica-polystyrene core-shell particles, *Colloid Polym. Sci.* **1999**, *277*, 1142–1151; (c) E. Bourgeat-Lami, J. Lang, Silica/Poly-styrene Composite Particles, in: *Polymers in Dispersed Media*, J. Claverie, M.-T. Charreyre, C. Pichot (Eds.), Macromolecular Symposia 151, Wiley VCH, Weinheim, **2000**, pp. 377–385; (d) S. Chalaye, E. Bourgeat-Lami, J.L. Putaux, J. Lang, Synthesis of composite latex particles filled with silica, in: *Fillers and Filled Polymer*, J.F. Gerard (Ed.), Macromolecular Symposia 169, Wiley VCH, Weinheim, **2000**, pp. 89–96.

155. I. Sondi, T.H. Fedynyshyn, R. Sinta, E. Matijevic, Encapsulation of nanosized silica by in-situ polymerization of *tert*-butyl acrylate monomer, *Langmuir* **2000**, *16*, 9031–9034.

156. R.E. Partch, S.G. Gangolli, D. Owen, C. Ljungqvist, E. Matijevic, Conducting Polymer Composites, Polypyrrole-Metal Oxide Latexes, in: *Polymer Latexes: preparation, characterization and applications*, Chap. 23, p. 369, E.S. Daniels, T. Sudol, M. El-Aasser (Eds.), ACS Symposium Series 492, Washington DC, **1992**.

157. (a) R.E. Partch, S.G. Gangolli, E. Matijevic, W. Cai, S. Arajs, Conducting polymer composites. I. Surface-induced polymerization of pyrrole on iron (III) and cerium (IV) oxide particles, *J. Colloid Interf. Sci.* **1991**, *144*, 27–35; (b) C.H. Huang, R.E. Partch, E. Matijevic, Coating of uniform inorganic particles with polymers: II. Polyaniline on copper oxide, *J. Colloid Interf. Sci.* **1995**, *170*, 275–283; (c) C.H. Huang, E. Matijevic, Coating of uniform inorganic particles with polymers: III. Polypyrrole on different metal oxides, *J. Mater. Res.* **1995**, *10*, 1327–1336.

158. S.P. Armes, Conducting polymer colloids, *Curr. Opin. Colloid Interf. Sci.* **1996**, *1*, 214–220 and references therein.

159. (a) J. Stejskal, Conducting Polymer Nanospheres and Nanocomposites. 1. Manufacturing Methods, in: *Dendrimers, Assemblies, Nanocomposites*, R. Arshady, A. Guyot (Eds.), MML Series 5, Citus Book, London, **2002**, pp. 195–243; (b) J. Stejskal, Conducting Polymer Nanospheres and Nanocomposites. 2. Properties and Applications, in: *Dendrimers, Assemblies, Nanocomposites*, pp. 245–281, R. Arshady, A. Guyot (Eds.), MML Series 5, Citus Book, London, **2002** and references therein.

160. M. Bakhshaee, R.A. Pethrick, H. Rashid, D.C. Sherrington, Encapsulation of carbon black in suspension polymerized copolymers, *Polym. Commun.* **1985**, *26*, 185–192.

161. E.C. Cooper, B. Vincent, Encapsulation of filler particles in poly(methyl methacrylate) beads by a double dispersion method, *J. Colloid Interf. Sci.* **1989**, *132*, 592–594.

162. (a) E. Duguet, M. Abboud, F. Morvan, P. Maheu, M. Fontanille, PMMA encapsulation of alumina particles through aqueous suspension polymerization processes, in: *Polymers in Dispersed Media*, J. Claverie, M.-T. Charreyre, C. Pichot (Eds.), Macromolecular Symposia 151, Wiley VCH, Weinheim, **2000**, pp. 365–370; (b) M. Abboud, L. Casaubieilh, F. Morvan, M. Fontanille, E. Duguet, PMMA-based composite materials with reactive ceramic fillers. 4. Radiopacifying particles embedded in PMMA beads for acrylic bone cements, *J. Biomed. Mater. Res. (Appl. Biomater.)* **2000**, *53*, 728.

163. (a) Q. Zhou, X. Fan, C. Xia, J. Mays, R. Advincula, Living anionic surface initiated polymerization (SIP) of styrene from clay surfaces, *Chem. Mater.* **2001**, *13*, 2465–2467; (b) X. Fan, C. Xia, R.C. Advincula, Grafting of polymers from clay nanoparticles via in-situ free radical surface-initiated polymerization: monocationic versus bicationic initiators, *Langmuir* **2003**, *19*, 4381–4389; (c) X. Fan, Q. Zhou, C. Xia, W. Cristofoli, J. Mays, R. Advincula, Living anionic surface initiated polymerization (LASIP) of styrene from clay nanoparticles using surface bound 1,1-diphenylethylene (DPE) initiators, *Langmuir* **2002**, *18*, 4511–4518; (d) Q. Zhou, S. Wang, X. Fan, R. Advincula, J. Mays, Living anionic surface-initiated polymerization (LASIP) of a polymer on silica, *Langmuir* **2002**, *18*, 3224–3331.

164. K.J. Warson, J. Zhu, S.T. Nguyen, C.A. Mirkin, Hybrid nanoparticles with block copolymer shell structures, *J. Am. Chem. Soc.* **1999**, *121*, 462–463.

165. (a) G. Carrot, D. Rutot-Houzé, A. Pottier, P. Degée, J. Hilborn, P. Dubois, Surface initiated ring-opening polymerization: a versatile method for nanoparticles ordering, *Macromolecules* **2002**, *35*, 8400–8404;

(b) M. Joubert, C. Delaite, E. Bourgeat-Lami, P. Dumas, Ring-opening polymerization of ε-caprolactone and L-lactide from silica nanoparticles surface, *J. Polym. Sci., Part A: Polym. Chem.* **2004**, *42*, 1976–1984.

166. M. Joubert, C. Delaite, E. Bourgeat-Lami, P. Dumas, Synthesis of poly(epsilon-caprolactone)-silica nanocomposites: from hairy colloids to core-shell nanoparticles. *New J. Chem.* **2005**, *29*, 1601–1609.

167. (a) N. Tsubokawa, K. Maruyama, Y. Sone, M. Shimomura, Grafting onto ultrafine ferrite particles: reaction of isocyanate-capped polypropylene glycol with hydroxyl groups on the surface, *Polym. J.* **1989**, *21*, 475–481; (b) N. Tsubokawa, A. Kogure, K. Maruyama, Y. Sone, M. Shimomura, Graft polymerization of vinyl monomers from inorganic ultrafine particles initiated by azo groups introduced onto the surface, *Polym. J.* **1990**, *22*, 827–833; (c) N. Tsubokawa, H. Ishida, Graft polymerization of vinyl monomers by peroxyester groups introduced onto the surface of inorganic ultrafine particles, *Polym. J.* **1992**, *24*, 809–816; (d) N. Tsubokawa, H. Ishida, Graft polymerization of methyl methacrylate from silica initiated by peroxide groups introduced onto the surface, *J. Polym. Sci., Part A: Polym. Chem.* **1992**, *30*, 2241–2246; (e) N. Tsubokawa, Y. Shirai, H. Tsuchida, S. Handa, Photografting of vinyl polymers onto ultrafine inorganic particles: photopolymerization of vinyl monomers initiated by azo groups introduced onto these surfaces, *J. Polym. Sci., Part A: Polym. Chem.* **1994**, *32*, 2327–2332; (f) N. Tsubokawa, Y. Shirai, K. Hashimoto, Effect of polymerization conditions on the molecular weight of polystyrene grafted onto silica in the radical graft polymerization initiated by azo or peroxyester groups introduced onto the surface, *Colloid Polym. Sci.* **1995**, *273*, 1049–1054.

168. (a) O. Prucker, J. Rühe, Imaging of polymer monolayers attached to silica surfaces by elemental specific transmission electron microscopy, *Polymer* **1996**, *37*, 1087–1093; (b) O. Prucker, J. Rühe, Mechanism of radical chain polymerizations initiated by azo compounds covalently bound to the surface of spherical particles, *Macromolecules* **1998**, *31*, 602–613; (c) O. Prucker, J. Rühe, Synthesis of poly(styrene) monolayers attached to high surface area silica gels through self-assembled monolayers of azo initiators, *Macromolecules* **1998**, *31*, 592–601.

169. (a) T. von Werne, T.E. Patten, Preparation of structurally well-defined polymer nanoparticle hybrids with controlled/living polymerizations, *J. Am. Chem. Soc.* **1999**, *121*, 7409–7410; (b) T. von Werne, T.E. Patten, M. Ellsworth, J. Goetz, Preparation of ordered hybrid nanoparticle/polymer films, *Polym. Mater. Sci. Eng.* **2000**, *82*, 233–234; (c) T. von Werne, T.E. Patten, Synthesis of hybrid organic/silica nanoparticles using atom transfer radical polymerization, *Polym. Mater. Sci. Eng.* **1999**, 80, 465–466; (d) T. von Werne, I.M. Suehiro, S. Farmer, T.E. Patten, Composite films of polymer-inorganic hybrid nanoparticles prepared using controlled/living radical polymerization, *Polym. Mater. Sci. Eng.* **2000**, *82*, 294–295; (e) T. von Werne, T.E. Patten, Atom transfer radical polymerization from nanoparticles: a tool for the preparation of well-defined hybrid nanostructures and for understanding the chemistry of controlled/living radical polymerizations from surfaces, *J. Am. Chem. Soc.* **2001**, *123*, 7497–7505.

170. J. Bai, J.-B. Pang, K.-Y. Qiu, Y. Wei, Synthesis and characterization of structurally well-defined polymer-inorganic hybrid nanoparticles via atom transfer radical polymerization, *Chin. J. Polym. Sci.* **2002**, *20*, 261–267.

171. H. Bottcher, M.L. Hallensleben, S. Nuβ, H. Wurm, ATRP grafting from silica surface to create first and second generation of grafts, *Polym. Bull.* **2000**, *44*, 223–229.

172. B. Gu, A. Sen, Synthesis of aluminium oxide/gradient copolymer composites by Atom Transfer Radical Polymerization, *Macromolecules* **2002**, *35*, 8913–8916.

173. C.R. Vestal, Z.J. Zhang, Atom transfer radical polymerization synthesis and magnetic characterization of $MnFe_2O_4$/polystyrene core/shell nanoparticles, *J. Am. Chem. Soc.* **2002**, *124*, 14312–14313.

174. Y. Wang, X. Teng, J.-S. Wang, H. Yang, Solvent-free atom transfer radical polymerization in the synthesis of Fe_2O_3 polystyrene core-shell nanoparticles, *NanoLetters* **2003**, *3*, 789–793.

175. (a) S. Nub, H. Böttcher, H. Wurm, M.L. Hallensleben, Gold nanoparticles with covalently attached polymer chains, *Angew. Chem. Int. Ed.* **2001**, *40*, 4016–4018; (b) K. Ohno, K.M. Koh, Y. Tsujii, T. Fukuda, Synthesis of gold nanoparticles coated with well-defined, high-density polymer brushes by surface-initiated living radical polymerization, *Macromolecules* **2002**, *35*, 8989–8993; (c) K. Ohno, K. Koh, Y. Tsujii, T. Fukuda, Fabrication of ordered arrays of gold nanoparticles coated with high-density polymer brushes, *Angew. Chem., Int. Ed.* **2003**, *42*, 2751–2754.

176. (a) S.C. Farmer, T.E. Patten, Synthesis of luminescent organic/inorganic polymer nanocomposites, *Polym. Mater. Sci. Eng.* **2000**, *82*, 237–238; (b) S.C. Farmer, T.E. Patten, Photoluminescent polymer/quantum dot composite nanoparticles, *Chem. Mater.* **2001**, *13*, 3920–3926.

177. C. Perruchot, M.A. Khan, A. Kamitsi, S.P. Armes, T. von Werne, T.E. Patten, Synthesis of well-defined polymer-grafted silica particles by aqueous ATRP, *Langmuir* **2001**, *17*, 4479–4481.

178. T.K. Mandal, M.S. Fleming, D.R. Walt, Production of hollow polymeric microspheres by surface-confined living radical polymerization on silica templates, *Chem. Mater.* **2000**, *12*, 3481–3487.

179. M.W. Weimer, H. Chen, E.P. Giannelis, D.Y. Sogah, Direct synthesis of dispersed nanocomposites by in situ living free radical polymerization using a silicate anchored initiator, *J. Am. Chem. Soc.* **1999**, *121*, 1615–1616.

180. S. Blomberg, S. Ostberg, E. Harth, A.W. Bosman, B. van Horn, C.J. Hawker, Production of crosslinked, hollow nanoparticles by surface-initiated living free-radical polymerization, *J. Polym. Sci., Part A: Polym. Chem.* **2002**, *40*, 1309–1320.

181. (a) J. Parvole, G. Laruelle, C. Guimon, J. François, L. Billon, Initiator-grafted silica particles for controlled free radical polymerization: influence of the initiator structure on the grafting density, *Macromol. Rapid Commun.* **2003**, *24*, 1074–1078; (b) G. Laruelle, J. Parvole, J. François, L. Billon, Block copolymer grafted-silica particles: a core/double shell hybrid inorganic/organic material, *Polymer* **2004**, *45*, 5013–5020.

182. (a) C. Bartholome, E. Beyou, E. Bourgeat-Lami, P. Chaumont, N. Zydowicz, Nitroxide-mediated polymerizations from silica nanoparticle surfaces. Graft-from polymerization of styrene using a triethoxysilyl-terminated alkoxyamine, *Macromolecules* **2003**, *36*, 7946–7952.

183. J. Parvole, L. Billon, J.P. Montfort, Formation of polyacrylate brushes on silica surfaces, *Polym. Int.* **2002**, *51*, 1111–1116.

184. A. Kasseh, A. Ait-Kadi, B. Riedl, J.F. Pierson, Organic/inorganic hybrid composites prepared by polymerization compounding and controlled free radical polymerization, *Polymer* **2003**, *44*, 1367–1375.

185. C. Bartholome, E. Beyou, E. Bourgeat-Lami, P. Chaumont, N. Zydowicz, Nitroxide-mediated polymerization of styrene initiated from the surface of silica nanoparticles. In-situ generation and grafting of alkoxyamine initiators, *Macromolecules* **2005**, *38*, 1099–1106.

186. (a) M. Okubo, H. Minami, K. Morikawa, Production of micron-sized, monodisperse, transformable rugby-ball-like-shaped polymer particles, *Colloid Polym. Sci.* **2001**, *279*, 931–935; (b) M. Okubo, T. Fujiwara, A. Yamaguchi, Morphology of anomalous polystyrene/polybutyl acrylate composite particles produced by seeded emulsion polymerization, *Colloid Polym. Sci.* **1998**, *276*, 186–189; (c) M. Okubo, T. Yamashita, H. Minami, Y. Konishi, Preparation of micron-sized monodispersed highly monomer-"adsorbed" polymer particles having snowman shape by utilizing the dynamic swelling method with tightly cross-linked seed particles, *Colloid Polym. Sci.* **1998**, *276*, 887–892; (d) M. Okubo, K. Kanaida, T. Matsumoto, Production of anomalously shaped carboxylated polymer particles by seeded emulsion polymerization, *Colloid Polym. Sci.* **1987**, *265*, 876–881; (e) Y.C. Chen, V. Dimonie, M.S. El-Aasser, Interfacial phenomena controlling parti-

cle morphology of composite latexes, *J. Appl. Polym. Sci.* **1991**, *42*, 1049–1063; (f) Y.G.J. Durant, J. Guillot, Some theoretical aspects on morphology development in seeded composite latexes. I. Batch conditions below monomer saturation, *Colloid Polym. Sci.* **1993**, *271*, 607–615; (g) H.R. Sheu, M.S. El-Aasser, J.W. Vanderhoff, Uniform nonspherical latex particles as model interpenetrating polymer networks, *J. Polym. Sci. Part A: Polym. Chem.* **1990**, *28*, 653–667; (h) I. Cho, K.W. Lee, Morphology of latex particles formed by poly(methyl methacrylate)-seeded emulsion polymerization of styrene, *J. Appl. Polym. Sci.* **1985**, *30*, 1903–1926; (i) A. Pfau, R. Sander, S. Kirsch, Orientational ordering of structured polymeric nanoparticles at interfaces, *Langmuir*, **2002**, *18*, 2880–2887.

187. (a) S. Reculusa, C. Poncet-Legrand, S. Ravaine, E. Duguet, E. Bourgeat-Lami, C. Mingotaud, Nanometric or mesoscopic dissymmetric particles, and method for preparing same, French Patent FR2846572 WO 2004/044061 (**2004**); (b) E. Duguet, S. Reculusa, A. Perro, C. Poncet-Legrand, S. Ravaine, E. Bourgeat-Lami, C. Mingotaud, From raspberry-like to dumbbell-like hybrid colloids through surface-assisted nucleation and growth of polystyrene nodules into macromonomer-modified silica nanoparticles, *Mater. Res. Soc., Symp. Proc.* **2005**, *847*, EE1.1.1–10.

188. S. Reculusa, C. Mingotaud, E. Bourgeat-Lami, E. Duguet, S. Ravaine, Synthesis of daisy-shaped and multipod-like silica/polystyrene nanocomposites, *NanoLetters* **2004**, *4*, 1677–1682.

189. C. Gauthier, G. Thollet, G. Vigier, E. Bourgeat-Lami, A. Guyot, Transmission electron microscopy observations of nucleated functionalized silica particles, *Polym. Adv. Technol.* **1995**, *66*, 345–348.

190. (a) M.J. Percy, J.I. Amalvy, C. Barthet, S.P. Armes, S.J. Grieves, J.S. Watts, H. Wiese, Surface characterization of vinyl polymer-silica colloidal nanocomposites using X-ray photoelectron spectroscopy, *J. Mater. Chem.* **2002**, *12*, 697–702; (b) J.I. Amalvy, M.J. Percy, S.P. Armes, C.A.P. Leite, F. Galembeck, Characterization of the nanomorphology of polymer-silica

colloidal nanocomposites using electron spectroscopy imaging, *Langmuir* **2005**, *21*, 1175–1179.

191. M. Hasegawa, K. Arai, S. Saito, The rate of heterogeneous polymerization in water for the encapsulation of inorganic powder with polymers, *J. Chem. Eng. Jpn.* **1988**, *21*, 30–35.

192. Y. Haga, S. Inoue, Copolymerization of vinyl monomers on the surface of CdS and photoconductivity properties of resulting composites, *Mater. Chem. Phys.* **1988**, *19*, 381–395.

193. Y. Haga, S. Inoue, T. Sato, R. Yosomiya, Photoconductivity properties of zinc oxide encapsulated in polymers, *Angew. Makromol. Chem.* **1986**, *139*, 49–61.

194. A. Bhattacharya, K.M. Ganguly, A. De, S. Sarkar, A new conducting nanocomposite – Ppy-zirconium (IV) oxide, *Mater. Res. Bull.* **1996**, *31*, 527–530.

195. C.F. Lee, W.Y. Chiu, Soap-free emulsion polymerization of methyl methacrylate in the presence of $CaSO_3$, *Polym. Int.* **1993**, *30*, 475–481.

5

Microencapsulation of Liquid Active Agents

Parshuram G. Shukla

5.1
Introduction

Microencapsulation has proved to be an effective method to protect different types of active agents in a variety of fields such as medicine, agrochemicals, food additives, perfumes, and industrial chemicals. In general, protection is required during storage and also at the application site, so that the active agent does not lose its activity and performs the required function more effectively. Many of these active agents are in liquid form, and consequently their handling in the conventional manner is often difficult. For example, many liquid pesticide formulations are available as emulsifiable concentrates (ECs) in a suitable solvent. The transportation of these solvent-based formulations may be difficult, with effective packaging systems being required because the agents are toxic and in liquid form. A typical example is that of farmers, who often handle harmful solvent-based formulations of pesticides.

One approach to overcoming these problems is to formulate materials such as liquid pesticides as an aqueous dispersion in microcapsules. This method is equally applicable to many other fields where a liquid active agent is required, and where problems of evaporation during storage and at the application site often occur. Evaporation during storage can be avoided with the correct packaging system, but the problem persists at the application site because the parameters responsible for evaporation are generally beyond our control. In such cases, it would be advantageous to protect a liquid in a convenient manner, and this can be achieved by enclosing the liquid active agent in a polymeric membrane. Chemical systems (e.g., resin + curing agent) composed of two liquids or one liquid and one solid can be placed in a single package, but a chemical reaction between the two is prevented by placing one constituent in a microcapsule form. Thus, a reaction cannot take place unless an external trigger occurs, for example, a change in temperature, pressure or pH. Other similar applications include perfumes, where release from a microcapsule can occur simply by a diffusion process, thus providing a longlasting fragrance. Alternatively, it may be

Functional Coatings. Edited by Swapan Kumar Ghosh
Copyright © 2006 WILEY-VCH Verlag GmbH & Co. KGaA, Weinheim
ISBN 3-527-31296-X

triggered by mechanisms used in applications such as cosmetics or laundry products.

5.2
Microencapsulation Methods

Several reviews on microencapsulation processes are available [1–5]. In this section, those methods that are novel and related to the microencapsulation of liquid active agents are discussed.

5.2.1
Microencapsulation by In-Situ Polymerization

Microencapsulation by a polycondensation process, which may be either normal dispersion polycondensation or interfacial polycondensation, is especially attractive for liquid active agents. The important advantage of this method is that in most cases very high active agent loadings can be obtained. These methods are well reviewed by Arshady [6], and are discussed in detail elsewhere in this book.

5.2.1.1 Dispersion Polycondensation

In dispersion polycondensation the monomers are initially soluble in the polymerization medium, while the active agent is in the emulsion form (in polymerization medium). As the polymerization reaction begins in homogeneous solution, oligo-condensate molecules collapse on the surface of the active agent droplets and grow into a polymer that encloses the active agent. Amino resin microcapsules prepared in aqueous medium (continuous phase) is an example of such a system. Amino resins used in microencapsulation are generally formed *in situ* in aqueous medium by the condensation of urea, melamine and other amino compounds with formaldehyde. Dietrich et al. have published a series of reports on amino resin microcapsules which include literature and patent reviews [7], preparation and morphology [8], release properties [9] and surface tension of the resins and mechanism of capsule formation [10].

Researchers at the National Chemical Laboratory, India, have prepared and evaluated amino resin microcapsules of solid pesticide (carbofuran), high-boiling liquid pesticide (quinolphos) and low-melting pesticide (chlorpyrifos) [11,12]. A typical procedure for this preparation of water-insoluble pesticides is as follows. Urea-formaldehyde (UF) methylol is initially prepared and mixed with an aqueous solution of polyvinylpyrrolidone (Luviscol K 90); this acts as a stabilizer for the dispersion and prevents agglomeration of the microcapsules. When UF-MF (melamine-formaldehyde) co-condensation is carried out, the individual methylols are prepared separately and then mixed together. Pesticide (e.g., technical quinalphos) is mixed with few drops of a suitable emulsifier and then dispersed in the above medium with high-speed stirring (ca. 1000 rpm) to obtain stable droplets of <15 μm diameter. A stepwise lowering of the pH by the addition of HCl, followed by raising of the

Table 5.1 Effect of crosslinking of urea-formaldehyde resin on density of matrix and release of carbofuran from microcapsules.

[F]/[U]	a.i. [%]	Density of matrix [g mL^{-1}]	Time for 30% release [h]
1.4	46.7	1.3535	0.5
1.6	50.5	1.3845	16.5
1.8	55.2	1.4321	30.5
2.0	55.5	1.4775	95.0
2.2	57.8	1.5039	187.0

temperature to 36 °C for a few hours and overnight agitation at ambient temperature was required to form stable microcapsules. Sodium chloride is added to methylol to aid in the formation of stable coacervate (colloidal precipitate) of the polymer, and finally fumed silica is added and the pH is adjusted to 5 to stabilize the microcapsular dispersion. By suitably adjusting the reaction parameters such as temperature and amino/formaldehyde ratio, with judicious combination of the amino compounds specific to each of the active agent and employing suitable surfactants, microcapsules of the desired strength and release properties can be prepared.

The structure–property relationships of amino microcapsules have been studied with UF microcapsules containing carbofuran [13]. An increase in formaldehyde:urea ratio increases the density of the UF resin matrix and decreases the release rate of carbofuran from the UF resin shell, thus indicating a corresponding increase in the extent of crosslinking (Table 5.1). This observation is confirmed by analyzing UF matrices (having different [F]:[U] ratios) using ^{13}C-CP/MAS NMR spectroscopy. A gradual disappearance of the unreacted methylol (-N-CH$_2$-OH) and the linear component (-NH-CH$_2$-NH) is observed when the [F]:[U] ratio is increased from 1.4 to 2.2.

Amino resins have been used to prepare microcapsules containing hydrophobic liquids for use in copying papers [14]. An emulsion of dibenzyltoluene, violet lactone and leuco *N*-benzoyl methylene blue is stirred at pH 7–8 while a 62–69% solution of etherified melamine resin is added in a specific proportion. When this emulsion is acidified to pH 4.2 and heated at 55 °C for 10 h, microcapsules with a size range of ca. 20 µm are obtained. The addition of pyrogenic Al$_2$O$_3$ (average particle size 10 nm) along with the dye has been shown to reduce the microcapsule size to 8 µm.

5.2.1.2 Interfacial Polycondensation in Nonaqueous Medium

Most of the interfacial methods reported employ oil-in-water or water-in-oil systems [6]. Polyurethane (PU) microcapsules containing monocrotophos (MCR) have been prepared by interfacial polymerization using oil-in-oil systems. MCR is both a systemic and contact pesticide which is useful for the treatment of a wide range of pests. Technical MCR (75% active ingredient content) is a semisolid at room temperature, and is freely soluble in water and in many organic solvents, except for aliphatic hydrocarbons. MCR is unstable in moist conditions at temperatures higher than 38 °C if exposed long-term, and also reacts with amine. By considering these chemical and physical properties of MCR, a practical method for its microencapsulation using polyurethane as a carrier polymer, and without any deleterious effect on its pesticide activity, has been demonstrated [15–17]. The diols used are insoluble in continuous phase (paraffin oil), whereas toluene diisocyanate (TDI) has limited solubility in paraffin oil (~10%). The active agent MCR is soluble in diol, and thus the microencapsulation method follows typical interfacial polymerization.

The microencapsulation method utilized is as follows. An oil phase containing a catalyst (1,4-diazobicyclo [2,2,2] octane), MCR, diol and crosslinker trimethylolpropane (TMP) is dispersed in paraffin oil containing a steric polymeric stabilizer (poly(butadiene-b-ethylene oxide)) in a reaction kettle maintained at 35 °C. The emulsion is stirred at 1200 rpm and TDI is added dropwise to the reaction mixture. After stirring the system at a specific stirring rate and temperature, PU microcapsules are formed which are then isolated by filtering the mixture, washing with hexane, and drying under vacuum.

Microcapsules prepared with ethylene glycol (EG) alone (without crosslinker) are found to be in the form of free-flowing powders, whereas those prepared with diethylene glycol (DEG) and 1,4-butane diol (BD) turn into lumps after one to two days, indicating leakage of MCR through the PU membrane [16]. The better holding capacity of MCR by the PU membrane with EG may be attributed to the relatively shorter chain length of the soft segments in PU and thus, less flexibility of PU chains. Microcapsules of DEG with 20% TMP showed better flowability, indicating that flexibility of PU chains can also be reduced by the incorporation of crosslinker (TMP).

Scanning electron microphotographs of microcapsules containing MCR, and those from which MCR has been extracted, are shown in Figure 5.1. The micro-

a

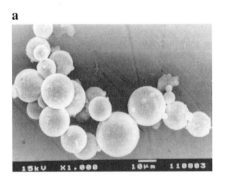

b

c

Figure 5.1 Scanning electron micrographs of microcapsules containing monocrotophos (a,b) and those from which monocrotophos has been extracted (c) [16].

capsules obtained have a typical reservoir-type structure (not monolithic type) (Fig. 5.1c), and the microencapsulation method follows capsule-forming interfacial polycondensation.

In a microencapsulation process conducted in a nonaqueous medium, the choice of surfactant is more critical. Surfactants, which prevent agglomeration of microcapsules, play an important role in the particle/capsule-forming polymerization process [18]. As the repulsive potential, between two particles, is directly proportional to the dielectric constant of the dispersion medium, stabilization of particles in nonaqueous medium – and especially in a nonpolar medium such as aliphatic hydrocarbon – becomes a difficult task. In such nonaqueous systems polymeric steric stabilizers are normally useful. These steric stabilizers include polymerizable surfmers (Fig. 5.2(1)), amphiphilic block (Fig. 5.2(2)), or graft copolymers (Fig. 5.2(3)) and homopolymers. Amphiphilic polymers such as conventional surfactants contain hydrophobic and hydrophilic moieties. Polymerizable stabilizers contain either hydrophobic or hydrophilic stabilizing polymer components, having reactive vinyl groups or two functional reactive groups such as hydroxyls through which the stabilizer becomes chemically anchored to main polymer chain.

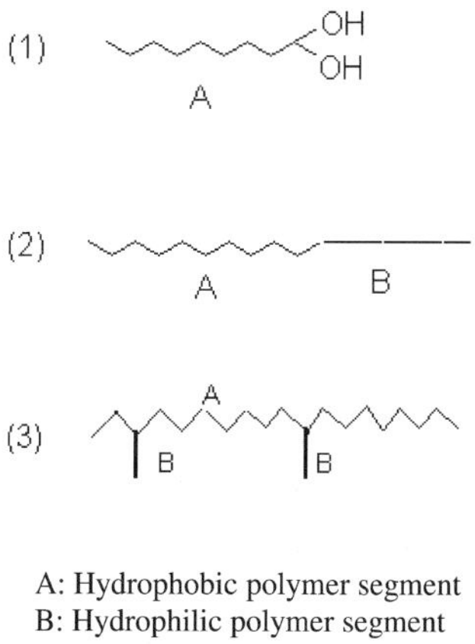

A: Hydrophobic polymer segment
B: Hydrophilic polymer segment

Figure 5.2 Polymeric steric stabilizers.

A variety of stabilizers, including reactive diol containing two primary hydroxyl groups with a long hydrophobic moiety (e.g., PLMA) (Fig. 5.3a), an amphiphilic block copolymer (Fig. 5.3b) and graft copolymer (Fig. 5.3c) have been used in the preparation of polyurethane microspheres or microcapsules in nonaqueous medium [15–17,19–24].

Frere et al. have reported the synthesis and characterization of PU microcapsules containing water, by interfacial polycondensation using three different diols, viz. 1,5- pentane diol, poly(ethylene glycol)s (PEG 600, PEG 1500, PEG 4200), and two different isocyanates, viz. diphenyl methylene diisocyanate (MDI) and poly(hexamethylene diisocyanate) [25]. The surfactant used is ABA block copolymer from

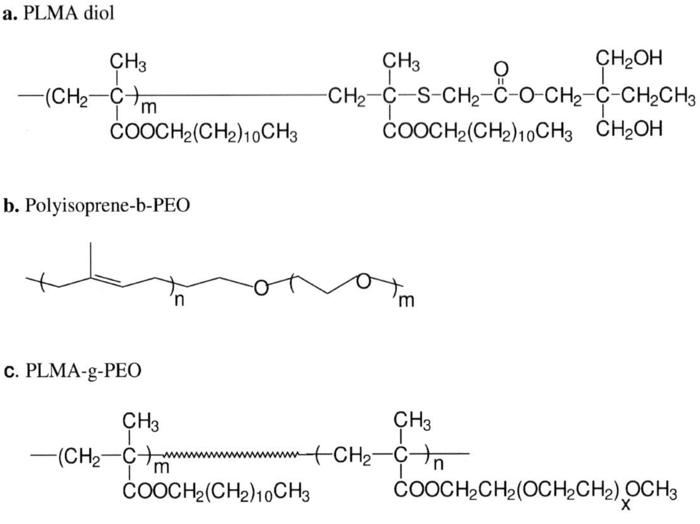

a. PLMA diol

$$-(CH_2-\underset{\underset{COOCH_2(CH_2)_{10}CH_3}{|}}{\overset{\overset{CH_3}{|}}{C}})_m \overline{} CH_2-\underset{\underset{COOCH_2(CH_2)_{10}CH_3}{|}}{\overset{\overset{CH_3}{|}}{C}}-S-CH_2-\overset{\overset{O}{\|}}{C}-O-CH_2-\underset{\underset{CH_2OH}{|}}{\overset{\overset{CH_2OH}{|}}{C}}-CH_2CH_3$$

b. Polyisoprene-b-PEO

c. PLMA-g-PEO

$$-(CH_2-\underset{\underset{COOCH_2(CH_2)_{10}CH_3}{|}}{\overset{\overset{CH_3}{|}}{C}})_m \text{\wwwwwwwwww} (CH_2-\underset{\underset{COOCH_2CH_2(OCH_2CH_2)_x OCH_3}{|}}{\overset{\overset{CH_3}{|}}{C}})_n$$

Figure 5.3 Polymeric stabilizers used in the preparation of polyurethane microcapsules.

poly(hydroxy stearic acid) and polyethylene oxide. The procedure involves the dispersion of solution containing diol, triol and water into polymerization medium – that is, toluene containing surfactant (under stirring) followed by the addition of mixture of two isocyanates (in different proportions) and catalyst dissolved in toluene surfactant solution. The microcapsules obtained are decanted, separated, and washed several times with water.

5.2.1.3 Microcapsules with High Concentrations

Interfacial polymerization to produce microcapsules has one drawback, namely that the process cannot be used to produce microcapsules at higher concentrations. Most of the conventional surfactants used in oil-in-water emulsions fail to maintain a stable suspension of microcapsules, especially when a high concentration of oil (water- immiscible material) is involved, and this results in large amounts of agglomerated microcapsules. The key parameter for obtaining high-concentration microcapsules is the choice of emulsifier which is used to prepare oil-in-water emulsion (prior to the actual encapsulation process). In the case of a microencapsulation process involving diisocyanate and diamine, when polyvinyl alcohol is employed to obtain a high-concentration emulsion, the emulsions become grainy in appearance following the addition of diamine to effect microencapsulation. Beestman and Deming have shown that lignosulfonates can produce concentrated oil-in-water emulsions that are stable to shell wall-forming chemical reactions at the oil/water interface [26]. Table 5.2 lists the generalized composition of a high-concentration microencapsulation formulation. It can be seen that the discontinuous phase (pesticide or solvent) to be microencapsulated constitutes a greater proportion of the emulsion than the continuous aqueous lignosulfonate phase into

Table 5.2   Generalized composition of a high-concentration microencapsulation formulation.

Ingredient	Weight [%]
Pesticide or solvent	49.0
Polyfunctional isocyanate[a]	3.7
1,6-Hexanediamine, 42.3% aqueous[b]	3.7
Emulsifier	1.0
Water	38.4
Formulation ingredient	4.2
Total	100

[a] Commercial products generically known as polymethylene polyphenyl isocyanate.
[b] Aqueous concentration such that stoichiometric equivalence was provided for equal weights of diamine solution and polyfunctional isocyanate. (From [26].)

which microcapsules are formed. The microcapsules obtained are unaggregated individual particles with a smooth continuous surface.

Microcapsules at higher concentration via interfacial polymerization have also been prepared using the salt of a partial ester of styrene-maleic anhydride copolymer as surfactant [27]. The process described is mostly suitable for water-immiscible active agents that are liquid at room temperature, or have a melting point <60 °C.

5.2.1.4   Microencapsulation Using a Static Mixer

Hirech et al. have described preparation of polyurea microcapsules containing liquid pesticide (e.g., diazinon) suspended in concentrated disinfectant solution, by use of a two-step microencapsulation process [28]. The first step is the liquid–liquid dispersion in a Sulzer static mixer (SMX); the second step is microencapsulation by interfacial polymerization in a stirred-tank reactor (Fig. 5.4). It has been shown that the

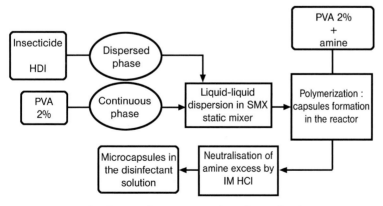

Figure 5.4   Principle of insecticide microencapsulation by interfacial polymerization [28].

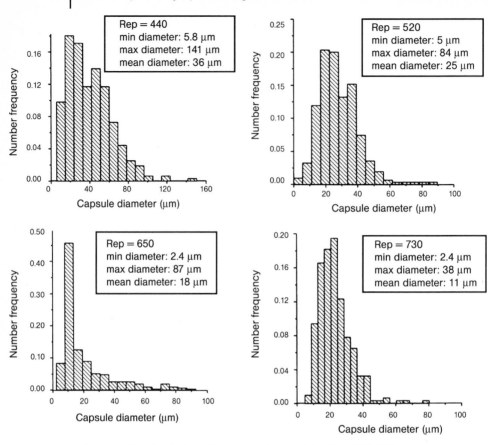

Figure 5.5 Size distributions of microcapsules containing insecticide [28].

size distribution of the microcapsules – which is an important factor to ensure the suspension of microcapsules in the disinfectant liquid – can be controlled by this two-step process. The droplet mean diameter of the oil/water dispersion produced in the static mixer depends on the flow rates of the dispersed phase (insecticide + HDI) and the continuous phase (2% PVA solution). In order to characterize the hydrodynamic conditions, a Reynolds number, Re_p, has been defined with the oil/water dispersion properties. An increase in the Reynolds number leads to a reduction in the mean diameter of microcapsules, and also narrow particle size distribution (Fig. 5.5).

5.2.2
Microencapsulation Using Liquid CO$_2$

Microcapsules containing dyes are usually prepared by dispersion polymerization. Dyes may be dissolved in the monomer, which is further polymerized [29]. This technique usually results in poor polymer conversion and particles having a broad size distribution. In another method, blank polymer particles are first produced as an aqueous dispersion in which an organic solvent containing a dissolved dye is emulsified [30]. The organic solvent swells the polymer particles and causes the transport of dye into the particles. This method, although effective, has certain drawbacks, including the use of toxic organic solvents and high energy demands for the multiple steps of freeze-drying and redispersion of particles in the aqueous medium.

Yates et al. reported a new microencapsulation technique in which liquid CO_2 is used to facilitate the transport of dyes into aqueous polymer colloids [31]. First, the polystyrene (PS) microspheres are prepared by dispersion polymerization, using ethanol as the dispersion medium and poly(N-vinyl pyrrolidone) (PVP) as surfactant. The PVP grafts to styrene during the dispersion polymerization, and thus the surface of the PS microspheres are coated with a layer of PVP. These PS microspheres (aqueous latex form) are dyed with Sudan Red 7B, using CO_2 at 25 °C at 310 bar in a stainless steel variable volume view cell. The CO_2 plasticizes the polymer particles and enhances the transfer of dye into the particle phase. Dye transfer into the polymer particles is greatly enhanced when the CO_2/water interfacial area is increased through emulsification, using both a fluorinated and a hydrocarbon-based surfactant (perfluro polyether ammonium carboxylate) (PFPE-NH$_4$) and poly(ethylene oxide)-b-poly(butylene oxide). It has been observed that when the dyeing experiment is carried out without surfactant, the incorporation of dye into polymer particles is 0.05%, whereas with PEO-b-PBO and PFPE-NH$_4$ it is 0.17 and 0.46%, respectively. There is no agglomeration or coalescence of the particles, although particles are highly plasticized during the dyeing process.

Further studies by Yates et al. with another surfactant (viz., poly(ethylene oxide)-b-poly(propylene oxide) and with various dyes having different solubilities in water, have shown that the partition coefficient of dyes into PS from water is the most important factor determining maximum dye loading [32]. Those dyes which are insoluble in water but have good solubility in benzene show maximum incorporation into the particles (Table 5.3). It has also been shown that kinetics of dye loading is improved by increasing the amount of added surfactant (Fig. 5.6), pressure, and amount of dye added (Fig. 5.7).

Studies on the microencapsulation of dyes using CO_2 as detailed above are carried out with PS microspheres prepared by dispersion polymerization. These sterically stabilized (with PVP) PS particles retain their original size and shape after the incorporation of dye. When PS microspheres are prepared by surfactant-free emulsion

Table 5.3 Dye loadings of different dyes[a]

Dye	Solubility in water	Solubility in benzene	Dye loading [wt.%]
Sudan Red 7B	Insoluble	3–10 wt.%	0.5613
Solvent Yellow 1	Slightly soluble	Soluble	0.1908
Solvent Violet 8	Very good	Slightly soluble	0.1193
Rose Bengal dye	Very good	Insoluble	0.0188

[a] Medium, water; dye amount, 0.050 g; F108, 0.60 wt.%; particle diameter, 2.3 μm; time, 24 h; pressure, 310 bar. (From [32].)

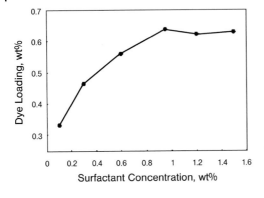

Figure 5.6 Effect of surfactant concentration on dye loading. Medium: water; dye: Sudan Red 7B, 0.050 g; particle diameter: 2.3 μm, time: 24 h, pressure: 310 bar [32].

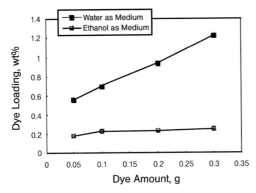

Figure 5.7 Effect of dye amount on dye loading. Dye, Sudan Red 7B; particle diameter, 2.3 μm; F108, 0.60 wt.%; time, 24 h; pressure, 310 bar [32].

polymerization, electrostatically stabilized particles are obtained. These electrostatically stabilized PS latexes, when exposed to liquid CO_2, lose their stability due to pH and ionic strength changes caused by carbonic acid formation in the aqueous phase. This problem has been solved by use of "dual function" surfactants [33]. Such surfactants are ionic surfactants (e.g., sodium dodecyl sulfate, SDS; hexadecyl trimethyl ammonium bromide, CTAB; cetyl pyridinium chloride monohydrate, CCM) and non-ionic surfactants (e.g., Pluronic F108, Tetronic 908, Triton X-100, Brij 78 and Brij 700). These surfactants, in being active at the CO_2/water interface, improve the microencapsulation process; in addition, being active at the polymer/water interface they enhance the colloidal stability of the PS particles. The nonionic surfactants provide a more stable emulsion of CO_2 into water, and this results in high dye loading.

5.2.3
Microencapsulation Using the Shirazu Porous Glass (SPG) Emulsification Technique

Microcapsules with a narrow size distribution containing oily core material can be prepared by a Shirazu porous glass (SPG) emulsification technique, followed by a suspension polymerization process. The SPG membrane is a special porous glass membrane with very uniform pore size. Guang Hui Ma et al. have reported the preparation of microcapsules containing hexadecane (oil core) using poly(styrene-

N,N-dimethylaminoethyl methacrylate) (DMAEMA) by an SPG emulsification technique [34]. The process consists of two steps – emulsification and polymerization. The oil phase containing monomers (styrene and DMAEMA), hexadecane (HD) and an initiator *N,N'*-azobis (2,4-dimethyl valeronitrile) (ADVN) is pressed by nitrogen gas through the SPG membrane into the aqueous phase. The aqueous phase contains stabilizer polyvinyl alcohol (PVA), surfactant sodium lauryl sulfate (SLS), electrolyte Na_2SO_4 and water-soluble inhibitor ($NaNO_2$ or diaminophenylene). The emulsion obtained is then transferred to a separate reaction kettle and polymerization is started by raising the reaction temperature to 70 °C. After 24 h, microcapsules with uniform size are obtained.

The effect of monomer conversion on the efficiency of encapsulation is observed by carrying out experiments using diaminophenylene as initiator in order to obtain a low conversion (53–76%). It has been observed that at lower conversion HD – irrespective of the loading amount – is not completely encapsulated by poly(St-*co*-DMAEMA).

Scanning electron microscopy (SEM) of microcapsules from which HD is extracted have been studied to understand the extent of encapsulation. One-hole, large-hole and half-moon morphologies imply that HD is not encapsulated completely by polymer, whereas a hollow morphology indicates complete encapsulation. Figure 5.8 shows SEM images of polymer microcapsules as a function of HD amount at lower monomer conversion (without DMAEMA). In the absence of DMAEMA, one-hole or

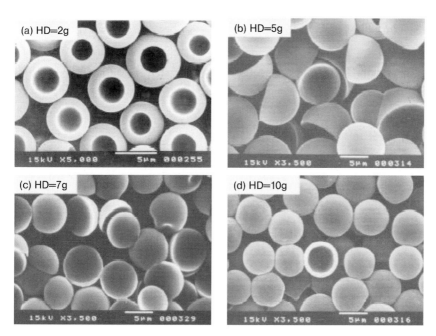

Figure 5.8 Scanning electron micrographs as a function of hexadecane quantity in the case of lower monomer conversion without addition of poly(styrene-*N,N*-dimethylaminoethyl methacrylate) (DMAEMA) [34].

half-moon morphology (indicating incomplete encapsulation) is also obtained after HD removal, except in the case where the HD amount was very high (10 g, 50 wt.% based on total oil phase) and conversion was also high (83.4%). When the experiment at highest loading of HD was carried out by reducing the polymerization time from 20 h to 10 h, a lower conversion (63%) was obtained, resulting in particles having incomplete encapsulation and showing half-moon morphology.

Higher conversions are obtained when $NaNO_2$ is employed as a water-soluble inhibitor in the presence of DMAEMA. In the absence of DMAEMA no inhibitor is required. In all of these experiments, with both lower and higher amounts of HD, complete encapsulation was observed irrespective of the addition of DMAEMA (2.5 wt.%, 0%)

Figure 5.9 shows SEM images of microcapsules (with higher monomer conversion) prepared without DMAEMA and with lower and higher amounts of HD. The hollow morphology indicates that complete encapsulation of HD is obtained. Thus, complete encapsulation of HD is achieved when the polymer conversion and HD loading are high. Incomplete encapsulation of HD at lower conversion is attributed to the fact that the interfacial tension of the HD phase with water and the polystyrene phase with water are closer (Table 5.4). Our understanding of such systems can be clarified when the morphology is simulated as functions of conversion and HD amount by employing the UNHLATEX- EQMURPH version 4 program, as developed by Sundberg's polymer research group at the University of New Hampshire (Fig. 5.10; Table 5.4).

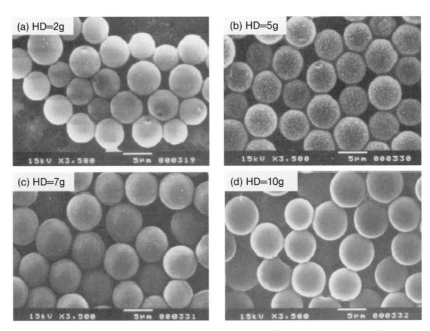

Figure 5.9 Scanning electron micrographs as a function of hexadecane quantity in the case of higher monomer conversion without addition of poly(styrene-N,N-dimethylaminoethyl methacrylate) (DMAEMA) [34].

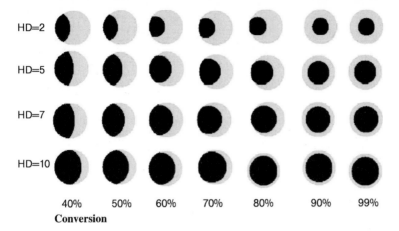

Conversion

Figure 5.10 Simulated morphology as functions of monomer conversion and hexadecane (HD) quantity when no poly(styrene-N,N-dimethylaminoethyl methacrylate) (DMAEMA) was added. Data processed using the Sundberg group's UNHLATEX_EQMORPH program. The black region represents the HD phase; the gray region represents the polymer phase [34].

Table 5.4 Simulated results of particle morphology by using UNHLATEX_EQMORPH version 4 program developed by Sundberg et al. (no DMAEMA added).

	Quantity of HD [g]			
	2	5	7	10
Conversion 50%				
Σ_{13}	1.87	2.18	2.37	2.65
σ_{12}	22.23	23.18	23.92	24.87
σ_{23}	21.30	21.72	22.09	22.56
Difference $(\sigma_{12} - \sigma_{23})$	0.93	1.46	1.83	2.31
Conversion 99%				
Σ_{13}	3.47	3.48	3.49	3.49
Σ_{12}	27.75	27.69	27.80	27.79
Σ_{23}	24.09	24.05	24.12	24.11
Difference $(\sigma_{12} - \sigma_{23})$	3.66	3.64	3.68	3.68

s_{13}: interfacial tension between HD (St/HD mixture) phase and PSt (St/PSt mixture) phase.
s_{12}: interfacial tension between HD (St/HD mixture) phase and aqueous phase;
s_{23}: interfacial tension between PSt (St/PSt mixture) phase and aqueous phase.
(Values taken from [34].)

5.2.4
Microencapsulation by Miniemulsion Polymerization

Oil-in-water miniemulsion can be obtained by dispersing an oil phase containing oil and a hydrophobe in the water phase containing an adequate amount of surfactant, using fast stirring and ultrasonication. Very small droplets of oil with interfacial tension larger than zero and in the size range of 50 to 500 nm with sufficient stability are obtained. The amount and type of surfactant and ultrasonication time governs the size of the droplets created [35,36].

For microencapsulation by miniemulsion polymerization, oil can be a monomer and the hydrophobe can be a liquid active agent in the above-mentioned miniemulsion system. The monomer and oil are chosen in such a way that these two components are miscible before the polymerization process. As soon as polymerization takes place, phase separation occurs and at the end of polymerization microcapsules containing active agent are obtained.

Tiarks et al. have described in detail the theory of droplets composed of binary mixtures in relation to the preparation of polymeric nanocapsules containing HD by miniemulsion polymerization using poly(methyl methacrylate) (PMMA) or PS polymer [37]. First, a miniemulsion of oil phase (monomer + HD + initiator) dispersed in aqueous phase and containing an appropriate surfactant is prepared by ultrasonication. The polymerization is initiated by heating the emulsion to 68 °C. When the polymer is formed it separates from the HD phase and, depending on the interfacial tension and spreading coefficients of monomer/polymer, HD (hydrophobe) and water, capsules with different morphologies are obtained.

5.2.5
Microencapsulation by Coacervation

5.2.5.1 **Simple Coacervation**
Simple coacervation is one of the oldest and widely used microencapsulation methods wherein active agent (liquid or solid) is dispersed in a homogeneous aqueous polymer solution. The formation of colloidal polymer aggregates (coacervates) is brought about by a change in temperature or pH. Thus, the polymer is deposited on an active agent surface, leading to the formation of microcapsules. Hydrophilic polymers such as gelatin, PVA, methyl cellulose, and cellulose acetate are mostly used in this process.

Tirkkonen et al. [38] developed an automatically controlled equipment for pilot processing of microcapsules formed by coacervation. It is possible to adjust the size of a batch of microcapsules manufactured by this equipment between 500 g and 5 kg [38].

The release of santosol oil from crosslinked PVA microcapsules prepared by coacervation, is shown to be mainly controlled by the crosslinking density of the PVA membrane and the size of the microcapsules [39]. As santosol oil used for encapsulation has very low solubility in water, release rate studies are carried out in water containing the surfactant SDS. It has been observed that, by increasing the

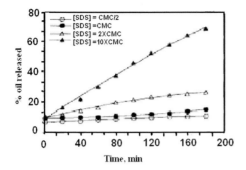

Figure 5.11 Effect of sodium dodecyl sulfate (SDS) concentration on release rate of santosol oil from polyvinyl alcohol microcapsules [39].

concentration of SDS in the release medium above its CMC, the solubilization of oil increases and this leads to faster release rates (Fig. 5.11).

Gelatin microcapsules containing essential oils, namely Rosmarinus and Thymus oil, with high encapsulation yields (over 98%), have been prepared using the coacervation process [40]. Microcapsules obtained by the conventional coacervation method are filtered, rinsed with cold water, and finally dehydrated by freeze-drying. Rosmarinus oil contains high levels of monoterpene hydrocarbons (64%), whereas carvacrol and thymol are the main constituents (68%) of Thymus oil. It has been observed that oil content of the wet microcapsules for both oils is almost the same, whereas in the dried product entrapment of Rosmarinus oil is more (65.2% oil content) than that of Thymus oil (55.6% oil content). This difference is attributed to the better hydrophilic characteristics of Thymus oil components compared to those of Rosmarinus oil, which favors aqueous phase entrapment during coacervation. Higher loss of essential oil occurred during the final dehydration stage, because oil volatility depends on the water concentration in the evaporating system. It has been suggested that these formulations must be stored in closed containers, since release rate studies of these microcapsules have shown that a loss of active agent (oil) occurs when the microcapsules are exposed to environmental humidity.

5.2.5.2 Complex Coacervation

Complex coacervation is similar to simple coacervation where another complimentary polyelectrolyte is used. Gelatin and gum arabic is a well-established system for microencapsulation by complex coacervation. Mayya et al. have reported a two-layer encapsulation of paraffin oil, based on a primary layer of interface active polyelectrolyte–surfactant complex, followed by a second layer of the conjugate polyelectrolyte–polyelectrolyte complex [41]. The procedure involves the dispersion of paraffin oil in 1% gelatin solution (pH adjusted to 6.5) containing SDS having concentration less than its CMC, followed by drop-wise addition of the solution of the other polyelectrolyte (1% gum arabic) into the dispersion. The pH is then adjusted to 4.5 to induce coacervation. After stirring for 1 h, the system is cooled to 8 °C, the crosslinking agent (glutaraldehyde) is added and, after adjusting the pH to 9, the system is stirred for 12 h. The microcapsules obtained are first washed with water, then with chilled isopropanol to dehydrate the walls of capsules, and then air-dried at room temperature.

As shown in Figure 5.12, these authors prepared a model for capsule formation. When SDS is added to the oil/water emulsion, complexation occurs between SDS and the oppositely charged polyelectrolyte (gelatin). This complex deposits at the oil droplet/water interface in the form of a primary layer (Fig. 5.12, B). Addition of the second polyelectrolyte (gum arabic) to the system induces further complexation between the two polyelectrolytes and covering of the primary layer surface. It is observed that the addition of surfactant (SDS) increases the encapsulation yield (Fig. 5.13).

The commonly used complex coacervation processes of microencapsulation have several drawbacks, including a need for constant attention and the adjustment of stirring conditions, solution viscosity, pH, and temperature. The process also often produces significant amounts of agglomerated microcapsules. A European

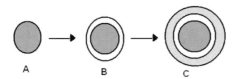

A   B   C

Figure 5.12 Schematic diagram of the model of capsule formation. The oil droplet (A) and oil droplet with primary layer (B) combine to form an oil droplet with secondary layer (C) [41].

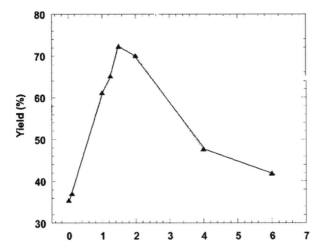

Figure 5.13 Micro-encapsulation yield (%) as a function of sodium dodecyl sulfate (SDS) concentration [41].

patent by Baker and Ninomiya provides an improved and simplified complex coacervation process, which produces high core-content microcapsules without agglomeration [42]. The process is suitable for a core ingredient which is a hydrophobic liquid or an emulsion of a hydrophilic liquid in a hydrophobic liquid. The two novel steps described for the process are: (1) mixing an ionizable colloid, ionic surfactant or ionizable long-chain organic compound with the core ingredient prior to emulsification of the core ingredient in a first aqueous colloid solution; and (2) adding a water-soluble wax derivative after gelation and prior to hardening.

5.2.6
Spray-Drying

Ninomiya et al. have patented a spray-drying technique useful for preparing microcapsules containing liquid active agent [43]. Polymer (polysulfone) solution prepared using organic solvent (dichloromethane), containing liquid active agent (e.g., insect attractant) is sprayed as liquid droplets (30–40 μm) through an aerosol-forming nozzle. The evaporation of organic solvent results in microcapsules of 1- to 10-μm size. It has been shown that microcapsules release 75% of the initial active agent in 30 days. Similar process producing microcapsules containing 40% insect attractant having 2- to 20-μm size range and 0.2- to 2-μm thickness film, has been described in one Japanese patent [44].

5.3
Characterization of Microcapsules

5.3.1
General Methods

Various methods are available to characterize microcapsules, and these mainly involve the active agent content, optical microscopy and SEM, particle size, and size-distribution and release-rate studies. The active agent in microcapsules is mostly determined by extracting an active agent in a suitable solvent, followed by quantitative analysis by UV spectroscopy or high-performance liquid chromatography or gas chromatography. The actual active agent content also provides an idea of encapsulation efficiency. Analysis of the extracted active agent by other spectral methods (e.g., infra-red and NMR) can be used to confirm that the active agent does not undergo any chemical change during the process of encapsulation. Optical microscopy and SEM are used to determine the size and morphology of microcapsules. Size distribution is monitored by using a suitable particle size analyzer instrument, or on occasion by measuring the size of several microcapsules by optical microscopy. Release-rate studies indicate the effect of particle size, active agent loading, the properties of polymer wall such as crosslink density, and the nature of the release medium. Since in the case of microcapsules containing liquid active agent it is important to know the physical strength of microcapsules, this aspect is described in detail in the following section.

5.3.2
Physical Strength of Microcapsules

In many applications, including copying paper, pest control, and self-healing composites (see Section 5.5), the release of active agent is triggered by employing external pressure in some way. Clearly, in such cases an understanding of the physical strength of microcapsules is required. The desired strength of microcapsules would be such that they do not rupture and/or release the active agent during storage, nor until they come into contact with an intended target. However, when such contact is made and pressure is applied to the microcapsule, its wall should break to release the active agent.

Ohtsubo et al. have described a method to identify those parameters which determine the physical strength of PU microcapsules containing liquid insecticide (viz., fenitrothion) [45]. These microcapsules have been found to be effective in controlling household pests such as cockroaches. As expected, the strength of the microcapsule depends mainly on the microcapsule mass mean diameter (D), the thickness of the membrane wall (T), and particle size distribution expressed as polydispersity (e.g., D_w/D_n). The coefficient of variation of the particle size distribution (CV) is first determined using any suitable instrument (particle size analyzer) and method. Wall thickness T is calculated from the following equation:

$$T = (W_w/W_c) \, (\rho_c/\rho_w) \, (D/6) \tag{1}$$

where W_w is the weight of wall material, W_c is the weight of core material, ρ_c is the density of core material, ρ_w is the density of wall material, and D is the mass median diameter. In experiments carried out with PU microcapsules containing fenitrothion, $\rho_c/\rho_w = 1$.

The physical breaking test of microcapsules (as described by Ohtsubo et al.) is conducted as follows. A prescribed number of microcapsules is applied to a glass plate. Another glass plate is gently attached to the surface of the treated plate and a rubber sheet is placed on the top and bottom sides (Fig. 5.14). The prescribed weight is softly loaded onto the rubber sheet for 1 min. After removing the loaded weight, the total amount of active agent (F_1) and the amount of active agent outside the microcapsules (F_0) are analyzed by previously established quantitative analytical methods. The broken ratio of microcapsules is calculated as follows:

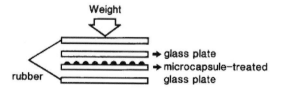

Figure 5.14 Method of breaking test for microcapsules [45].

Broken ratio (%) = $(F_0/F_1) \times 100$ (2)

The weight (W) per milligram of fenitrothion applied to the plate has been used as a parameter of the weight loaded onto the sample. Following the theory of destruction of an empty sphere with a thin wall, when introduced to the breaking of microcapsules, the pressure (P) required to break a microcapsule is calculated as follows:

$P = W_s/[\pi (D/2)^2]$ (3)

where W_s, the weight loaded onto a single microcapsule, is calculated as follows:

$W_s = W/\{(1/\sigma_{mc})/[4/3 \, \pi \, (D/2)^3]\}$ (4)

where σ_{mc} is the density of the microcapsule.

Figure 5.15 shows plot of percentage broken microcapsules versus P (log scale) for two microcapsule samples having the same D/T ratio but with different size distribution (CV = 17.9 and 4.9%). The slope of the for microcapsules with a small CV is steeper than that of the sample with a large CV, indicating that the particle-size distribution of microcapsules affects the breaking behavior. However, the pressure required to break 50% of microcapsules (P_{50}) is independent of CV (when D/T is constant). Table 5.5 shows P_{50} values for microcapsules having different D/T ratios. When the D/T ratio is decreased (i.e., when the microcapsule diameter decreases and thickness of membrane wall increases), the P_{50} will also increase. It has been further shown that when P = 20 μg μm^{-2} , the physical breaking behavior of the microcapsules agrees best with the biological breaking by German cockroaches [45].

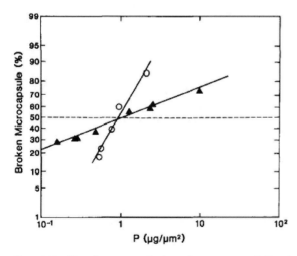

Figure 5.15 Plot of percentage broken microcapsules versus pressure (P) (log scale) for two microcapsule samples having the same D/T ratio but with different size distribution. CV = 17.9% (▲) and 4.9% (○) [45].

Table 5.5 Relationship between mass mean diameter/wall thickness ratio (D/T) and P_{50}. (From [45].)

	D/T						
	697	699	351	279	176	141	71
P_{50} ($\mu g\ \mu m^{-2}$)	9.2×10^{-1}	1.0	4.8	3.5×10	6.8×10	1.2×10^2	2.1×10^2

Taguchi and Tanaka have described a similar method to measure the mechanical strength of microcapsules [46]. The instrument, a "destruction tester" (Aikoh Co. Ltd., Model 1307), which is used for such studies is shown in Figure 5.16.

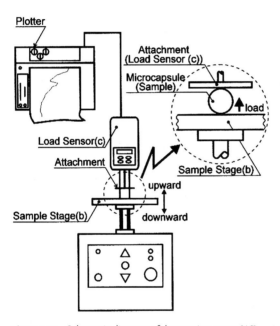

Figure 5.16 Schematic diagram of destruction tester [46].

5.4
Microcapsules with Desired Release Patterns

Microcapsules with desired release rate and patterns can be obtained by carrying out either chemical or physical modifications to the microcapsules. In the case of chemical modifications, the polymeric wall material is modified or, along with active agent, another additive is added. This allows the release of active agent to occur in response to physical/chemical changes in the additive, due in turn to change(s) in the microcapsule environment. Physical modifications involve controlling porosity of the capsule wall, or double encapsulation. The adsorption of an additive such as a surfactant or a wax material that may either enhance or retard the release rate is also feasible. The following sections describe some of these novel modifications.

5.4.1
Chemical Modifications

5.4.1.1 Incorporation of Disulfide Linkages

It has been shown that microcapsules containing a liquid active agent, with variable release rates, can be prepared using *in-situ* polymerization methods [47]. The polymeric shell wall is formed by oxidative coupling of a polythiol compound, or a combination of oxidative coupling and interfacial polymerization and condensation of a mixture of a polythiol compound and butylated urea formaldehyde prepolymer. Thus, the incorporation of disulfide links in the polymeric wall results in microcapsules that show either gradual controlled release or rapid triggered release, depending upon the environmental conditions, such as the presence of base and/or reductive systems. The rate of evaporative weight loss of butylate (active agent) from microcapsules *in vacuo* has been studied using the Cahn RH electrobalance. The release of active agent due to diffusion mechanisms can be modified mainly by the amounts of crosslinker and oxidant added to form disulfide linkages. It has also been shown that, under alkaline release conditions, microcapsules having disulfide linkages have a faster release of encapsulated active agent compared to non-triggered diffusion-controlled conditions. This rapid release under alkaline conditions is attributed to cleavage of the disulfide bonds.

Microcapsules composed only of aminoplast due to an absence of disulfide linkages do not break down under alkaline conditions. Biological evaluations of these microcapsules have been carried out to confirm the results of *in-vitro* release studies. Mortality assessments for microcapsules containing chlorpyrifos, made at different days after treatment (DAT) for *Lygus hesperus* (a sucking pest), showed the LC_{50} values of microencapsulated chlorpyrifos to be higher than those of commercial chlorpyrifos (Lorsban 4E) (Table 5.6). These results indicate that microcapsules with >90% disulfide linkages (see example 13 in Table 5.6) exhibit good barrier properties, and thus provide improved beneficial (non-foliar feeding) insect protection compared to the standard chlorpyrifos formulation (Lorsban 4E). A decrease in LC_{50} values over a period of time was attributed to the slow diffusion-controlled release of chlorpyrifos from the microcapsules.

Table 5.6 LC_{50} values (ppm) at different days after treatment (DAT).

Formulation	1 DAT	2 DAT	3 DAT	4 DAT	5 DAT	6 DAT
Lorsban 4E[a]	262	253	252	258	260	257
Example 13	2118	1433	1245	1253	1218	1199

[a] Chlorpyrifos emulsion concentrate produced by Dow Chemical containing 4 lb chlorpyrifos per gallon (2060 g L^{-1}) [47].

When mortality assessment experiments were carried out with *Helicoverpa zea* (a foliar-feeding lepidopteran with an alkaline gut), the LC_{50} values at 2 DAT for Lorsban 4E, standard aminoplast microcapsules and microcapsules with >90% disulfide linkages were 14.5, 96.4 and 14.7 ppm, respectively. The results indicated that microcapsules made only of aminoplast resin did not have disulfide linkages and thus did not break down in the gut of the insect, as indicated by the higher LC_{50} value. However, those microcapsules with disulfide linkages ruptured in the insect gut and provided comparable control to that with Lorsban 4E.

5.4.1.2 Side-Chain Crystalline Polymers

Side-chain crystalline (SSC) polymers are useful in the preparation of microcapsules from which release of active agent is required to occur in response to change in temperature. A research group at Landec Labs, Inc. (USA) has prepared microcapsules containing diazinon, a liquid organophosphate insecticide using SSC polymer (viz., poly(hexadecyl acrylate) by a standard emulsion encapsulation process [48]. Figure 5.17 illustrates the comparative release of diazinon from microcapsules prepared from a polymer with a melt temperature (T_m) of 30 °C and conventional polyurethane capsules. A sharp increase in diazinon release is seen when the temperature is switched from 20 to 30 °C. PU capsules showed little change in release profile over the same temperature range.

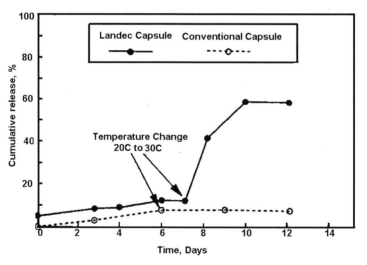

Figure 5.17 Release of insecticide from Landec and conventional microcapsules [48].

5.4.1.3 Photochemical Rupture of Microcapsules

Microcapsules can be made to release their contents on exposure to light by incorporating into the core of the microcapsule a material that photochemically eliminates a gaseous product. Mathiowitz and Raziel have described such system, which involves polyamide microcapsules prepared by interfacial polycondensation of terephthaloyl chloride and diethylenetriamine (DETA) [49]. Microcapsules contain the core material (solvent, e.g., benzene/xylene, 2:1, v/v) such that it has a good solving power for photoeliminator (azo-bis-isobutyronitrile, AIBN) and terephthaloyl chloride, and is immiscible with water. AIBN decomposes both thermally and photochemically, with production of nitrogen and tetramethyl succinodinitrile. The reason for choosing polyamide as the encapsulating polymer is that it has low permeability to nitrogen, is transparent to near-UV light, and is chemically stable to the photoeliminator used. The mechanism of microcapsule rupture is due to AIBN, on UV irradiation, producing N_2 gas which leads to a build-up of pressure within the capsule.

5.4.2
Physical Modifications

5.4.2.1 The Capsule-in-Capsule Approach
Microcapsules containing liquid pesticide have certain drawbacks. One example is when the pesticide is itself both volatile and toxic and has a high vapor pressure. A second example is when the capsule shell is strong and thick. In the first case, the pesticide diffuses very rapidly from the capsules and its odor initially repels the pest. Diffusion from the capsules is rapid, however, and when they are empty the pests return to the site (e.g., crops). In the second case, the capsules do not release the pesticide to produce a minimum effective level at the application site, and so pesticidal action is not achieved. In order to overcome these problems, a WO patent disclosed the preparation of microcapsules of pesticides containing pest attractant using a capsule-in-capsule approach [50]. As shown in Figure 5.18, the outer capsule contains pest attractant or food, in which the inner capsule containing the pesticide, is encapsulated. Such a product would bait or lure a pest to a capsule, but release the pesticide only when the pest digests or attacks the capsule. Using this method, UF microcapsules containing pesticide were prepared such that the capsule size was very small (~15 μm). The second wall material is formed by mixing appropriate amounts of gelatin, gum arabic, and ethyl cellulose in water. UF microcapsules mixed with peanut oil (as attractant) are added to the second wall material solution which, after stirring for 1 h at 65 °C, was cooled to room temperature and stirred for another hour. The resultant gelatin-based shell material encapsulated both peanut oil and UF microcapsules.

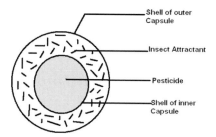

Figure 5.18 Capsule-in-capsule arrangement.

5.4.2.2 **Porosity Control**

Microcapsules with desired release rates according to the application involved can be prepared by incorporating a selected fluid (solvent), while keeping the other capsule parameters constant. These fluids include dearomatized and isoparaffinic hydrocarbons, aromatic hydrocarbons, acetate derivatives, and blends of these [51].

In the preparation of polyurethane-polyurea microcapsules containing heptenophos (liquid pesticide, b.p. 64 °C), it has been reported that the porosity of microcapsules can be controlled by the addition of 2-ethoxy ethyl acetate or ethyl acetate in the organic phase containing isocyanate prepolymer and pesticide [52].

5.4.2.3 **Incorporation of Fibers in the Microcapsule Wall**

In many applications, release is triggered by the application of external or internal pressure, preferably at an optimal value. Consequently it is necessary to adjust the mechanical strength of the microcapsule wall. Taguchi and Tanaka reported the preparation of microcapsules containing water (core material) using waste paper fibers and expanded polystyrene (EPS) which is used for packing electrical appliances [46]. The microencapsulation method involved the preparation of EPS solution in CH_2Cl_2, followed by the addition of pretreated fibers. To this solution, water (as core material) was added to obtain a water/oil dispersion that was then added to an aqueous PVA solution to produce a water/oil/water dispersion. The microcapsules were prepared using the drying-in-liquid method. The fibers, before being added to the EPS solution, were treated with different agents such as acetone, two types of styrene maleic acid (S1 and S2) and surfactant Span 80 in order to adjust the affinity of the fibers to the solution dissolving EPS.

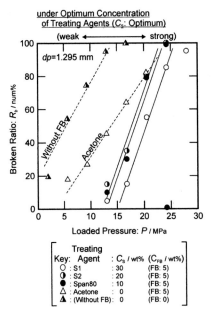

Figure 5.19 Dependencies of broken ratio (R_b) (in numerical %) on loaded pressure (P) [46].

The mechanical strength of microcapsules has been measured using the method described in Section 5.3.2. Loaded pressure is calculated from Eq. (4) (see Section 5.3.2), and the broken ratio is defined as the ratio of the number of microcapsules broken at a given pressure to that of all microcapsules measured. From Figure 5.19 it is clear that the broken ratios for microcapsules with fiber incorporated into the capsule wall are smaller than for microcapsules without fiber. This mechanical strength of microcapsules is increased by the incorporation of fiber into the EPS. The microcapsules also become stronger following the addition of fibers treated with acetone, 20 wt.% S2, 10 wt.% Span 80, and 30 wt.% S1.

5.5
Applications

5.5.1
Fragrances

Microcapsules containing fragrances are utilized in many areas, including air-fresheners, cosmetics, textiles, and laundry detergents. As most perfumes are volatile substances, their effect is lost very rapidly from the application site. Perfumes added to detergents are lost during the laundering operation due to the relative warmth of the water, as well as during the washing and drying processes, and consequently laundered fabric will have only a very faint desired odor. However, when perfume is added to the detergent in the form of microcapsules, the microcapsules penetrate the fabric during laundering and release the perfume during drying, and/or when laundered fabric is being used [53,54]. Epoxy resin microcapsules containing lipophilic fragrances in cosmetic composition do not cause any "clammy" feeling, even at high fragrance contents [55]. Microcapsules which release the fragrance when exposed to a solution at predetermined pH have also been reported [56]. These microcapsules range from 100 to 250 μm in size, and are prepared by a suspension polymerization technique using an acrylic acid monomer and a second monomer such as acrylates, methacrylates or styrene.

UF microcapsules containing orange or rose extracts are reported to be most suitable for nonwoven fabrics due to their open structures [57]. Microcapsules containing perfume prepared by coacervation methods using gelatin and gum arabic have been reported to be useful in solid soap composition. The perfume is released when the soap is placed in contact with water during use [58].

5.5.2
Lubricating Oils

Microencapsulation is also useful in enhancing the performance of lubricating oils when used at the application site. Pressure-ruptured microcapsules containing lubricating oil are used in powder metallurgy. The requirements for the microcapsule shell wall is that it should have a high abrasion resistance, be able to withstand the high temperatures used for mixing operations (with powdered metal), and that it should rupture when subjected to the pressure exerted during compaction. The preferred microcapsule shape is spherical, and size should be as small as possible. A US patent by Blachford discloses the preparation of such microcapsules using di-isocyanate and diamine by an interfacial polymerization method [59]. The microcapsules containing lubricating oil are superior in injection force, apparent density and shrinkage, and are similar to the commercial lubricants in terms of tensile strength and transverse rupture strength. Microcapsules containing synthetic lubricating oil have been shown to be useful in the electrochemical deposition of nickel from $NiSO_4$ solution on steel substrates [60].

5.5.3
Edible Oils

The oxidative stability of seal blubber oil (SBO) is improved by microencapsulation. Microcapsules containing SBO prepared with β-cyclodextrins as the wall material are easily handled and incorporated into food formulations [61].

Microcapsules of Perilla seed oil (PSO) are prepared using soyabean protein and maltodextrin as the wall material, using a spray-drying method [62]. Moreover, the linolenic acid present in microencapsulated PSO is not lost during processing.

The microencapsulation of fish oil protects it against oxidation, and this has led to the successful incorporation of these oils into normal food components. Freeze-drying techniques have been shown to produce microcapsules containing fish oils of high quality and oxidation stability [63]. Emulsions containing 10% fish oil, 10% sodium caseinate, 10% carbohydrate and 70% water are frozen to obtain the encapsulated product. The best shelf-life is obtained by freeze-drying the product at a slow freezing rate.

5.5.4
Dental Composites

Precise dosage of reactive components is essential for the reproducible hardening of dental cements, without adversely affecting the physical properties of the hardened product. Cellulose acetate butyrate microcapsules containing poly(acrylic acid) prepared by phase separation when mixed with glass ionomer powder result in single-phase, free-flowing powders [64]. The contents of microcapsules can be released and become mixed with the solid phase by mechanical stress, vibration microwaves, and/or sonication.

5.5.5
Self-Healing Agents

On many occasions, cracks occur in polymer materials used in a wide variety of materials, including those used in transportation (automobiles, airplanes, space-craft), sporting goods (tennis rackets, helmets, skis), medical devices (pacemakers,

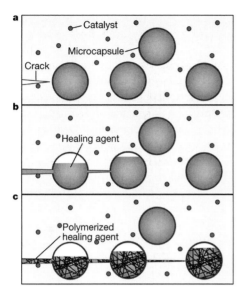

Figure 5.20 The autonomic healing concept. A microencapsulated healing agent is embedded in a structural composite matrix containing a catalyst capable of polymerizing the healing agent. (a) Cracks form in the matrix wherever damage occurs. (b) The crack rup- tures the microcapsules, releasing the heal- ing agent into the crack plane through capil- lary action. (c) The healing agent contacts the catalyst, triggering polymerization that bonds the crack faces closed [65].

body part replacements, dental materials), electronics (circuit boards, electronic components) and paints and coatings. White et al. have reported a structural poly- meric material which has the ability autonomically to heal cracks within its struc- ture [65]. Microcapsules containing a healing agent are embedded in a structural composite matrix, which contains a catalyst capable of polymerizing the healing agent. The autonomic healing mechanism is illustrated in Figure 5.20.

When cracks develop they rupture the embedded microcapsules, and this results in a release of the healing agent into the crack plane through capillary action. Poly- merization of the healing agent is triggered by contact with the embedded catalyst, and this leads to bonding of the crack faces. This autonomic control of repair has been demonstrated with an epoxy matrix composite containing Grubb's catalyst and UF microcapsules containing a healing agent (viz., dicyclopentadiene, DCPD). The polymerization of DCPD takes place by ring-opening metathesis polymeriza- tion (ROMP).

UF microcapsules containing DCPD have been prepared by *in-situ* polymeriza- tion in an oil-in-water emulsion [66]. Different experimental parameters such as agitation speed, temperature and pH have been studied in order to obtain micro- capsules with a long shelf-life – that is, microcapsules which are impervious to leakage and diffusion of the encapsulated (liquid) healing agent for a considerable time.

5.5.6
Pesticides

Herbert has reviewed the microencapsulation of pesticides by interfacial polymerization with reference to process and performance considerations [67].

In another report, Scher discussed pesticide microcapsules and reported that these capsules could be used to reduce mammalian toxicity, to extend activity, to reduce phytotoxicity, to protect pesticides from rapid environmental degradation, and to reduce pesticide levels in the environment [68]. Thus, many advantages are obtainable with microcapsular pesticide formulations. In an overview, Gimeno has discussed the definition of the criteria to select an active agent for microencapsulation, together with the techniques used in commercial microencapsulated formulations [69].

Spherical UF microcapsules of quinolphos, prepared with different degrees of crosslinking and varying in terms of the weight ratios of urea to melamine, have been found to be quite stable as dry powders [11,12]. Quinolphos is commercially

Table 5.7 Bioefficacy of quinalphos microcapsular dispersion on the aphid, Aphis gossypii in okra cultivation. (Values denote average numbers of aphids on five random plants.)

Days after treatment	Microcapsular quinalphos [250 g a.i. ha^{-1}]	Ecalux [250 g a.i. ha^{-1}]	Untreated control
Pretreatment count			
−1	27.6	24.0	24.3
First spray			
1	12.3[b]	17.3[b]	28.3[a]
5	8.6[b]	8.3[b]	32.3[a]
10	8.6[b]	12.6[b]	37.6[a]
15	8.0[b]	14.6[b]	40.3[a]
Second spray			
16	7.0[b]	12.6[b]	44.3[a]
20	5.3[c]	14.3[b]	52.6[a]
25	0.0[c]	17.3[b]	58.3[a]
30	0.0[c]	19.0[b]	58.0[a]
Third spray			
31	0.0[c]	15.6[b]	60.6[a]
35	0.0[c]	12.6[b]	64.6[a]
40	0.0[c]	16.0[b]	66.3[a]
45	0.0[c]	17.3[b]	67.6[a]

Note: The same alphabetic superscript in a row indicates no significant difference.

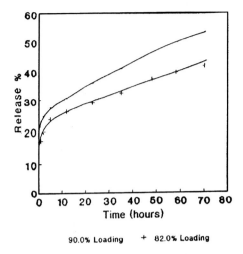

90.0% Loading   + 82.0% Loading

Figure 5.21 Release profiles of chlor-pyrifos from amino microcapsules at different core loadings [11].

available as an EC (Ecalux), and is used as a foliar spray against pests on rice, veg-etables, fruit trees, and cotton. It has been shown that microcapsules can show drastic decreases in release rate below a threshold level of 80% loading, and an in-verse relationship between release rate and melamine component of resin has been identified. Field evaluations of quinolphos in okra cultivation, as indicated by aphid (pest) counts (Table 5.7), showed that while the commercial EC formulation was ef-fective for 7–10 days, the microencapsulated product – even at lower doses – was ef-fective for about 20 days [70].

Chlorpyrifos, a well-known pesticide that is effective against subterranean pests such as white grubs and termites, has been microencapsulated as its melt in an amino resin matrix [11,12]. Its release profile at loadings of 90% and 82% showed that, unlike quinolphos, stable capsules could be obtained at high loadings (Fig. 5.21). The bioefficacy of microcapsular and commercial chlorpyrifos seed-coat treatments against white grub in the cultivation of groundnuts showed plant mor-tality of ~5% and ~16% respectively, indicating a clear advantage of microencapsu-lation (Table 5.8). Further benefit/cost analyses showed spectacular performance

Table 5.8 Bioefficacy of microcapsular and commercial chlorpyriphos seed-coat treatments against white grub in cultivation of groundnut crop.

Study no.	Treatment	Dose [g a.i. kg⁻¹ seed]	Plant mortality [%]	Pod yield [q ha⁻¹]
1	CRCH I 90 SC[a]	5.0	5.26	25.49
2	CRCH II 82 SC[a]	5.0	5.51	25.16
3	Chlorpyrifos 20 EC[b]	5.0	16.26	20.97
4	Untreated control	–	88.32	2.47

[a] Microcapsules of chlorpyriphos.
[b] Commercial chlorpyriphos (20 EC).

by these microcapsular chlorpyrifos formulations (Table 5.9). The results indicated that a 1-Rupee value of commercial chlorpyrifos formulation provided a benefit of 29.6 Rupees; in contrast, a 1-Rupee value of microcapsular formulation provided a benefit of 82–91 Rupees.

Aminoplast (MF) has been used to prepare microcapsules of liquid pesticides, namely methyl parathion [71] and metachlor [72]. Other selected examples of microcapsules containing agrochemicals such as pesticides, herbicides, fertilizers and insecticides have been reported and are summarized in Table 5.10.

Table 5.9 Benefit-cost (B/C) ratios of seed-coat microcapsular and commercial chlorpyriphos formulations in groundnut cultivation.

Treatment	Dose [g a.i. kg^{-1} seed]	Cost of pesticide [Rs]	Yield over control [q ha^{-1}]	Cost of additional produce over control [Rs]	B/C ratio
CRCH-90	5.6	302	23.02	27 624	91.47
CRCH-80	6.1	329	22.69	27 228	82.76
Chlorpyrifos 20 EC	25 mL	750	18.50	22 200	29.60

Notes: a) Application charges not considered.
b) Groundnut seed rate 100 kg ha^{-1}.
c) Price of groundnut pods is Rs 1200 q^{-1}.
d) Price of chlorpyrifos 20 EC is Rs 300 L^{-1}.
e) Price of microcapsular chlorpyrifos-90 and -80 SC is Rs 540 kg^{-1}.

Table 5.10 Microcapsules containing agrochemicals.

Polymer	Active agent	Reference(s)
Gum arabic-gelatin	Biocide	73
Poly (lactic acid)	Pesticide	74
Gelatin	Pesticide	75
Acrylic polymer (Carboset 525)	Insecticide, herbicide	76
Ethyl cellulose (containing photodegradation-enhancing agent)	N-P-K fertilizer	77
Polyurethane	Pesticide	78–80
Kraft lignin	Pesticide	81
Polyurea	Herbicide, insecticide	82,83
Amino resin	Pesticide	84,85
Polyamide	Pesticide	86
Styrene-divinylbenzene copolymer	Pesticide	87

5.6
Conclusions

The great need to protect a liquid active agent from environmental factors such as temperature and humidity, or to protect it against other chemicals in a one-pack system, can be fulfilled by using microencapsulation. In addition to conventional microencapsulation, new techniques such as the use of liquid CO_2, miniemulsion, the use of SPG membranes, and the use of novel polymeric surfactants are being employed to produce microcapsules of desired size, size distribution, and performance. Interfacial polymerization methods are more attractive and advantageous as they can be used to prepare microcapsules with higher concentrations and higher core loadings. Modifications to microcapsules which are either chemical or physical can be carried out in order to obtain microcapsules with desired release rates and patterns. As our understanding of the physical strength of microcapsules – especially of those containing liquid agents – has expanded, methods have been developed for such characterization. Field evaluations of microcapsules containing pesticides have shown that advantages such as reductions in mammalian toxicity, in phytotoxicity, and in the number of applications can be obtained. Moreover, microencapsulation has proved to be an effective and powerful technique that can be used in a wide variety of areas including fragrances, lubricating oils, edible oils and self-healing agents.

Abbreviations

ADVN	*N,N*'-azobis (2,4-dimethyl valeronitrile)
AIBN	azo-bis-isobutyronitrile
BD	1,4-butane diol
CCM	cetyl pyridinium chloride monohydrate
CMC	critical micelle concentration
CTAB	hexadecyl trimethyl ammonium bromide
DCPD	dicyclopentadiene
DEG	diethylene glycol
DETA	diethylenetriamine
DMAEMA	poly(styrene-*N,N*-dimethylaminoethyl methacrylate)
EC	emulsifiable concentrate
EG	ethylene glycol
EPS	expanded polystyrene
HD	hexadecane
MCR	monocrotophos
MF	melamine-formaldehyde
PFPE-NH$_4$	perfluro polyether ammonium carboxylate
PMMA	poly(methyl methacrylate)
PS	polystyrene
PSO	perilla seed oil

PU	polyurethane
PVA	polyvinyl alcohol
PVP	poly(*N*-vinyl pyrrolidone)
ROMP	ring-opening metathesis polymerization
SBO	seal blubber oil
SDS	sodium dodecyl sulfate
SLS	sodium lauryl sulfate
SPG	Shirazu porous glass
SSC	side-chain crystalline
TDI	toluene diisocyanate
TMP	trimethylolpropane
UF	urea-formaldehyde

References

1. P.L. Madan, Microencapsulation (in the pharmaceutical industry), *Asian J. Pharm. Sci.* **1979**, *1*, 1–46.
2. C. Thies, Microencapsulation, *Encyclopedia of Polymer Science and Engineering*, Vol. 9. Wiley & Sons, New York., **1987**.
3. H. Porte, G. Couarraze, *Handbook of Powder Technology, Vol. 9 (Powder Technology and Pharmaceutical Processes)*, **1994**, pp. 513–543.
4. S. Benita (Ed.), *Microencapsulation-methods and industrial applications*. Marcel Dekker Inc., New York, **1996**,
5. R. Arshady (Ed.), *Microspheres Microcapsules and Liposomes*, Vols. 1 & 2. Citus Books, UK, **1999**.
6. R. Arshady, *J. Microencapsulation* **1989**, *6*(1), 13–28.
7. K. Dietrich, H. Herma, R. Nastke, E. Bonatz, W. Teige, *Acta Polymerica* **1989**, *40*, 243–251.
8. K. Dietrich, E. Bonatz, H. Geistlinger, H. Herma, R. Nastke, H.J. Purz, W. Teige, *Acta Polym.* **1989**, *40*, 325–331.
9. E. Bonatz, K. Dietrich, H. Herma, R. Nastke, M. Walter, W. Teige, *Acta Polym.* **1989**, *40*, 683–690.
10. K. Dietrich, E. Bonatz, R. Nastke, H. Herma, M. Walter, W. Teige, *Acta Polym.* **1990**, *41*, 91–95.
11. N. Rajagopalan, C. Bhaskar, P.G. Shukla, N. Amarnath, in: *Proceedings, FAO/IAEA International Seminar on "Research and Development and Controlled Release Formu-*lations of Pesticides*, Vienna, Austria, 1993, Vol. I , **1994**, pp. 91–110.
12. C. Bhaskar, P.G. Shukla, N. Rajagopalan, *Indian Patent IN184975*, **2000**.
13. C. Bhaskar, N. Rajagopalan, P.G. Shukla, R.B. Mitra, in: *Polymer Science: Recent Advances*, **1994**, Vol. I, pp. 437–442.
14. D.C. Creech, R. Curtis, *PCT WO 9104661 A2*, **1991** (CA 115:273482).
15. P.G. Shukla, N. Rajagopalan, S. Sivaram, *U.S. Patent 5,962,003*, **1999**.
16. P.G. Shukla, S. Sivaram, *J. Microencapsulation* **1999**, *16*(4), 517–521.
17. P.G. Shukla, B. Kalidhass, A. Shah, D.V. Palaskar, *J. Microencapsulation* **2002**, *19*(3), 293–304.
18. K.E.J. Barrett, *Dispersion polymerization in organic media*. Wiley, London, **1975**.
19. P.G. Shukla, S. Sivaram, *U.S. Patent 5,814,675*, **1998**.
20. P.G. Shukla, S. Sivaram, *U.S. Patent 5,859,075*, **1999**.
21. L.S. Ramanathan, P.G. Shukla, S. Sivaram, *Pure Appl. Chem.* **1998**, *70*(6), 1295–1299.
22. L.S. Ramanathan, S. Sivaram, *U.S. Patent 6,022,930*, **2000**.
23. L.S. Ramanathan, S. Sivaram, *U.S. Patent 6,123,988*, **2000**.
24. L.S. Ramanathan, P.G. Shukla, D. Bhaskaran, S. Sivaram, *Macromol. Chem. Phys.*, **2002**, *203*(7), 998–1002.
25. Y. Frere, L. Danicher, P. Gramain, *Eur. Polym. J.* **1998**, *34*(2), 193–199.

26. G.B. Beestman, J.M. Deming, *Pesticide Formulations and Application Systems*, **1988**, 8th Vol., ASTM STP 980, pp. 25–35.

27. Lo Chien-Cho, *European Patent 0551796*, **1993**.

28. K. Hirech, S. Payan, G. Carnelle, L. Brujes, J. Legrand, *Powder Technol.* **2003**, *130*, 324–330.

29. D. Horak, F. Svec, J. Frechet, *J. Polym. Sci. Part A: Polym. Chem.* **1995**, *33*, 2961–2968.

30. C.K. Ober, K.P. Lok, M.L. Hair, R.E. Branston, *U.S. Patent 4,613,559*, 1986.

31. M.Z. Yates, E.R. Birnbaum, T.M. McCleskey, *Langmuir* **2000**, *16*(11), 4757–4760.

32. H. Liu, M.Z. Yates, *Langmuir* **2002**, *18*, 6066–6070.

33. H. Liu, M.Z. Yates, *Langmuir*, **2003**, *19*, 1106–1113

34. G.H. Ma, Z.G. Su, S. Omi, D. Sundberg, J. Stubbs, *J. Colloid Interface Sci.* **2003**, *266*, 282–294.

35. K. Landfester, *Macromol. Symp.* **2000**, *150*, 171–178.

36. K. Landfester, *Adv. Mater.* **2001**, *13*(10), 765–768.

37. F. Tiarks, K. Landfester, M. Antonietti, *Langmuir* **2001**, *17*, 908–918.

38. S.M. Trikkonen, P. Puumalainen, M.J. Juslin, T.P. Paronen, *Chim. Oggi.* **1986**, *10*, 35–36.

39. A.R. Bachtsi, C. Kiparissides, *J. Control. Release* **1996**, *38*, 49–58.

40. M.D.L. Moretti, G. Sanna-Passino, S. Demontis, E. Bazzoni, *AAPS PharmSciTech*, **2002**, *3*(2) , Article 13, URL: http://www.aapspharmscitech.org.

41. K.S. Mayya, A. Bhattacharyya, J.F. Argillier, *Polymer Int.* **2003**, *52*, 644–647.

42. R. Baker, Y. Ninomiya, *European Patent EP 126583 A1*, **1984**. (CA 102:64293).

43. Y. Ninomiya, C. Komamura, Y. Musa, *German Patent DE 3417200*, **1985** (CA 103:100429).

44. *Japanese Patent JP 59095928*, **1984** (CA101:186163).

45. T. Ohtsubo, S. Tsuda, K. Tsuji, *Polymer* **1991**, *32*(13), 2395–2399.

46. Y. Taguchi, M. Tanaka, *J. Appl. Poly. Sci.* **2001**, *80*, 2662–2669.

47. I.M. Shirley, K. Van, E. Juanita, H.B. Scher, R. Follows, P. Wade, F.G.P. Earley, D.B. Shirley, *U.S. Patent US 6485736*, **2002**.

48. V.Y. Yoon, D.H. Carter, D.K. Brandom, R.F. Stewart, S.P. Bitler, *Polym. Prepr.* (ACS, Div. Polym Chem), **1991**, *32*(2), 225–226.

49. E. Mathiowitz, A. Raziel, M.D. Cohen, E. Fishcher, *J. Appl. Poly. Sci.* **1981**, *26*, 809–822.

50. B.K. Redding, Jr., *PCT WO 90/00005*, 1990.

51. R.A. Verbelen, S. Lemoine, *Basic Chemicals and Intermediates technology*, **1999**, pp. 65–80 (CA 132:147992).

52. R. Heinrich, H. Frensch, K. Albrecht, *German Patent DE 2757017*, (CA 91:108634).

53. D.K. Brain, M.T. Cummins, *U.S. Patent 4,145,184*, **1979**.

54. H. Uchiyama, J.R. Cetti, M. Alonso, D.L. Montezinos, D.S. Cobb, *U.S. Patent 0215417A1*, **2003**.

55. K. Oka, Y. Ehata, *Japanese Patent 62084127 A2*, **1987**, (CA 107:155760).

56. M.L. Kantor, E. Barantsevitch, S.J. Milstein, *PCT WO 9747288 A1*, **1997**, (CA 128:79828).

57. M. Haghighatkish, F. Mazaheri, M. Norouzzadeh, *J. Polym. Sci. Technol.* **1998**, *11*(3), 179–186, (CA 131:59881).

58. A.L.F. Baptista, P.J.G. Coutinho, M.E.C.D. Real Oliveira, J.I.N. Rocha Gomes, *J. Liposome Res.* **2003**, *13*(2), 111–121, (CA 140:183187).

59. J. Blachford, *U.S. Patent 4002474*, **1977**.

60. J.P. Celis, J. Fransaer, J.R. Roos, *Mater. Tech. (Paris)* **1995**, *83*(1-2) (5-8), (CA 123:345197).

61. U.N. Wanasundara, F. Shahidi, *J. Food Lipids* **1995**, *2*(2), 73–86 , (CA 23:284102).

62. C. Kai, D. Xiaolin, *Zhongguo Liangyou Xuebao* **1997**, *12*(6), 36–39, (CA 128:101314).

63. K. Heinzelmann, K. Franke, *Colloids Surf. B.* **1999**, *12*(3-6), 223–229 (CA 130:295756).

64. H.P. Hollwege, M.T. Plaumann, *German Patent DE 19802644 A1*, **1999** (CA 131:120918).

65. S.R. White, N.R. Sottos, P.H. Geubelle, J.S. Moore, M.R. Kessler, S.R. Sriram, E.N. Brown, S. Viswanathan, *Nature* **2001**, *409*, 794–797.

66. E.N. Brown, M.R. Kessler, N.R. Sottos, S.R. White, *J. Microencapsulation* **2003**, *20*(6), 719–730.

67. H. Scher, *Proc. Int. Congr. Pestic. Chem.* **1982**, *4*, 295–300.

68. H.B. Scher, M. Rodson, K.S. Lee, *Pesticide Sci.* **1998**, *54*(4), 394–400.

69. M. Gimeno, *J. Environ. Sci. Health, Part B* **1996**, *B31*(3), 407–420.

70. C. Bhaskar, P.G. Shukla, in: *Proceedings, 23rd International Symposium on 'Controlled Release of bioactive Materials"*, **1996**, 371.

71. T. Dreyfus, H. Laurent, *European Patent EP 463926 A1*, **1992**. (CA 116:108516).

72. K. Dietrich, R. Nastke, H. Geistlinger, E. Lewandowski, W. Teige, H.J. Koerner, W. Wildgrube, W. Grasshoff, D. Otto, *Ger. (East) Patent DD 240843 A1*, **1986**. (CA 107:213619).

73. B.M. Nunn, *U.S. Patent 5164096*, **1992**.

74. H. Jaffe, *U.S. Patent 4272398*, 1981.

75. O. Yoshio, I. Yuriko, W. Koju, *Ger. Offen DE*, 2815139, **1978** (CA 90:1711).

76. N.F. Cardarelli, C.M. Himel, *Fr. Demande FR*, 2430259, **1980** (CA 93:48075).

77. T. Norio, *Japanese Patent JP 54109078*, **1979** (CA 92:40649).

78. T. Norio, *Japanese Patent JP 54004282,*, **1979** (CA 90:198862).

79. K.Y. Choi, K.S. Min, I.H. Park, K.S. Kim, T. Chang, *Pollimo* **1990**, *14*(4), 392–400.

80. T. Otsubo, S. Tsuda, Y. Manabe, T. Ito, H. Kawada, G. Shinjo, K. Tsuji, *Japanese Patent 62161706*, 1987 (CA 107:149243).

81. G.W. Smith, *Eur. Pat. Appl. EP 140548*, **1985**.

82. G.B. Beestman, J.M. Deming, *Romanian Patent RO 85310 L*, **1984**, (CA 104:16550).

83. J.L. Chen, K.S. Lee, M. Rodson, H.B. Scher, *PCT Int. Appl. WO 9744125*, **1997**.

84. R. Nastke, A. Loenhardt, E. Neuenschwander, *Eur Pat. Appl. EP 532463*, **1993** (CA 118:249829).

85. K.Y. Choi, K.S. Min, T. Chang, *Pollimo* **1991**, *15*(5), 548–555.

86. S. Kiyoyama, Y. Kono, Y. Hatade, A. Mizuno, M. Suzuki, *Japanese Patent JP 2004196718*, **2004** (CA 141:101555).

6

Core-Shell Systems Based on Intrinsically
Conductive Polymers and their Coating Applications

Swapan Kumar Ghosh

6.1
Introduction

Until the 1960s, organic polymers were generally considered to be insulators, and were always used extensively by the electronics industries for this very property. The concept of conductivity in polymers was first noted by Hatano and co-workers [1]. These authors' studies revealed that polyacetylene had conductivities in the order of 10^{-5} Siemen (S) cm^{-1}. Later in 1977, Shirakawa, MacDiarmid, and Heeger et al. found that the conductivity of polyacetylene could be increased by several orders of magnitude (up to 10^5 S cm^{-1}) when exposed to halogen vapor [2]. Since this initial discovery, numerous investigations have been carried out in this field, leading to a new class of polymers referred to as intrinsically (inherently) conducting polymers (ICPs). These ICPs are also referred to as "organic metals", as they can achieve electrical conductivities close to that of metals such as copper.

Depending on the type of charge transport carrier, conductive polymers can be classified into two main groups of ionically and electronically conductive polymers (Fig. 6.1). Polyethylene oxide with lithium perchlorate (LiClO$_4$) is an example of an

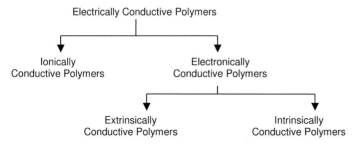

Figure 6.1 Classification of electrically conductive polymers.

Functional Coatings. Edited by Swapan Kumar Ghosh
Copyright © 2006 WILEY-VCH Verlag GmbH & Co. KGaA, Weinheim
ISBN 3-527-31296-X

ionically conductive polymer where the ions are responsible for the conduction of electricity. These polymers are mainly used as an electrolyte in solid-state batteries. Graphite-containing coatings and polypyrrole (PPy) are typical examples of electronically conductive polymers. This class of polymer can be further divided into extrinsically and intrinsically conductive polymers.

Extrinsically conductive polymers conduct electricity due to the presence of conductive fillers such as graphite, carbon fiber, and metal particles, whereas ICPs conduct electricity due to their property of conjugation. The latter is a term used for the presence of alternate single (σ) and double (σ and π) bonds in polymer molecules. The carbon atoms along the ICP backbone have sp2 hybridizations (three sp2 orbitals in the plane and pz orbital out of the plane). In sp2 hybridization, the pz orbital of the adjacent carbon atoms overlap each other and form additional pi (π) bonds, and the sp2 orbitals form the sigma (σ) bonds. The energy difference between the highest occupied molecular orbital (HOMO) and the lowest unoccupied molecular orbital (LUMO) is smaller in case of π bonds compared to the σ bonds that provide the semiconducting behavior, and the optical absorption properties are in or near the visible spectral range. This π-π overlap causes the delocalization of electrons that provide the pathway for the free electron charge carriers to move along the polymer chain. However, the presence of conjugation in ICPs leads to poor processability that hinders their industrial applications. The following section will emphasize the various possibilities of enhancing the processability of ICPs, and will also include an extensive discussion on ICP-based core-shell systems.

6.2
Intrinsically Conducting Polymers

The initial discovery of ICPs attracted the scientific community, and this in turn led to the development of several conductive polymers. Examples of the most prominent and widely studied ICPs include polyacetylene, polyaniline (PANI), PPy, polythiophene (PT), and poly(phenylene vinylene) (PPV) (Fig. 6.2).

The conductivity of undoped ICP is low, with a band gap in the range of 1 to 4 eV (electron volt). The band gap (E_g) is the difference in the energy between HOMO

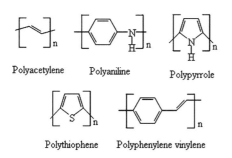

Polyacetylene   Polyaniline

Polypyrrole

Polythiophene   Polyphenylene vinylene

Figure 6.2 Repeat units of intrinsically conductive polymers (ICPs).

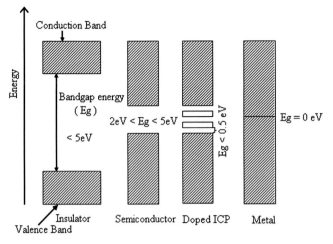

Figure 6.3 Schematics depicting the band model of ICPs.

and LUMO – that is, the energy required to transfer an electron from the top of the valance band to the bottom of the conduction band. When ICPs are exposed to oxidants or reductants, a change in the band gap is observed (Fig. 6.3).

The doping of ICPs is accomplished by the addition (reduction) or removal (oxidation) of electrons to/from the π bonds of the polymer backbone. A number of methods including chemical, electrochemical, charge injection or photo doping are available to enhance the conductivity of the ICPs. The most two important types of doping are the redox and the acid–base type. The mechanism of doping for the conjugated polymer having a degenerate ground state (e.g., polyacetylene) differs from that for PT or PPy. Polyacetylene can be doped by either chemical or anodic electrochemical oxidation of the polymer chain. The removal of an electron from the polyacetylene chain produces polycarbonium cations (Fig. 6.4a), while removal of the second electron gives rise to a second radical cation which combines to form a spinless di-cation. Further oxidation of these di-cations produces charge carriers which are known as "solitons". The charge spin relation of a soliton is the opposite of free charge. A charge soliton is spinless, while a neutral soliton has spin of $^1/_2$ (Fig. 6.4b).

Solitons produced in polyacetylene are delocalized over approximately 12 CH units, with the maximum charge density to the dopant counterion. Soliton formation results in the creation of a new localized electronic state which is in the middle of the energy gap. At a high level of doping the charged solitons produce soliton bands that can merge to behave like a metallic conductor.

For heterocyclic polymers such as PPy and PT, different charge configurations are formed (Fig. 6.5). The removal of an electron from the π system of the polymer backbone creates a radical cation known as a "polaron" (half-oxidized state) (Fig. 6.5b). The creation of a polaron induces the presence of a quininoids structure sequence in the polymer chain. Further removal of electrons leads to the formation

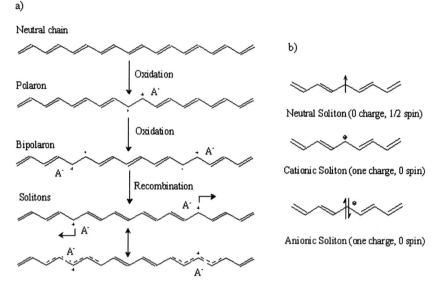

a)

Neutral chain

Polaron

Oxidation
. A˙

Oxidation

Bipolaron

A˙⁻

+ A˙

Recombination
A˙⁻ ↱

Solitons

↲ A˙⁻

A˙⁻

A˙⁻

b)

Neutral Soliton (0 charge, 1/2 spin)

Cationic Soliton (one charge, 0 spin)

Anionic Soliton (one charge, 0 spin)

Figure 6.4 (a) Doping (p-type) of polyacetylene by an oxidator. (b) Spin–charge relationship of solitons.

of new polaron or spinless bipolarons (dications) that separate the quininoid structure from the aromatic type of bonds in the ICP chain (Fig. 6.5c).

PANI can undergo redox as well as acid–base type doping. The most reduced form of PANI (i.e., the leucoemeraldine base; LB) undergoes oxidative doping to form radical cations that are bipolarons. The semioxidized form of PANI (i.e., the

a)

$+e^-$ ‖ $-e^-$

b)

$+e^-$ ‖ $-e^-$

c)

Figure 6.5 p-Type doping of polypyrrole (X = NH) and polythiophene (X = S). (a) neutral polymer; (b) polarons; (c) bipolarons.

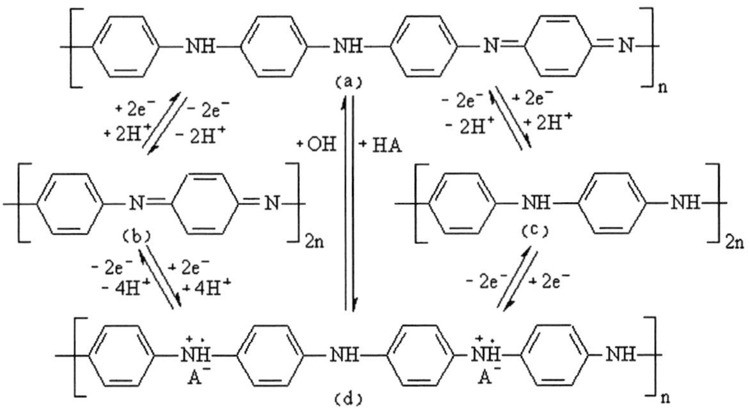

Figure 6.6 Oxidative and acid/base doping of polyaniline.
(a) Emeraldine base; (b) pernigraniline base; (c) leuco base;
(d) emeraldine salt.

emeraldine base; EB) can be protonated to give corresponding emeraldine salts
(ES). The different redox forms of PANI are summarized in Figure 6.6.

The conductivity of ICPs can be tuned in several ways, including modification of
the polymer chain, variation of the type of dopants, the level of doping, and blending with other polymers. Usually, anions are incorporated during the synthesis of
ICPs that neutralize the charge of the polymer. Unlike the inorganic semi-conduc-
tors, ICPs can be reversibly doped [Eq. (1)].

$$(ICP)_n^+ A^- \underset{-e}{\overset{+e}{\rightleftharpoons}} (ICP)_n^0 + A^- \tag{1}$$

Equation (1) shows that the dopant anion leaves the polymer upon reduction.
However, it was later proved that if the dopant anion is large and immobile (e.g.,
polystyrene sulfonate), then the cation from the supporting environment (electrolyte) is incorporated into the polymer [Eq. (2)]. Thus, the dopant anions have
greater influence on the physico-chemical properties of the conductive polymer.

$$(ICP)_n^+ A^- + M^+ \underset{-e}{\overset{+e}{\rightleftharpoons}} (ICP)_n^0 A^- M^+ \tag{2}$$

Similar to p doping, n doping can also be carried out for ICPs. However, the n
doping of ICPs is difficult as the obtained redox forms are less stable in air. The
doping of ICPs does not interfere with the σ bonds. Thus, although the carbon–carbon skeleton remains intact, it changes the electrical, spectral and other properties
of the polymer.

6.3
Synthesis of ICPs

ICPs can be prepared using various polymerization techniques such as step-growth condensation, oxidative polymerization, and photochemical synthesis. Among these methods, oxidative polymerization – using either a chemical or an electrochemical approach – is the most widely used polymerization method for ICP synthesis [3]. Numerous laboratory research investigations have been carried out on ICP synthesis using electrochemical techniques. This technique offers the advantages of direct synthesis of conductive film on the electrode, the ability to control the film properties by adjusting the applied potential, the flexibility of incorporating different types of anions (dopants), and characterization of the film by changing the electrolyte [4]. Despite the above-mentioned benefits, the technique has several drawbacks, such as difficulties in removing the deposited ICP film from the electrode, and the process is not cost-effective. On the other hand, although chemical methods of polymerization have certain limitations, they have gained commercial success. The synthesis of few well-known ICPs is outlined in the following sections.

6.3.1
Synthesis of Polyaniline

Polyaniline can be synthesized by either chemical or electrochemical oxidative polymerization of aniline [5]. Chemical synthesis of PANI is the most popular method for large-scale production. PANI powders are usually obtained by oxidizing aniline monomer with an oxidant in a nonoxidizing protonic acid medium such as HCl or H_2SO_4 [6–8]. Several oxidizing agents such as $FeCl_3$, $KMnO_4$, and $K_2Cr_2O_7$ have been used for PANI synthesis [9], but most commonly $(NH_4)_2S_2O_8$ (ammonium persulfate) is used as an oxidant. A schematic representation of the mechanism of oxidative polymerization of aniline is shown in Figure 6.7.

The different steps involved in the formation of PANI are as follows:
Oxidation of aniline to radical cation.

- Dimerization of radical cation, loss of proton and formation of neutral dimer.
- Oxidation of dimer to form dimmer radical cation.
- Combination of dimer radical cation with another radical cation to form larger species such as trimer, tetramer and finally the polymer.

Para coupling of the radical cations is preferred over ortho coupling, as it does not lead to disrupted polymer. The use of an ortho-substituted monomer can enhance para coupling. The properties of the end products are greatly influenced by the polymerization conditions such as temperature and pH.

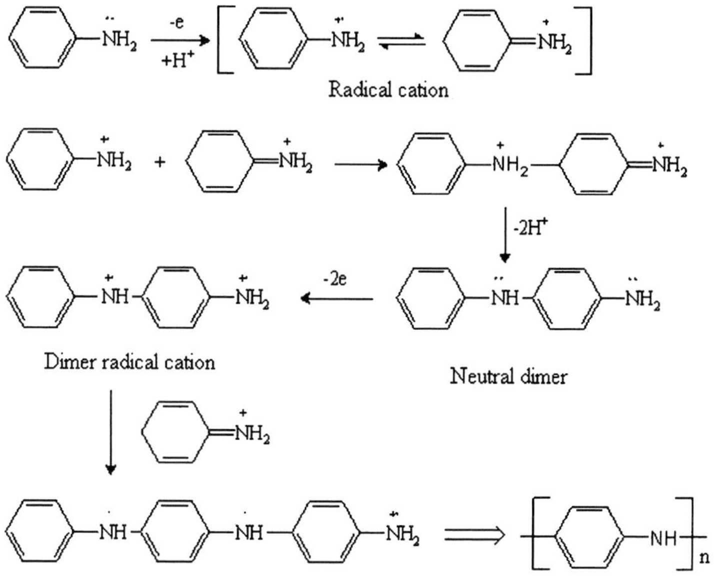

Figure 6.7   The mechanism of oxidative polymerization of aniline.

6.3.2
Synthesis of Polypyrrole

Electrochemical polymerization of pyrrole is the most versatile and widely used method for PPy synthesis. PPy can be polymerized both from an aqueous or organic solvent medium, and at neutral pH [4,10]. A general mechanism of oxidative polymerization of pyrrole is shown in Figure 6.8.

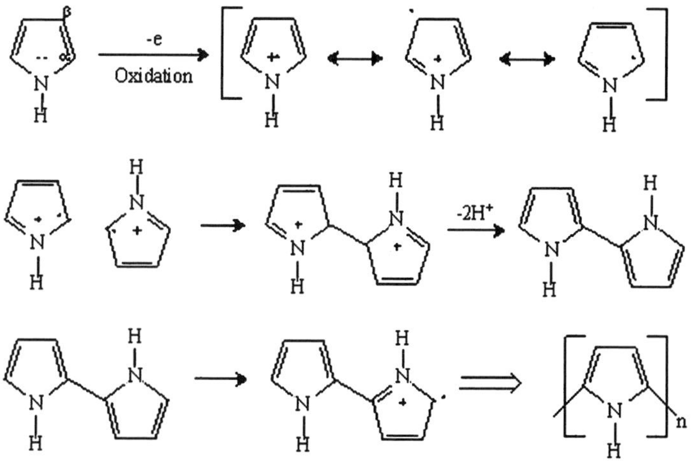

Figure 6.8   The mechanism of oxidative polymerization of pyrrole.

Figure 6.8 shows that the radical cationic species generated by the oxidation of pyrrole monomer is coupled to form the dimer. This undergoes further oxidation due to its lower oxidation potential compared to pyrrole monomer, and produces cationic radical species. In this way, the oligomer is formed and further oxidation and coupling with active cationic species increases the chain length and finally the formed polymer.

The synthesis of PPy via chemical oxidative polymerization is well known. Chemical syntheses are especially preferred by the oxidation of pyrrole monomer with appropriate oxidants such as $Fe^{(III)}Cl_3$, H_2O_2, or $(NH_4)_2S_2O_8$. If organic acid or peroxides are used as oxidants for the synthesis of PPy, the polymer obtained exhibits lower conductivity. From the reaction scheme shown in Figure 6.8 it can be deduced that α-α coupling is the most preferred one as it gives rise to a greater conjugation length. Moreover, α-β or β-β' coupling leads to branched structures. To overcome this difficulty, the synthesis of an exclusive α-α linked PPy by organometallic coupling reactions have been developed [11]. The nature of the counterions incorporated during synthesis has a strong influence on the final properties of the prepared PPy.

6.3.3
Synthesis of Polythiophene

Thiophene and its derivatives can undergo both chemical and electrochemical oxidation to produce PTs [12,13]. The electrochemical synthesis of PT is more difficult as the oxidation potential of thiophene monomer is relatively high compared to that of other monomers such as pyrrole or aniline. Thus, strong oxidants are required for the polymerization of thiophene. The reaction mechanism is similar to PPy synthesis.

Chemical synthesis of PT is relatively simple compared to the electrochemical route. The most widely used chemical polymerization techniques for PT derivatives are based on organometallic coupling reactions. Among the different derivatives of thiophene, 3,4-ethylenedioxythiophene (EDT or EDOT) is the most studied and the corresponding polymer poly (3,4-ethylenedioxythiophene) (PEDT; also known as PEDOT) have found several industrial applications. The polymer itself is not soluble or dispersible due to the formation of undesired products through α-β or β-β' coupling of the cation radicals in the polymerization reactions. The solubility was improved by polymerizing EDT monomer in an aqueous solution of polystyrene sulfonate (PSS) using $Na_2S_2O_8/Fe_2(SO_4)_3$ oxidant/catalyst systems. The chemical structure of PEDOT/PSS blends is shown in Figure 6.9.

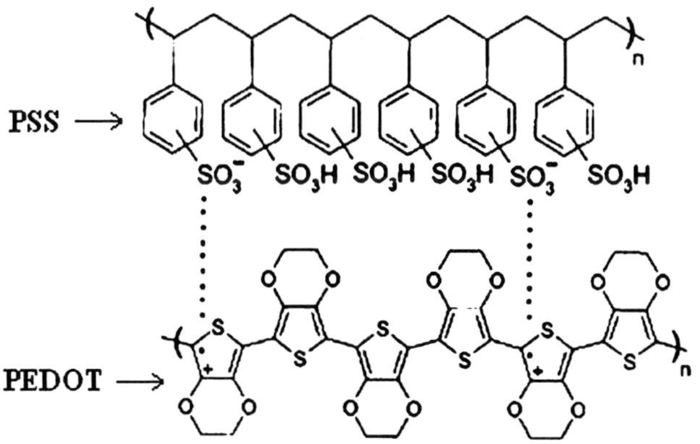

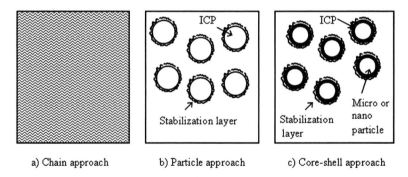

Figure 6.9 The chemical structure of the PEDOT-PSS complex (adapted from [12]).

6.4
Processing of ICPs

The unique feature of ICPs is their electrical and mechanical properties. Due to the presence of the conjugated backbone, they tend to be insoluble and infusible, and this results in poor processability. To overcome this problem, extensive research has been carried out into the synthesis of ICPs that are easily processable. The different approaches that have been considered to improve processability of ICPs may be summarized as chain, particle, or core-shell models (Fig. 6.10).

a) Chain approach	b) Particle approach	c) Core-shell approach

Figure 6.10 Different strategies for improving the processability of ICPs.

6.4.1
Chain Model

In this approach, functionalized monomers are employed to synthesize processable ICPs. During the past few years, some progress has been made in the chemical modification of the parent ICP structures to obtain more soluble polymer derivatives. For example, the introduction of an allyl chain or alkyl group into the 3-position of pyrrole, or an alkyl group into the *ortho* position of aniline, yields processable PPy or PANI upon polymerization [4,14–16]. If the substituent contains soluble groups in its side chain (e.g., sulfonate), then the polymer shows water dispersibility. The N-group substitution in PPy by alkyl sulfonate yields processable PPys [17]. Similarly, PT or PANI can be rendered processable by the above-mentioned approach [18,19]. The other route is to dope ICPs with functional organic acids such as camphorsulfonic acid or dodecyl benzene sulfonic acid [20,21]. However, modification of the monomer can alter the physical, chemical or electrical properties. Very often it is found that modification of monomer decreases the electrical conductivity of the prepared polymer. Another possibility for the chain model is to blend ICPs with other processable polymers such as polyvinyl chloride (PVC), polyethylene, polypropylene, or polystyrene [22,23]. The blends of ICPs with processable polymers can be produced by using solution or melt processing techniques. Another approach is to copolymerize different monomers to produce ICPs with unique properties compared to the homopolymer [24,25]. Copolymerization of pyrrole in the presence of formaldehyde resins have been investigated by Kizilcan and Ustamehmetoglu [26]. These authors have investigated the role of the type and amount of resin, the addition order of the reactants, and the concentration of pyrrole on the conductivity and solubility of the obtained copolymer. Papila et al. have synthesized graft copolymers of thiophene-functionalized polystyrene with PPy by using electrochemical techniques [27]. More recently, block copolymers of PEDOT with flexible polymers such as polyethers, polysiloxanes or polyacrylates have also become available commercially [28].

6.4.2
Particle Model

The second approach for improving the processability of ICPs is to prepare their colloidal dispersions in water or an appropriate solvent. The colloid dispersions of ICPs can be obtained by chemical or electrochemical oxidation of the monomer in the presence of a steric stabilizer [29–31].The key parameter for such synthesis is the choice of an appropriate steric stabilizer which adsorbs or grafts onto the polymer colloidal particles to prevent their aggregation or precipitation. Several polymers such as poly(ethylene oxide) [32], poly(vinyl pyrrolidone) [33,34], poly(vinyl alcohol) [35], ethyl hydroxy cellulose [36], poly(vinyl alcohol-*co*-acetate) [37], poly(vinyl methyl ether) [38,39] and block copolymer stabilizer [40] have been used as steric stabilizers to produce PPy colloidal dispersions. Surfactants are also employed for the synthesis of ICP colloidal dispersions [41,42]. Very recently, stable PPy dispersions were prepared by Lu et al. by polymerizing pyrrole in an aqueous medium containing different anionic salts such as sodium benzoate, potassium hydrogen phthalate, and sodium succinate [43]. These authors also reported that the conductivity of PPy dispersions was enhanced when sodium benzoate was used as dopant. Chemical oxidative polymerization in the presence of PSS in aqueous medium produces colloidal dispersions and improves processability [44]. Colloidal dispersions of PANI have also been prepared successfully [45]. Synthesis of ICP colloidal dispersions definitely improves the processability. However, ICP film formation is difficult as the particles are composed of pure conducting polymer. PANI and PEDOT are commercially available as dispersions in solvent or water [46–48]. A recent report revealed the possibility of using polymeric dopants to synthesize double-stranded ICPs such as PANI [49]. The double-stranded PANI is a noncovalently bonded molecular complex of two linear polymers such as PANI and PSS, PANI and polyacrylic acid, PANI and poly(methacrylate-co-acrylic acid) (Fig. 6.11).

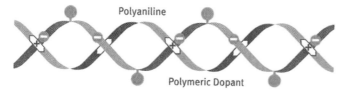

Figure 6.11 Double-stranded polyaniline (the green strand represents the conductive polymer backbone, the red circles are the dopant anions, and the purple strand is the polyelectrolyte). (Reprinted with permission from [49].)

These types of double-stranded polymers are synthesized by using a template method. For example, an adduct of aniline and PSS is formed by adsorbing aniline onto a PSS template which, upon polymerization, leads to the formation of PANI: PSS double-stranded ICP. The second polymer chain (other than ICP) provides the opportunity to prepare water- or solvent-dispersible ICPs. It also offers the possibility of improving the compatibility of ICP strands with common organic coating resins. Dispersions of polyindole were reported by Sudharani et al. [50]. A new approach was reported for the synthesis of aqueous dispersions of PPy, PANI, and PEDOT by polymerizing the corresponding monomer in the presence of a polymeric ionic liquid (PIL) stabilizer such as poly(1-vinyl-3-ethylimidazolium bromide) [51]. In this approach, the prepared submicrometer ICP particles can be transferred between water and different organic solvents. Recently, the synthesis of ICP particles in supercritical carbon dioxide has also been explored [52].

6.4.3
Core-Shell Model

The third approach is the most promising one, where a thin layer of conducting polymer is applied to sterically stabilize latex particles to form colloidal dispersions of ICPs with core-shell morphology. In this approach, the conducting polymer is the minor component. Thus, by choosing a soft core material it is possible to prepare composite particles with good film-forming properties. The advantages of core-shell approach are as follows:

• Particles with useful properties can be obtained with lesser amounts of conducting polymer.

- Easy film formation when a low T_g (glass transition temperature) polymer is employed as core material.
- Electrical conductivity can be tuned.

6.5
ICP-Based Core-Shell Particles

Polymer-coated particles offer interesting properties within a broad spectrum of applications, ranging from catalysts to additives and pigments, where they are exploited in the manufacture of cosmetics, inks, and paints. In some cases the conductivity of the particles can be induced by coating the core material with an intrinsically conductive polymer. In principle, the shell can be produced by adsorbing preformed macromolecules or monomers onto the core particles, followed by its polymerization. It is important to bind sufficient amounts of monomers onto the core and the subsequent polymerization by an appropriate initiator to increase the encapsulation efficiency. The adsorption of monomers onto dispersed particles may occur directly, but modification of the core surface provides better solid–solute interaction. The modification can be carried out either by depositing preformed polymer or by binding coupling agents onto the core surface.

There is technological interest concerning the use of composite particles in functional coatings, and a number of excellent reviews have been prepared on conductive nanocomposites [53–56]. Although the synthesis of composites always demands some entrapment (encapsulation) of polymers, the following sections will illustrate mainly the core-shell composite particles. These composite particles can be divided broadly into either organic-ICP or inorganic-ICP.

6.5.1
Organic-ICP Composite Particles

The chemical synthesis of ICP-coated particles where nonconducting polymeric materials are used was first claimed by Yassar et al. [57], though no experimental evidence was produced at the time regarding the colloidal stability of such dispersions. In 1995, Wiersma et al. first demonstrated the synthesis of stable colloidal ICP-coated dispersions with core-shell morphology [58]. These authors reported that sterically stabilized latex particles can be encapsulated either by PPy or PANI in aqueous media to form conductive composite latexes with good colloidal stability. The colloidal stability of the hybrid particles produced was achieved by adsorbing polyethylene oxide or hydroxymethylcellulose onto the polymer particles prior to polymerization. The conducting polymer layer was formed without significant interference with the steric stabilization layer. The same authors utilized different polymer particles such as nonionic surfactant-stabilized aqueous polyurethane dispersions or alkyd resins to produce colloidal particles with good film-forming properties. Evidence of the core-shell morphology of the hybrid particles was produced using transmission electron microscopy, together with measurements of electrophoretic mobility and dielectric constant. It was also shown that the stability of such dispersions could be further enhanced by the use of aromatic sulfonate dopants. In another publication, Bremer et al. studied the synthesis of composite particles with a benzidine-free polyurethane core and thin ICP shell. The use of a low-T_g polymer core offers the advantage of film-forming properties for the composite colloidal dispersions [59]. Later, these products were commercialized by DSM under the trade name of ConQuest™. A similar protocol was used by Armes et al. for the synthesis of ICP-coated organic polymer core-containing composite particles in the presence of suitable steric stabilizers [60–65]. The general scheme of organic-ICP composite latex particle synthesis is shown in Figure 6.12.

Since then, a number of composite particles have been developed [66–70]. Lu et al. have demonstrated the effect of surfactants on the morphology of ICP-coated polystyrene (PS) or poly(styrene-co-butyl acrylate) [PST-co-BuA] particles [71]. These authors showed that when the core particles are stabilized with ionic surfactants, a raspberry-type of morphology is obtained, whereas a homogeneous core-shell morphology is achieved in the case of nonionic surfactant-stabilized particles. It was also shown that the core materials influence the conductivity of the composite particles. Nano-sized PS–PPy composites have been reported by Xu et al. [72], who also noted that when the cationic, nonpolymerizable surfactant cetyltriemethylammonium bromide (CTAB) was used as stabilizer, the latex particles prepared showed poor colloidal stability and low electrical conductivity (Fig. 6.13) compared to that of latexes stabilized by nonionic polymerizable surfactant ω-methoxy[poly(ethylene oxide)$_{40}$] undecyl α-methacrylate (PEO-R-MA-40) and cationic polymerizable surfactant ω-acryloyloxyundecyltrimethylammonium bromide (AUTMAB). These authors also proposed that the poor colloidal stability and low conductivity of the composite particles were due to the desorption of CTAB

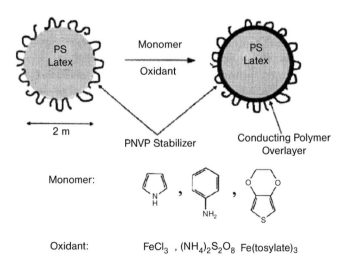

Figure 6.12 Schematic representation of the synthesis of organic-ICP composite particles (reprinted from [64]).

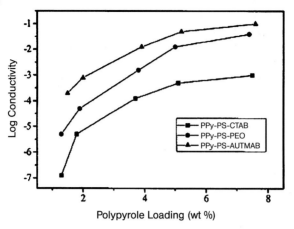

Figure 6.13 Plot of log conductivity against PPy loading (wt.%) for PPy-PS composite particles stabilized by CTAB, PEO macromonomer and AUTMAB, respectively (reprinted from [72]).

during deposition of PPy on PS particles, and the lower surface coverage of PPy. On the other hand, the chemical binding of a polymerizable surfactant onto the PS particle produced a stable dispersion with a distinct PPy phase.

Sapurina et al. have reported the synthesis of PANI-coated waterborne polyurethane latexes in the presence of a polymeric stabilizer, polyvinylpyrrolidone (PVP). The composite particles showed a conductivity of 10^{-2} S cm^{-1}, with good colloidal and mechanical properties [73]. Recently, Huang et al. synthesized a series of PPy-coated styrene-butyl acrylate (SBA) core-shell latex particles [74]. The T_g of the composite particles was shown to be determined mainly by the core material. It was reported that the conductivity of the hybrid particles could be tuned by varying the butyl acrylate content in the SBA copolymer. The composite particles showed a conductivity of 0.17 S cm^{-1}. Composite particles having a photochromic dye as the core and ICP as shell, and possessing properties of photoluminescence, have been studied by Jang et al. [75]. Thin PEDOT-coated PS particles (Fig. 6.14) were reported for self-assembled crystalline colloidal arrays with a stop band in the visible regime by Han and coworkers [76].

Currently, carbon nanotubes (CNT) are of major interest for their applications in nanomaterials and nano devices, and composite particles based on a CNT core and ICP shell have been reported recently [77,78]. Yu and coworkers successfully synthesized multi-walled carbon nanotubes (MWNT) coated with a PPy shell with core-shell morphology [78]; the composite particles obtained showed an improvement of conductivity and the redox behavior of PPy.

Another strategy that allows the combination of electronic, electrical and optical properties of ICP with tunable properties of the matrix has also been explored. Nanostructure materials with an ICP core and an insulating polymer shell have also been reported [79–81]. Highly transparent conductive coatings with core-shell nanoparticles have been reported by Jang et al. [82]. Poly(methyl methacrylate)

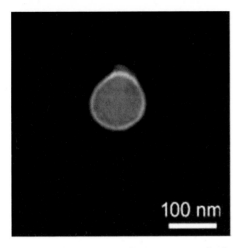

Figure 6.14 Transmission electron micrograph of a poly (3,4-ethyl-enedioxythiophene)-coated polystyrene particle (reprinted from [76]).

(PMMA)-coated PPy nanoparticles were prepared by microemulsion polymerization, and the composite particles were incorporated into a PMMA matrix to obtain transparent conductive coatings. The prepared particle showed core-shell morphology (Fig. 6.15).

Coatings based on these particles exhibit high optical transparency (>90% at 10 wt.% of particle loading) in the entire visible light range. The percolation threshold in the coating film was observed at a higher particle loading (15–20 wt.%) compared to PPy- coated PMMA composite particles, due mainly to the insulating shell.

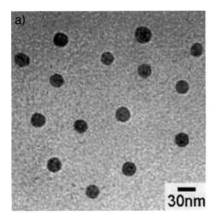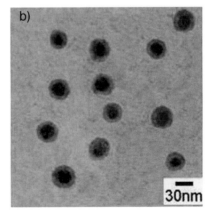

Figure 6.15 Transmission electron micrographs of (a) PPy nanoparticles and (b) PPy/PMMA core-shell nanospheres (reprinted from [82]).

In another approach, core-shell particles with multilayer shells have been prepared using a layer-by-layer assembly technique [83]. Multicore composite particles have also been reported by Shin et al. [84].

6.5.2
Inorganic-ICP Composite Particles

Polymer encapsulation of inorganic particles offers the advantage of a combination of the properties of the core and the polymer shell materials. Thus, it is possible to alter the solubility, surface charge, catalytic activity, conductivity, etc., by covering the particle surface with a thin polymer layer. The choice of ICPs as shell material provides a route for inducing conductivity to the composite particles. Inorganic particles such as SiO_2 [85], TiO_2 [86], CeO_2 [87], Fe_2O_3 [88], and MnO_2 [89] have been mixed with, or incorporated into, the conducting polymer for many applications. However, the poor compatibility of inorganic particles with the ICPs is an issue for optimization of properties for final applications. The synthesis of ICP-coated particles ensures the means to improve conductive coating performances. A number of methods are available for the encapsulation of inorganic particles by organic polymers to produce core-shell composite particles. One of the most frequently employed means of producing composite particles involves the adsorption of pre-formed polymers or monomers, followed by subsequent polymerization on the solid surface by an initiator. A sufficient quantity of monomer adsorption onto the core particles, in addition to an appropriate choice of initiator, are required to obtain good coatings. The use of initiators can be avoided in the polymerization process by using active cores [87,90]. Other methods include heterocoagulation-polymerization and emulsion polymerizations.

Seeded emulsion polymerization is a widely accepted technique for the encapsulation of inorganic particles (see Chapter 4). Polymerization onto the particle surface is favored over the formation of new particles in the bulk. This is usually achieved by adsorbing monomers onto the surface or in a bilayer, or by localizing the initiator near the surface. Surfactants are employed for the colloidal stability of the particles. Modification of the inorganic particle surface enhances the encapsulation efficiency.

Stable colloidal dispersions of ICPs are obtained via dispersion polymerizations in the presence of a polymeric stabilizer (see Section 6.4.2). Armes et al. have reported the synthesis of inorganic-ICP nanocomposites in the absence of a polymeric stabilizer [91–96]. In this approach, PPy or PANI is chemically polymerized in presence of ultrafine inorganic colloidal particles such as silica or tin (IV) oxides. The prepared composites show a raspberry-type morphology, and the surface of these nanocomposites are rich in inorganic phase. In another study, Huang et al. prepared PANI-coated CuO particles by emulsion polymerization without employing water-soluble oxidants [97]. Although a quite large number of ICP-coated silica particles have been reported, very few of them show good colloidal stability with core-shell morphology. One such report by Hao et al. demonstrated the successful encapsulation of silica latexes by a PPy shell with controllable wall thickness [98].

Lee et al. have synthesized multilayer (SiO_2/PS/PANI) conductive composite particles by seeded emulsion polymerization [99], the particles having been synthesized by several sequential steps. Nanosize SiO_2 particles were synthesized and surface-modified by silane for better polystyrene grafting; the PS-grafted SiO_2 nanoparticles were then polymerized using an initiator that led to the formation of SiO_2/PS core-shell composite particles. These particles were used as seed particles to produce SiO_2/PS/PANI conductive composite particles by emulsion polymerization of aniline monomer in the presence of sodium dodecyl sulfate (SDS) as surfactant. Morphological studies revealed the core-shell type of composite particles (Fig. 6.16).

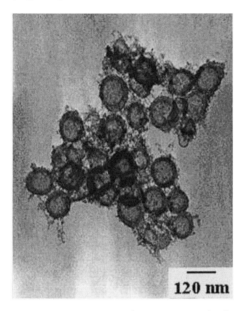

120 nm

Figure 6.16 Transmission electron micrograph of SiO_2/PS/PANI conductive composite particle prepared in the presence of sodium dodecyl sulfate (SDS) (reprinted from [99]).

Core-shell particles with PANI shell or PEDOT shell have also been synthesized recently [100,101]. Composite particles having both conductive and ferromagnetic properties have attracted much attention due to their potential applications in electrochemical display devices, microelectronics, and electromagnetic interference (EMI) shielding. Nanocomposites based on composite particles of Fe_3O_4 and ICPs have been successfully prepared using several methods such as polymerization by chemical or electrochemical techniques, layer-by-layer assembly, or simultaneous gelation and polymerization [102–105]. In most cases the obtained composites Fe_2O_3-PPy showed low room temperature conductivity and low coercive force due to complex synthetic methods involved. A relatively simple method to synthesize a core-shell Fe_3O_4-PPy composite particle was described by Chen et al. [106]. The composites were prepared by chemical polymerization of pyrrole monomer with

Fenton's reagent in the presence of polyethylene glycol (PEG) as surfactant and Fe_3O_4 nanoparticles. The prepared composites showed a magnetic coercive force ranging from 98.4 to 116.3 Oe, and electrical conductivity from 10^{-5} to 10^{-2} S cm^{-1}, depending on the Fe_3O_4 content. These authors proposed that the low Fe_3O_4 particle size (10–20 nm) might be responsible for the low coercive force. Deng et al. have also reported the synthesis of Fe_3O_4-PPy composite particles via *in-situ* emulsion polymerization in aqueous solution containing Fe_3O_4 nanoparticles and NaDS (sodium dodecylbenzene sulfonate) as a surfactant and dopant [107]. The prepared composite particles showed core-shell morphology, with both the magnetic and electrical properties being dependent upon the amount of Fe_3O_4 used. In another publication, the same authors reported the synthesis of core-shell (where Fe_3O_4 is the core and cross-linked PANI is the shell) composite particles by emulsion polymerization [108].

6.6
Conductive Coatings with Core-Shell Particles

Organic coatings are nonconductive in nature, so in order to produce conductive coatings conductive pigments such as carbon black, graphite, or metal particles are usually added to the organic resins. The different possibilities for the design of conductive coatings can be summarized as follows:

- To use a conductive polymer as a continuous matrix.
- The addition of conductive pigments into the organic resin.

ICPs can be used either as a continuous matrix or they can be incorporated into nonconductive organic resins to design conductive coatings. Conductive coatings based on core-shell latex particles can be used to design conductive thin films by evaporating the solvent, or by mixing these particles into organic resins. The film-formation process of ICP-based core-shell particles is illustrated in Figure 6.17.

When soft polymers are used as core it is possible to obtain good quality conductive films by drying. Core-shell-based coatings can be applied by spraying or roll coating. The electrolytic co-deposition of ICP-based core-shell particles with metallic ions such as zinc or copper on metallic substrates can also be used to design composite coatings with functional properties. Pfleger and coworkers have deposited PT-coated TiO_2 particles electrophoretically onto conductive surfaces [109].

Figure 6.17   Film-formation process of the ICP-coated latex particles.

6.7
Applications of ICP-Based Materials

The number of potential applications of ICPs is almost unlimited due to their special features such as light weight, flexibility, mechanical properties, and tuneable conductivities that lie between insulators and metallic conductors [110,111]. However, difficulties with the processability of ICPs has hindered their commercial success. During the past 10 years, a wide variety of processable ICPs has been developed, with some becoming commercially available (Table 6.1).

Table 6.1 Examples of commercially available ICPs.

Product	Company	Product type	Applications
Panipol	Panipol Ltd. (Finland)	Polyaniline powders and dispersions	Corrosion protection, antistatic, electro-chromic windows, EMI shielding
Ormecon™	Ormecon Chemie, Germany	Polyaniline dispersions	Corrosion protection, antistatic coatings
Eeonomer®	Eeonyx Corporation, USA	Polyaniline/polypyrrole-coated carbon powders	Static dissipative applications
ConQuest	DSM, The Netherlands	Polypyrrole-coated latexes, powders	Corrosion protection, antistatic, electrostatic discharge
Orgacon™ ICP Baytron®	Agfa Gevaert, Belgium and HC Starck/Bayer Material Science, Germany	PEDOT : PSS	Photographic films, electrode materials, antistatic coatings
	Sigma-Aldrich	PANI/PPy/PPV/PEDOT dispersions and powders, doped and undoped versions	Electronic applications

Conductive polymers can be regarded as semiconductors when their conductivity is below 10^{-5} S cm^{-1}. The band gap of undoped ICPs depends not only on the chemical composition but also on the substituents attached to the main chain. Undoped conjugated polymers have found extensive applications for the fabrication of electroluminescent devices. Conductive polymers behave like metallic conductors when they are doped. Both doped and undoped forms of ICPs have an array of potential applications (Table 6.2).

Table 6.2 Application potential of ICPs.

Applications of ICPs	
As semiconductors	As conductors
Light-emitting diodes (LEDs)	Antistatic applications
Transistors and integrated circuits	Corrosion protection
Solar cells	Batteries
Sensors	Electromagnetic shielding
	Electrochromic screens or windows
	Lithography
	Conductive textiles
	Microwave welding of plastics
	Electrolytic capacitors

6.7.1
ICPs as Semiconductors

Organic semiconducting polymers combine the properties of plastics and semiconductors. They have outstanding potential applications in electronics, optoelectronics, in the field of information technologies, and in biomedical sciences [112,113]. This combination of properties has opened new possibilities for the design of electronic devices such as light-emitting diodes (LEDs), transistors, and plastic solar cells. These materials offer the advantage of easy manufacturing processes, higher flexibility, low cost, and new market possibilities.

6.7.1.1 Polymer Light-Emitting Diodes (PLEDs)
Semiconducting ICP films emit light upon the application of an external voltage to the film. This property of light emission (electroluminescence) made ICPs a suitable candidate for fabricating display devices, and the first LED was produced commercially in 1997 by the Japanese electronics company, Pioneer Corporation, for use in electronic equipment for automobiles. Organic light-emitting diodes (OLEDs) can be made by using organic molecules or polymers; likewise, ICPs are sandwiched between two electrodes to design PLEDs. Under an applied voltage, positive (h^+) and negative (e^-) charge carriers are ejected from the anode and cathode, respectively. These charge carriers combine in the emissive ICP layer to produce light. One of the electrodes (mainly the anode) is made transparent so that the light produced may be transmitted. Indium tin oxide (ITO) is usually used as anode material to design electroluminescence devices. A schematic diagram of a PLED is shown in Figure 6.18.

Electrodes are chosen to facilitate the charge injections into the luminescent polymer layer. Thus, the use of additional polymer layers (i.e., making multilayer electroluminescence devices) raises the device efficiency. For example, using an electron-transporting (hole-blocking) layer between the cathode and the electroluminescent polymer moves the recombination of holes and electrons away from the

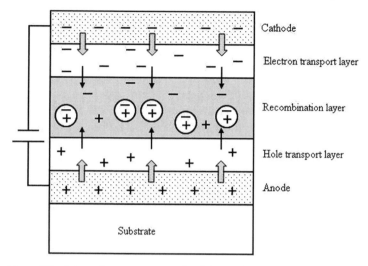

Figure 6.18 Schematic representation of multilayer polymer light-emitting diode (PLED).

polymer/electrode interface to the bulk of the polymer [114]. Derivatives of PPVs such as MEH-CN-PPV (alkoxy and cyano-modified PPV) can be used for this purpose. Similarly, the use of hole transport layers between the anode and the luminescent polymer enhances the rate of hole injection. Multilayer devices shows better performance compared to single-layer devices, but the cost of designing multilayer devices is significantly higher. Polymers such as poly(*p*-phenyl vinyl), poly(alkyl thiophenes) and their derivatives are used as an electroluminescence layer. ICPs such as a PEDOT:PSS complex or PANI are used as electron transport and hole transport layers respectively to design multilayer PLEDs [115,116]. ICP-based PLEDs offer certain advantages, such as higher contrast, larger viewing angle, lower power consumption, flexibility and good processability compared to the well-known liquid crystal displays (LCDs).

6.7.1.2 Polymer Solar Cells

There is a growing need for energy in the world, and the quest for economically competitive renewable energy sources such as hydroelectric, wind, and solar power are increasing due to high oil prices and environmental concerns. The conversion of sunlight into electricity using solar cells is one of the efficient ways of utilizing renewable energy. Silicon-based solar cells are most widely used for this purpose and, although such cells are promising in terms of efficiency, they are expensive. The manufacturing process is quite complex, and simple wet application techniques cannot be used to prepare large-scale silicon solar cells.

These problems may be overcome by the use of organic photovoltaic solar cells [117–120], which consist of a donor and an acceptor material, each possessing its individual HOMO and LUMO energy levels. For an efficient device configuration, the HOMO and LUMO energy levels of the donor material must be higher com-

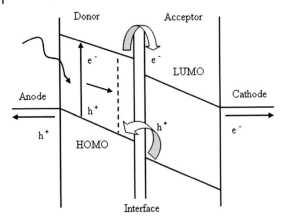

Figure 6.19   Working principle of a bilayer heterojunction organic solar cell.

pared to that of the acceptor molecules. When light energy strikes either on donor or acceptor molecules, coupled electron-hole pairs called "excitons" are generated (Fig. 6.19). The excitons are formed as the band gap (energy difference between HOMO and LUMO) energy of the molecules is smaller compared to the required energy for generating free electron-hole pairs. When the band gap energy is higher than the binding energy of the excitons, they tend to move towards the donor/acceptor or organic molecules/electrode interface. At these interfaces, the excited electrons jump from the LUMO of the molecule to the LUMO of the acceptor, or the holes in the HOMO jumps into the HOMO of the donor. These free electrons and holes then migrate to their respective electrodes.

Organic solar cells can be designed as either single-layer, bilayer heterojunction, or bulk heterojunction cells [121]. The device configuration of a single-layer solar cell consists of metal-polymer-metal; these types of cell are not popular as the photo-induced charge generation is limited. The excitons created in a single-layer cell recombine, and this decreases the efficiency of the device. In order to overcome these difficulties, bilayer heterojunction (donor/acceptor) cells have been developed in which the individual layers of donor and acceptor molecules are sandwiched between the anode and cathode. The efficiency of such systems is further enhanced by the use of duel (donor/acceptor) molecule (composite) structures instead of separate layers of donor and acceptors. For example, a composite film of ICP and fullerene can be used to fabricate bulk heterojunction photovoltaic devices. Solar cells with higher efficiency of photoelectrical conversion were achieved by using dye-sensitized semiconductor particles in combination with an electrolyte. These cells, which are known as "Grätzel-type" solar cells [122], have one main disadvantage in that insert, they use liquid electrolytes. Much research has been focused on using solid electrolyte or organic hole transport materials to design more efficient solar cells [123–127]. Core-shell composite particles can be used to design hybrid-type solar cells, and photosensitive layers containing core-shell

(TiO$_2$ core and PT shell) particles have been reported recently [109]. Cathodic electrophoretic deposition was used to fabricate solar cells with these hybrid particles.

6.7.1.3 Organic Thin-Film Transistors

One of the most exclusively studied applications of undoped ICPs is their use in organic thin-film transistors (OTFTs) [112]. OTFTs are used in a number of low-cost, large-area electronic applications such as liquid crystal displays, smart cards, inventory tags, and large-area sensors. OTFTs have the advantage of low-temperature fabrication and significantly low costs compared to the traditional inorganic semiconductor-based thin-film transistors. The low-temperature fabrication allows OTFTs to be integrated into inexpensive plastic substrates rather than onto glass substrates. Most research into OTFTs is carried out using oligomers or low molecular-weight systems. Field-effect transistors (FETs) are composed of two metal electrodes (source and drain) deposited on a semiconducting polymer layer. The third electrode (the gate) is placed onto the dielectric layer. The layout of OTFTs are of two major types, namely "top contact" and "bottom contact". For the top contact layout, the semiconductor layer is deposited onto the dielectric layer, after which the source and drain contacts are placed. In contrast, the bottom contact transistor layout is obtained when the semiconducting polymer layer is deposited onto the source and drain contacts (Fig. 6.20).

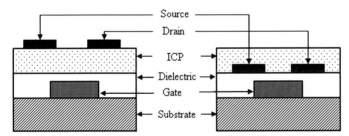

Figure 6.20 Layouts of organic thin film transistors (OTFTs).
(Left) Top contact. (Right) bottom contact.

When the transistor is in its "off" state (gate voltage = 0), the current flow between the drain and the source electrode is low. As the gate voltage is applied, charges are induced at the semiconducting and dielectric interface, and this results in a flow of current between the drain and the source electrode. Thus, a higher gate voltage results in more charge carriers and higher current. The crucial parameters for the performance of TFTs are the field-effect mobilities and the "on"/"off" ratio. The level of conductivity in the "off" state is highly dependent on the amount of doping in the polymer. In the "on" state, it is proportional to the concentration of the charge carriers induced by the gate voltage and their mobilities. Thus, ICPs used for OTFT applications must have a highly ordered structure and less impurity in the polymer.

6.7.2
Applications as Conductors

6.7.2.1   Antistatic Coatings

Static charges may cause sparks and damage on nonconductive surfaces, and in order to overcome these problems antistatic agents such as ionic salts or polyionic polymers are used as surface-active components. The antistatic agents provide conductivity to the polymers in the presence of moisture. These agents consist mainly of a hydrophilic portion that is able to attract moisture, and a hydrophobic portion that increases the compatability with polymer coatings. External antistatic agents (antistats) are usually applied from solvent or aqueous solutions; thus, they suffer from being easily removed by rinsing or wiping of the substrate. In contrast, internal antistats are incorporated into the bulk of the polymer film, and must migrate to the surface to provide antistatic properties; hence, their efficiency decreases with time. Antistatic coatings requires very high ($\sim 10^{10}$ Ohms) surface resistivity to promote static charge dissipation by grounding.

Electronically conductive materials offer the advantage of being active without the presence of moisture. Doped metallic oxides are used to provide antistatic effects; for example, ITO-sputtered films are used for display applications. Unfortunately, these products are expensive and less flexible, and consequently transparent conductive layers based on ICPs as antistatic coatings are gaining more industrial applications. The latter coatings cannot be washed from the surface, and are more flexible. Important examples where ICPs are used as antistatic coating include antistatic films, electronic packaging, and cathode ray tubes. Core-shell ICP-based systems can be applied for antistatic coating developments. These particles can be incorporated into organic coating formulations with an adhesion promoter, cross linkers and additional additives to obtain suitable conductivities for antistatic applications.

6.7.1.2   Electromagnetic Shielding

Modern electronic equipment such as laptop computers must operate in smaller packages with higher frequencies, and consequently face the problem of EMI. Plastics are insulatory by nature and are, therefore, more vulnerable to EMI. The latter is defined as the unwanted degradation of systems or equipment by electrical signals ranging from low to high frequencies. The solution to this problem can be achieved in various ways, such as making insulatory plastic conductive by adding fillers, using thin metal sheets, and coating surfaces with conducting films. In general, the shielding effectiveness (in decibels, dB) is expressed as:

SE (Shielding effectiveness) = 20 log (E_1/E_2)

where E_1 and E_2 represent the incident and transmitted electric/magnetic field strengths, respectively. Usually, the lower the coating sheet resistance, the better the shielding effect. Beside the traditional approach of using conductive fillers for EMI shielding, ICPs offer an alternative material to solve EMI problems. ICPs are preferred over the use of metals due to good conductivity, low density, better corrosion resistance and shielding by absorption mechanism. A more detailed discussion on the application potentials of ICPs such as PANI and PPy for EMI shielding can be found in Ref. [128].

6.7.1.3 Corrosion Protection of Metals

Both ferrous and nonferrous metals are susceptible to corrosion, and several strategies have been developed to tackle this problem. For example, the use of inorganic films, particularly noble metals, as a protective coating layer over a less noble metal provides anodic galvanic protection and minimizes corrosion. The other most widely used strategy is the use of cross-linked organic films, with or without corrosion inhibitors. An electrochemical corrosion cell required the presence of oxidizing species and ion movement from the cathodic to anodic sites of the corrosion cell. Application of organic coatings hinders the transport of oxidizing species to be available in the metal/coating interface and, at the same time, provides a resistance to ion movement at the interface. However, the presence of defects in the coating (either natural or accidental) expedites the transport process of cathodic reactants, thereby enhancing the corrosion rate. Therefore, organic coating systems are typically composed of a primer and a barrier layer (top coat) possessing functional properties. The primer coating is chosen as it provides good adhesion to the metal substrate while simultaneously containing ingredients that react with the oxides to form a barrier layer that prevents corrosion when the latter is breached. In general, a phosphate treatment followed by chromate treatment is applied to achieve adhesion and good anticorrosion properties. Although a wide variety of primer coating compositions exists, the most widely used primers contain heavy metals, but these suffer the disadvantage of being toxic and carcinogenic. Conductive polymer coatings offer an alternative route to the design of environmentally friendly anticorrosive primers for metals.

Mechanism of corrosion protection by conductive coatings

A great deal of research has been carried out to develop anticorrosion primers. Reviews of studies of the mechanism of corrosion protection by conductive polymer coatings often yield quite confusing information that is not yet fully understood. In general, it can be said that conductive polymers offer anticorrosive properties by a combination of the following hypothesis:

- Barrier effect: ICPs in their insulating (undoped) states act as barriers because in these states they have poor electronic or ionic conductivities. They are preferred over other insulating polymers due to their ability to form dense and less porous adherent films.
- Galvanic protection: ICPs in their conductive (doped) state act as efficient oxidizers and maintain the metals in the passive domain.
- Electronic effect: When doped ICPs come into contact with metals, an electrical field is generated that prevents the flow of electrons from the metal to the oxidizing species, thereby reducing the rate of corrosion.

Various anticorrosion mechanisms have been described, the most common being ennobling, dislocation of the oxygen reduction site, and the release of dopant anions.

Ennobling mechanism

The principle of this mechanism is based on the fact that conductive (p-doped) coatings could provide anodic galvanic protection to the metal substrate and act as an oxidizer to maintain the metal in the passive domain. This mechanism can lead to passivation of the exposed metal surface at small defects in the passive layer. A considerable number of reports have been made claiming the formation of passive films at the exposed site, with subsequent inhibition of further metal dissolution [129].

Dislocation of the oxygen reduction site

When a conductive polymer-coated metallic substrate is exposed to an electrolyte, a galvanic coupling may occur between the metallic substrate and the electrolyte due to the electronic conductivity of ICPs (Fig. 6.21).

Figure 6.21 shows that the metal oxidation [Eq. (2)] at the metal/ICP interface in the initial stage of exposure triggers the ICP reduction [Eq. (3)], and it decreases the passivating power of the ICP film.

$$M \rightarrow M^{n+} + ne^- \tag{2}$$

$$ICP^{n+} + ne^- \rightarrow ICP^0 \tag{3}$$

In the case of insulating coatings, oxygen reduction takes place at the metal–ICP interface, which causes delamination/debonding of the organic coatings [130]. However, due to their electronic conductivity, ICPs could transfer electrons produced in the exposed surface into the coating. Thus, the presence of ICPs triggers the reduction of dissolved oxygen [Eq. (4)] at the ICP–electrolyte interface that accelerates reoxidation [Eq. (5)] of the ICP film.

$$ICP^0 \rightarrow ICP^{n+} + ne^- \tag{4}$$

$$n/4\, O_2 + n/2\, H_2O + n\, e^- \rightarrow nOH^- \tag{5}$$

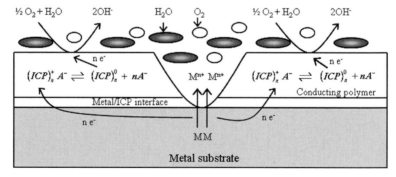

Figure 6.21 Schematic representation depicting the dislocation of oxygen reduction site.

Displacement of the O_2 reduction site causes retardation of the cathodic delamination effect. A model study has been carried out by Michalik et al., using the delamination test of PPy coatings in an atmosphere containing isotopic oxygen ($^{18}O_2$), to establish the dislocation site of oxygen reduction [131]. The hydroxyl ions ($^{18}OH^-$) generated by reduction of the isotopic oxygen was recognized by ToF-SIMS measurements. The study results confirmed dislocation of the oxygen reduction from the metal/ICP interface to the top of the conductive coatings. It was reported that this effect is not good for corrosion protection, as a conductive polymer only shifts high concentrations of OH^- ions from the ICP/metal interface to the top of ICP surface, thereby enhancing delamination of the topcoat. Smear-out of the oxygen reduction site into the bulk of the ICP film could be achieved by using ICP as pigments in a nonconductive organic matrix. Even then, the conductivity of such systems might be enough to dislocate the oxygen reduction site to the surface of the coating.

Release of dopant anions

In the presence of any defect, ICP may polarize through galvanic coupling of the metal at the defects, while the anions of the p-dopants would be forced out of the reduced ICP film in accordance with the electroneutrality principle. The release mechanism is shown schematically in Figure 6.22.

The release of anions can take place due to cathodic reduction of the conducting polymer, or ion exchange with the generated OH^-, or by both. If the dopant anion has some corrosion-inhibiting properties, then damage-responsive corrosion protection will be shown. Additionally, an organic oxygen reduction reaction inhibitor may also be incorporated in the form of a dopant to protect scratches. Several studies have suggested the use of different dopants such as phosphates, phosphonates, borates, and nitrates as anionic corrosion inhibitors in ICP coatings [132–134]. As the release of anions is dependent upon the limited diffusion of the dopant anion to the damaged site, this effect is prominent in the vicinity of a defect (exposed site). This interesting feature of ICPs makes them a suitable candidate for the design of self-healing coatings. In this context, it should be noted that a true self-healing coating for corrosion protection should release corrosion-inhibitive anions only in the

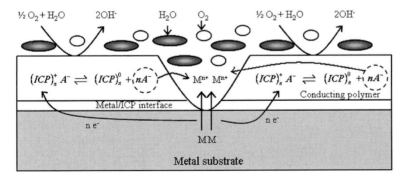

Figure 6.22 Schematic representation of the release mechanism for corrosion protection.

case of corrosive attack. This means that the inhibitor anions must not be leached out or released through ion-exchange processes. Very recently, self-healing anti-corrosive coatings based on corrosion-inhibitive dopant release as a consequence of electrochemical ICP reduction were reported for the first time by Porebska et al. [135]. The coatings were prepared by mixing stable PPy dispersions doped with either MoO_4^- or $[PMo_{12}O_{40}]^{3-}$ into a nonconducting organic resin. The authors claimed that the designed coating is even capable of passivating large defects. The dopant anions appear when there is a decrease in potential at the metal/coating interface during the delamination. The same authors also mentioned that the release mechanism is negatively influenced by presence of small cations and high pH at the metal/coating interface.

6.7.1.4 Corrosion Protection by ICP-Based Core-Shell Latexes

Corrosion protection by ICPs has been well documented in several reviews [136–141]. The use of ICP-based core-shell latexes for the corrosion protection of steel is a relatively new field of research, and a recent review has proposed the exploitation of such possibilities [142]. In a recent study, we found that ICP-based core-shell latexes offer the possibility of achieving anticorrosion properties [143], though the presence of pinholes or scratches in the coatings enhances the corrosion rate of metals. This problem may be due to poor dispersion of the composite particles in the insulating resin systems, and future research will need to develop conductive polymer-based anticorrosive primers for metals. The anticorrosive properties of PANI-coated polystyrene latex microspheres has been reported recently [144]. PS-PANI composite particles with core-shell structure were prepared by chemical oxidative polymerization of aniline monomer in the presence of a PVP-stabilized PS latex suspension. The reduced form of the particle was obtained by adding hydrazine monohydrate to the suspension. Both oxidized and reduced PANI-PS particles were used to obtain a PANI-PS-coated iron electrode (PANI-PS-Fe). Pure PANI

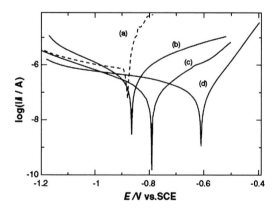

Figure 6.23 Tafel plots of: (a) Fe; (b) reduced PANI-PS-Fe; (c) PANI-Fe; (d) oxidized PANI-PS-Fe, electrodes in 3% (w/v) NaCl aqueous solutions at a scan rate of 1 mV s^{-1} (reprinted from [144]).

film was obtained in two ways: (i) by dissolving the PS cores of PANI-PS-Fe film in tetrahydrofuran; and (ii) by electropolymerizing aniline on Fe [PANI-Fe]. The tafel plots of the above-mentioned samples are presented in Figure 6.23.

The data in Figure 6.23 show that the films are stable up to −1.2 V. Polarization resistance of the films (calculated from the tafel plots) was found to be enhanced by increasing the oxidized form of the particles in the film. The corrosion current obtained was smaller compared to the uncoated Fe.

6.7.3
Other Applications

Several other highly promising applications involving ICPs have been suggested, but none is described in detail here. For example, intrinsically conductive polymers can be used in sensors, while chemical transducers made from ICPs such as PT have been used to determine precisely the chemical composition of a formulation. Another widespread use of ICPs is in the manufacture of lightweight, rechargeable batteries; such polymer-based batteries are superior to conventional nickel-cadmium batteries due to the phenomenon of doping and dedoping of polymer molecules during charging and discharging. The highly redox properties of ICPs are also utilized in the design of electrochromic smart devices [145]. Conducting polymers have also shown promise in "smart" structures; an example is that of the recently developed "smart skis", which avoid vibration during skiing by generating a force opposite to that vibratory force. ICP-based functionalized core-shell systems are also widely used in biomedical applications [146].

6.8
Conclusions

The growth potential of this field is clearly illustrated not only by reviewing the available literature, but also by the number of patents on conducting polymer research which have been published during the past few years. Nonetheless, the questions that arise always from an industrial perspective are:

- What are the commercial possibilities for conductive polymer-based products?
- What are the potentials for return of stakes (the pay-back of investments)?
  And finally
- Are conductive polymer-based products ready for commercialization?

Though several conductive polymer-based products are available commercially, many possibilities exist and are yet to come to market. The main disadvantages for the commercialization of these conductive polymer-based products are their ill-defined and nonoptimized properties. At present, the traditionally used chromium- or lead-based products are losing market share due to public concerns over their adverse effects on the environment and to their potential carcinogenic properties.

However, this should pave the way for conducting polymers to be used as an alternative solution. Today, the major challenge for scientists working in the field of conductive polymer-based products is to overcome any existing problems such that they may offer easily processed, environmentally friendly and stable conductive polymer-based functional coatings for general use.

Abbreviations

AUTMAB	ω-acryloyloxyundecyltrimethylammonium bromide
CNT	carbon nanotube
CTAB	cetyltriemethylammonium bromide
EB	emeraldine base
EDT (EDOT)	3,4-ethylenedioxythiophene
Eg	band gap
EMI	electromagnetic interference
ES	emeraldine salt
eV	electron volt
FET	field-effect transistor
HOMO	highest occupied molecular orbital
ICP	intrinsically (inherently) conducting polymer
ITO	indium tin oxide
LB	leucoemarlidine base
LCD	liquid crystal display
LED	light-emitting diode
LUMO	lowest unoccupied molecular orbital
MWNT	multi-walled carbon nanotubes
OLED	organic light-emitting diode
OTFT	organic thin-film transistor
PANI	polyaniline
PEDT (PEDOT)	poly (3,4-ethylenedioxythiophene)
PEO-R-MA-40	ω-methoxy[poly(ethylene oxide)$_{40}$] undecyl α-methacrylate
PLED	polymer light-emitting diode
PMMA	poly(methyl methacrylate)
PPV	poly(phenylene vinylene)
PPy	polypyrrole
PSS	polystyrene sulfonate
PT	polythiophene
PVC	polyvinyl chloride
PVP	polyvinylpyrrolidone
S	Siemen
SBA	styrene-butyl acrylate
SDS	sodium dodecyl sulfate
T_g	glass transition temperature
ToF-SIMS	time-of-flight-secondary ion mass spectrometry

References

1. M. Hatano, S. Kambara, S. Okamoto, *J. Polym. Sci.* **1961**, *51(156)*, S26–S29.
2. H. Shirakawa, E.J. Louis, A.G. MacDiarmid, C.K. Chiang, A.J. Heeger, *J. Chem. Soc. Chem. Commun.* **1977**, *16*, 578–580.
3. Y. Wei, J. Wang, X. Jia, J.M. Yeh, P. Spellane, *Polym. Mater. Sci. Eng.* **1995**, *72*, 563–564.
4. J. Rodriguez, H.J. Grande, T.F. Otero, in: H.S. Nalwa (Ed.), *Handbook of Organic Conductive Molecules and Polymers.* John Wiley & Sons: New York, **1997**, Vol. 2, Chapter 10, pp. 415–468.
5. M. Angelopoulos, G.E. Asturias, S.P. Ermer, A. Ray, E.M. Scherr, A.G. MacDiarmid, *Mol. Cryst. Liq. Cryst.* **1988**, *160*, 151–163.
6. D.C. Trivedi, in: H.S. Nalwa (Ed.), *Handbook of Organic Conductive Molecules and Polymers.* John Wiley & Sons: New York, **1997**, Vol. 2, Chapter 12, pp. 505–572.
7. M. Vecino, J. Gonzalez, M.E. Munoz, A. Santmaria, E. Ochoteco, J.A. Pomposo, *Polym. Adv. Technol.* **2004**, *15*, 560–563.
8. J. Stejskal, P. Kratochvil, A.D. Jenkins, *Polymer* **1996**, *37(2)*, 367–369.
9. Y. Cao, A. Andreatta, A.J. Heeger, P. Smith, *Polymer* **1989**, *30 (12)*, 2305–2311.
10. T.V. Vernitskaya, O.N. Efimov, *Russ. Chem. Rev.* **1997**, *66 (5)*, 443–457.
11. S. Martina, V. Enkelmann, G. Wegner, A.D. Schlüter, *Synth. Metals* **1992**, *51*, 299–305.
12. L. Groenendaal, F. Jonas, D. Freitag, H. Pielartzik, J.R. Reynolds, *Adv. Mater.* **2000**, *12(7)*, 481–494.
13. E.E. Sheina, S.M. Khersonsky, E.G. Jones, R.D. McCullough, *Chem. Mater.* **2005**, *17(13)*, 3317–3319.
14. R.C. Foitzik, A. Kaynak, J. Beckmann, F.M. Pfeffer, *Synth. Metals* **2005**, *155*, 185–190.
15. J. Rühe, T.A. Ezquerra, G. Wegner: (a) *Makromol. Chem. Rapid. Commun.* **1989**, *10*, 103–108; (b) *Synth. Metals* **1989**, *20*, 177–181.
16. F.R. Díaz, C.O. S nchez, M.A. del Valle, J.L. Torres, L.H. Tagle, *Synth. Metals* **2001**, *118*, 25–31.
17. B. Wessling, J. Posdorfer, *Synth. Metals* **1999**, *102*, 1400–1401.
18. P.J. Kinlen, D.C. Silverman, C.R. Jeffreys, *Synth. Metals* **1997**, *85*, 1327–1332.
19. J.L. Camalat, J.C. Lacroix, S. Aeiyach, K. Chaneching, P.C. Lacaze, *Synth. Metals* **1998**, *93*, 133–142.
20. C.Y. Yang, Y. Cao, P. Smith, A.J. Heeger, *Synth. Metals* **1993**, *53*, 293–301.
21. E. Virtanen, J. Laakso, K. Väkiparta, H. Ruohonen, H. Järvinen, M. Jussila, P. Passiniemi, J.E. Österholm, *Synth. Metals* **1997**, *84*, 113–114.
22. C.F. Liu, T. Maruyama, T. Yahamoto, *Polym. J.* **1993**, *25 (4)*, 362–372.
23. A.W. Rinaldi, R. Matos, A.F. Rubira, O.P. Ferreira, E.M. Girotto, *J. Appl. Polym. Sci.* **2005**, *96(5)*, 1710–1715.
24. G. Cakmak, Z. Küçükyavuz, S. Küçükyavuz, *Synth. Metals* **2005**, *151*, 10–18.
25. M.K. Ram, N. Sarkar, H. Ding, C. Nicolini, *Synth. Metals* **2001**, *123 (2)*, 197–206.
26. N. Kizilcan, B. Ustamehmetoglu, *J. Appl. Polym. Sci.* **2005**, *96 (3)*, 618–624.
27. Ö. Papila, L. Toppare, *Int. J. Polym. Anal. Charact.* **2004**, *9*, 13–28.
28. http://www.tda.com/eMatls/icp.html.
29. S.P. Armes, M. Aldissi, *Polymer* **1990**, *31(3)*, 569–574.
30. S.P. Armes, M. Aldissi, G.C. Idzorek, P.W. Keaton, L.J. Rowton, G.L. Stradling, M.T. Collopy, D.B. McColl, *J. Colloid Interface Sci.* **1991**, *141(1)*, 119–126.
31. J.N. Barisci, J. Mansouri, G.M. Spinks, G.G. Wallace, C.Y. Kim, D.Y. Kim, J.Y. Kim, *Colloids. Surf. A: Physicochem. Eng. Aspects* **1997**, *126*, 129–135.
32. R. Odegard, T.A. Skotheim, H.S. Lee, *J. Electrochem. Soc.* **1991**, *138(10)*, 2930–2934.
33. G. Markham, T.M. Obey, B. Vincent, *Colloids. Surf.* **1990**, *51*, 239–253.
34. S.P. Armes, B. Vincent, *J. Chem. Soc., Chem. Commun.* **1987**, 288–290.
35. S.P. Armes, J.F. Miller, B. Vincent, *J. Colloid Interface Sci.* **1987**, *118 (2)*, 410–416.
36. T.K. Mandal, B.M. Mandal, *J. Polym. Sci. Part A: Polym. Chem.* **1999**, *37*, 3723–3729.
37. N. Cawdery, T.M. Obey, B. Vincent, *J. Chem. Soc., Chem. Commun.* **1988**, 1189–1190.

38. M.L. Digar, S.N. Bhattacharyya, B.M. Mandal, *Polymer* **1994**, *35 (2)*, 377–382.

39. A. Pich, Y. Lu, H.J.P. Adler, T. Schmidt, K.F. Arndt, *Polymer* **2002**, *43 (21)*, 5723–5729.

40. P.M. Beadle, L. Rowan, J. Mykytiuk, N.C. Billingham, S.P. Armes, *Polymer* **1993**, *34 (7)*, 1561–1563.

41. C. DeArmitt, S.P. Armes, *Langmuir* **1993**, *9*, 652–654.

42. Y.H. Geng, Z.C. Sun, J. Li, X.B. Jing, X.H. Wang, F.S. Wang, *Polymer* **1999**, *40*, 5723–5727.

43. Y. Lu, A. Pich, H.J.P. Adler, *Macromol. Symp.* **2004**, *210*, 411–417.

44. C. Arribas, D. Rueda, *Synth. Metals* **1996**, *79*, 23–26.

45. J. Stejskal, I. Sapurina, *J. Colloid Interface Sci.* **2004**, *274*, 489–495.

46. http://www.panipol.com/.

47. http://www.hcstarck.de/.

48. http://www.ormecon.de/.

49. S. Yang, R. Brown, J. Sinko, *Eur. Coat. J.* **2005**, *11*, 48–54.

50. G. Rajasudha, D. Rajeswari, B. Lavanya, R. Saraswathi, S. Annapoorni, N.C. Mehra, *Colloid. Polym. Sci.* **2005**, *283*, 575–582.

51. R. Marcilla, E. Ochoteco, C. Pozo-Gonzalo, H. Grande, J.A. Pomposo, D. Mecerreyes, *Macromol. Rapid Commun.* **2005**, *26*, 1122–1126.

52. K.T. Lim, G.H. Subban, H.S. Hwang, J.T. Kim, C.S. Ju, K.P. Johnston, *Macromol. Rapid Commun.* **2005**, *26*, 1779–1783.

53. R. Gangopadhyay, A. De, *Chem. Mater.* **2000**, *12*, 608–622.

54. J. Njuguna, K. Pielichowski, *J. Mater. Sci.* **2004**, *39*, 4081–4094.

55. A. Malinauskas, *Polymer* **2001**, *42*, 3957–3972.

56. M.L. Cantú, P.G. Romero, in: P.G. Romero, C. Sanchez (Eds.), *Functional Hybrid Materials*. Wiley-VCH, Weinheim, **2003**, Chapter 7, pp. 210–270.

57. A. Yassar, J. Roncali, F. Garnier, *Polym. Commun.* **1987**, *28*, 103–104.

58. A.E. Wiersma, L.M.A. vd Steeg, T.J.M. Jongeling, *Synth. Metals* **1995**, *71*, 2269–2270.

59. L.G.B. Bremer, M.W.C.G. Verbong, M.A.M. Webers, M.A.M.M. van Doorn, *Synth. Metals* **1997**, *84*, 355–356.

60. S.F. Lascelles, S.P. Armes, *J. Mater. Chem.* **1997**, *7(8)*, 1339–1347.

61. C. Barthet, S.P. Armes, S.F. Lascelles, S.Y. Luk, H.M.E. Stanley, *Langmuir* **1998**, *14*, 2032–2041.

62. M.A. Khan, S.P. Armes, *Langmuir* **1999**, *15*, 3469, 3475.

63. D.B. Cairns, S.P. Armes, L.G.B. Bremer, *Langmuir* **1999**, *15*, 8052–8058.

64. M.A. Khan, S.P. Armes, *Adv. Mater.* **2000**, *12 (9)*, 671–674.

65. D.B. Cairns, M.A. Khan, C. Perruchot, A. Riede, S.P. Armes, *Chem. Mater.* **2003**, *15*, 233–239.

66. B.J. Kim, S.G. Oh, M.G. Han, S.S. Im, *Polymer* **2002**, *43*, 111–116.

67. F.M. Huijs, F.F. Vercauteren, G. Hadziioannou, *Synth. Metals* **2002**, *125*, 395–400.

68. M. Okubo, S. Fugi, H. Minami, *Colloid. Polym. Sci.* **2001**, *279*, 139–145.

69. H.J. Choi, M.S. Cho, S.Y. Park, C.H. Cho, M.S. John, *Designed Monomers and Polymers* **2004**, *7(1-2)*, 111–117.

70. C.S. Kuo, L.A. Samuelson, S.P. McCarthy, S.K. Tripathy, J. Kumar, *J. Macromol. Sci. Part A: Pure. Appl. Chem.* **2003**, *40(12)*, 1383–1396.

71. Y. Lu, A. Pich, H. Adler, *Synth. Metals* **2003**, *135-136*, 37–38.

72. X.J. Xu, L.M. Gan, K.S. Siow, M.K. Wong, *J. Appl. Polym. Sci.* **2004**, *91*, 1360–1367.

73. I. Sapurina, J. Stejskal, M. ?pirkpva, J. Kotek, J. Proke?, *Synth. Metals* **2005**, *151*, 93–99.

74. L.Y. Haung, W.B. Hou, Z.P. Liu, Q.Y. Zhang, *Chin. Sci. Bull.* **2005**, *50(10)*, 971–975.

75. J. Jang, J.H. Oh, *Adv. Mater.* **2003**, *15(12)*, 977–980.

76. M.G. Han, S.H. Foulger, *Adv. Mater.* **2004**, *16(3)*, 231–234.

77. J. Fan, M. Wan, D. Zhu, B. Chang, Z. Pan, S. Xie, *J. Appl. Polym. Sci.* **1999**, *74*, 2605–2610.

78. Y. Tu, C. Ouyang, Y. Gao, Z. Si, W. Chen, Z. Wang, G. Xue, *J. Polym. Sci. Part A: Polym. Chem.* **2005**, *43*, 6105–6115.

79. H.J. Choi, Y.H. Lee, C.A. Kim, M.S. Jhon, *J. Mater. Sci. Lett.* **2000**, *19*, 533–535.

80. H. Li, E. Kumacheva, *Colloid. Polym. Sci.* **2003**, *281*, 1–9.

81. J.H. Sungi, Y.H. Lee, I.B. Jangi, H.J. Choi, M.S. Jhon, *Designed Monomers and Polymers* **2004**, *7*,101– 110.

82. J. Jang, J.H. Oh, *Adv. Funct. Mater.* **2005**, *15(3)*, 494–502.

83. X.Y. Shi, A.L. Briseno, R.J. Sanedrin, F.M. Zhou, *Macromolecules* **2003**, *36(11)*, 4093–4098.

84. J.S. Shin, J.H. Kim, I.W. Cheong, *Synth. Metals* **2005**, *151*, 246–255.

85. S.P. Armes, *Curr. Opin. Colloid Interface Sci.* **1996**, *1*, 214–221.

86. C.A. Ferreira, S.C. Doménech, P.C. Lacaze, *J. Appl. Electrochem.* **2001**, *31*, 49–56.

87. R. Partch, S.G. Gangolli, E. Matijevic, W. Ca?, S. Arjas, *J. Colloid Interface Sci.* **1991**, *144*, 27–35.

88. R. Gangopadhyay, A. De, S. Das, *J. Appl. Phys.* **2000**, *87*, 2363–2372.

89. S. Kuwabata, A. Kishimoto, T. Tanaka, H. Yoneyama, *J. Electrochem. Soc.* **1994**, *141*, 10–14.

90. C.L. Huang, E. Matijevic, *J. Mater. Res.* **1995**, *10(5)*, 1327–1344.

91. M.D. Butterworth, R. Corradi, J. Johal, S.F. Lascelles, S. Maeda, S.P. Armes, *J. Colloid Interface Sci.* **1995**, *174*, 510–517.

92. J. Stejskal, P. Kratochvíl, S. P. Armes, S. F. Lascelles, A. Riede, M. Helmstedt, J. Proke?, I. Krivka , *Macromolecules* **1996**, *29*, 6814–6819.

93. G.P. McCarthy, S.P. Armes, *Langmuir* **1997**, *13*, 3686–3692.

94. M.I. Goller, C. Barthet, G.P. McCarthy, R. Corradi, B.P. Newby, S.A. Wilson, S.P. Armes, S.Y. Luk, *Colloid Polym. Sci.* **1998**, *276*, 1010–1018.

95. M.G. Han, S.P. Armes, *J. Colloid Interface Sci.* **2003**, *262*, 418–427.

96. S. Maeda, S.P. Armes, *Synth. Metals* **1995**, *69*, 499–500.

97. C.L. Huang, R.E. Partch, E. Matijevic, *J. Colloid Interface Sci.* **1995**, *170*, 275–283.

98. L. Hao, C. Zhu, C. Chen, P. Kang, Y. Hu, W. Fan, Z. Chen, *Synth. Metals* **2003**, *139*, 391–396.

99. C.F. Lee, H.H. Tsai, L.Y. Wang, C.F. Chen, W.Y. Chiu, *J. Polym. Sci. Part A: Polym. Chem.* **2005**, *43*, 342–354.

100. L. Zhang, M. Wan, Y. Wei, *Synth. Metals* **2005**, *151*, 1–5.

101. M.G. Han, S.H. Foulger, *Chem. Commun.*, **2004**, *19*, 2154–2155.

102. F. Yan, G. Xue, J. Chen, Y. Lu. *Synth. Metals* **2001**, *123*, 17–20.

103. H.S. Kim, B.H. Sohn, W. Lee, J.K. Lee, S.J. Choi, S.J. Kwon, *Thin Solid Films* **2002**, *419*, 173–177.

104. K. Suri, S. Annapoorni, R.P. Tandon, N.C. Mehra, *Synth. Metals* **2002**, *126*, 137–142.

105. O. Jarjayes, P.H. Fries, G. Bidan, *Synth. Metals* **1995**, *69*, 343–344.

106. W. Chen, X. Li, G. Xue, Z. Wang, W. Zou, *Appl. Surf. Sci.* **2003**, *218*, 215–221.

107. J. Deng, Y. Peng, C. He, X. Long, P. Li, A.S.C. Chan, *Polym. Int.* **2003**, *52*, 1182–1187.

108. J. Deng, X. Ding, W. Zhang, Y. Peng, J. Wang, X. Long, P. Li, A.S.C. Chan, *Polymer* **2002**, *43*, 2179–2184.

109. J. Pfleger, M. Pavlik, N. Hebestreit, W. Plieth, *Macromol. Symp.* **2004**, *212*, 539–548.

110. J.W. Schultze, H. Karabulut, *Electrochimica* **2005**, *50*, 1739–1745.

111. A. Pron, P. Rannou, *Prog. Polym. Sci.* **2002**, *27*, 135–190.

112. M. Angelopoulos, *IBM J. Res. Dev.* **2001**, *45*, 57–75.

113. S. Kirchmeyer, L. Brassat, *Kunststoffe Plast. Europe* **2005**, *10*, 202–208.

114. Y. Yang, Q. Pei, *J. Appl. Phys.* **1995**, *77(9)*, 4807–4809.

115. Y. Yang, A.J. Hegger, *Appl. Phys. Lett.* **1994**, *64(10)*, 1245–1247.

116. S. Karg, J.C. Scoot, J.R. Salem, M. Angelopoulos, *Synth. Metals* **1996**, *80(2)*, 111–117.

117. G.G. Wallace, P.C. Dastoor, D.L. Officer, C.O. Too, *Chem. Innov.* **2000**, *30(1)*, 14–22.

118. C.J. Brabec, N.S. Sariciftci, J.C. Hummelen, *Adv. Funct. Mater.* **2001**, *11(1)*, 15–26.

119. J. Nelson, *Curr. Opin. Solid State Mater. Sci.* **2002**, *6*, 87–95.

120. K.M. Coakley, M.D. McGehee, *Chem. Mater.* **2004**, *16(23)*, 4533–4542.

121. H. Hoppe, N.S. Sariciftci, *J. Mater. Res.* **2004**, *19*, 1924–1945.

122. B. O'Regan, M. Grätzel, *Nature* **1991**, *353*, 737–740.

123. F. Cao, G. Oskam, P.C. Searson, *J. Phys. Chem.* **1995**, *99(47)*, 17071–17073.

124. U. Bach, D. Lupo, P. Comte, J.E. Moser, F. Weissortel, J. Salbeck, H. Spreitzer, M. Grätzel, *Nature* **1998**, *395*, 583–585.

125. S.X. Tan, J. Zhai, M.X. Wan, L. Jiang, D.B. Zhu, *Synth. Metals* **2003**, *137*, 1511–1512.

126. M.E. Rincon, R.A. Guirado-Lopez, J.G. Rodríguez, M.C. Arenas-Arrocena, *Solar Energy Materials & Solar Cells* **2005**, *87*, 33–47.

127. J. Kois, S. Bereznev, J. Raudoja, E. Mellikov, A. Öpik, *Solar Energy Materials & Solar Cells* **2005**, *87*, 657–665.

128. Y. Wang, X. Jing, *Polym. Adv. Technol.* **2005**, *(16)*, 344–351.

129. J. Reut, A. Öpik, K. Idla, *Synth. Metals* **1999**, *102*, 1392–1393.

130. P.J. Kinlen, D.C. Silverman, C.R. Jeffreys, *Synth. Metals* **1997**, *85*, 1327–1332.

131. A. Michalik, M. Rohwerder, *Z. Physikal. Chem.* **2005**, *219(11)*, 1547–1559.

132. M. Kendig, M. Hon, L. Warren, *Prog. Org. Coat.* **2003**, *47*, 183–189.

133. P.J. Kinlen, Y. Ding, D.C. Silverman, *Corrosion* **2002**, *58(6)*, 490–497.

134. P.J. Kinlen, V. Menon, Y. Ding, *J. Electrochem. Soc.* **1999**, *146(10)*, 3690–3695.

135. G. Paliwoda-Porebska, M. Stratmann, M. Rohwerder, K. Potje-Kamloth, Y. Lu, A.Z. Pich, H.-J. Adler, *Corrosion Sci.* **2005**, *47*, 3216–3233.

136. D.E. Tallman, G. Spinks, A. Dominis, G.G. Wallace, *J. Solid. State Electrochem.* **2002**, *6*, 73–84 and 85–100.

137. H.N. Thi Le, B. Garcia, C. Deslouis, Q.L. Xuan, *Electrochim. Acta* **2001**, *46*, 4259–4272.

138. R. Rajagopalan, J.O. Iroh, *Surf. Eng.* **2002**, *18(1)*, 59–63.

139. P. Zarras, N. Anderson, C. Webber, D.J. Irvin, J.A. Irvin, A. Guenthner, J.D. Stenger-Smith, *Radiat. Phys. Chem.* **2003**, *68*, 387–394.

140. A.B. Samui, S.W. Phadnic, *Prog. Org. Coat.* **2005**, *54*, 263–267.

141. U. Rammelt, P.T. Nguyen, W. Plieth, *Electrochim. Acta* **2003**, *48*, 1257–1262.

142. W.J. Hamer, Ph. D. Thesis: *Polypyrrole Electrochemistry; Environmental Friendly corrosion protection of steel, (im) possibilities*, **2005**, Delft University of Technology, The Netherlands, Chapter 6.

143. C. Celsing, M. Rohwerder, A. Michalik, C. Ostwaid, S.K. Ghosh, Final ECSC project report, Number: 7210-PR-259, **2004**.

144. Y.M. Abu, K. Aoki, *J. Electroanal. Chem.* **2005**, *583*, 133–139.

145. H.W. Heuer, R. Wehrmann, S. Kirchmeyer, *Adv. Funct. Mater.* **2002**, *12(2)*, 89–94.

146. S. Bousalem, C. Mangeney, M.M. Chehimi, T. Basinska, B. Miksa, S. Slomkowski, *Colloid. Polym. Sci.* **2004**, *282*, 1301–1307.

7
Smart Textiles Using Microencapsulation Technology

Marc Van Parys

7.1
Introduction

A spherical particle with a solid shell is commonly denominated as a microcapsule. Microencapsulation technology, which originated from the paper and the pharmaceutical industries, has now attracted the attention of a wide number of other industries, including that of textiles. Microencapsulation has great potential for improving processing ability over a wide range of applications, as well as product performance, particularly in terms of controlled release and enhanced stability against external factors such as light and oxidizing agents.

The textile industry has reacted slowly to the many possible opportunities that microencapsulation can offer. Until now, its exploitation in textiles has been limited owing to a lack of awareness across the industry, and the relatively high costs of the global processing. Other issues which have limited exploitation are related to the low durability of the effect, as conventional washing and other high-input thermal processes such as ironing and tumble-drying might cause dramatic reductions in the desired effects.

During the past few years a growing number of applications of microcapsules on textiles have emerged, although most of these have been restricted to fabrics utilized in clothing and interior textiles [1]. The huge potential of microcapsules has not yet been fully tapped, with only a handful of microencapsulated finishes currently in use. Indeed, microcapsules, when firmly anchored within the fabric or coating of a textile, can introduce new smart functionalities without affecting the look and feel of the textile.

In particular, the combination of microcapsules and coatings or laminates allows the introduction of new smart functionalities that often are not possible with any other existing technology. The combination of these technologies holds great promise for the textile industry in the future, not only for clothing and interior textiles but also in technically advanced textiles. The possibilities are virtually unlimited.

Functional Coatings. Edited by Swapan Kumar Ghosh
Copyright © 2006 WILEY-VCH Verlag GmbH & Co. KGaA, Weinheim
ISBN 3-527-31296-X

This chapter provides an overview of the general frame and the constraints of microcapsules in textiles, and exemplifies promising routes for smart (responsive) coatings, laminates and other end-uses.

7.2
Why Use Microencapsulation Technology?

Depending on the end use of the textile materials, encapsulation technology is used for one of the following purposes:

• Controlled release
• Protection
• Compatibility

7.2.1
Controlled Release

Controlled or targeted release is the most common application of microcapsules. Microencapsulation is an ideal tool to effect and achieve a desired delay of the active core until the correct stimulus is encountered. Most microcapsules are extraordinarily stable and show a great retardation of the encapsulated substances. Fragrances, aromas, drugs, enzymes, cosmetics and even pharmaceutical formulations (vitamins, drugs) are encapsulated for delivery from textiles at a specific time, rate, or situation. The encapsulated substances can either be released at once – for example, if pressure is applied to the capsule or the shell material is dissolved – or little by little – an effect which is especially beneficial when long-term effects are required, for example the administration of an anti-allergic drug.

7.2.2
Protection

Microcapsules are also used to store volatile or unstable substances on a temporary basis. The active agents in these capsules are usually not suitable for their direct use, due to a variety of reasons including low solubility, reactivity (may be too low or too high) and stability (low). On the other hand, it may be advantageous to optimize the properties of the active agents, for example by using sustained or controlled release functions.

Microencapsulation for protective purposes is equally important as the controlled release uses, as microcapsules may prevent the possible premature loss of an active ingredients, which is reflected in a decline of its activity.

Microencapsulation is an effective and longlasting method of protecting unstable or reactive products against external influences such as oxidation, alkalinity, acidity, heat, polluting gases, moisture and evaporation, and in this way it can be used to increase the shelf life of a product. Microencapsulation prevents unwanted interactions between the active ingredients and other components in the system or fibers, and thereby increases their washing durability. For example, the encapsulation of a fragrance involves surrounding the active molecules with a layer of material that prevents its release, as well as delaying its penetration into the environment until desired. The aim is to maintain the fragrance in perfect condition until it is liberated from the product and subsequently enjoyed by the customer. Microencapsulation may also be used in a masking role to prevent the release of unpleasant flavors or malodors during manufacture.

Soluble products can be rendered temporarily insoluble by encapsulation, and this in return improves their washing durability. In addition to single products, complex mixtures – where a combination of ingredients in exact ratios yields a specific functionality – can be added to other formulations without loss of functionality. Key examples of the latter approach are light- and temperature-sensitive dyes (Leucodye®) and flame-retardant formulations (e.g., diamide methylphosphonic acids in polyolefines). Microencapsulation is also involved in health issues during production processes such as textile coating, enabling both workers and end-users to be protected against exposure to hazardous or other toxic substances.

7.2.3
Compatibility

Finally, active products are permanently encapsulated in order to permit the mixing of incompatible products. Compatibility and protection are somewhat inter-related, and for a variety of applications both will apply. Classic examples are the phase-change materials (PCMs); these are permanently entrapped in their encapsulated forms, which means that the effect offered is not diminished throughout its period of use.

Other interesting compatibility features include:

- The conversion of liquids into powders in order to prevent clumping and improve mixing and compounding.
- The improved handling of active species before processing.
- Health issues with regard to protecting workers or end-users against exposure to hazardous or other toxic substances.

7.3
Preparation and Application of Microcapsules

Currently, many techniques are available for the preparation of microcapsules, though no single technique can be considered "ideal". In view of the diversity and size of the microencapsulation field, the relevance of some techniques is outlined in Table 7.1 [2].

Table 7.1 Relevance of some production techniques [1].

Technique	Advantages	Disadvantages
PAN coating	Low-cost equipment	Difficult to control, high skill level required
Spray-drying	Equipment and know-how widely available, versatile	High equipment and operating costs, high temperatures, uses organic solvents
Spray-cooling	Solvent-free	Low production volume
Spray-coating	Low cost, higher production volume	Difficult to control
Pressing/grinding	Easy to control	High temperature, wastage of material
Coacervation	Versatile	Organic solvents, aldehyde as hardener
Emulsification	Well established	Organic solvents
Extrusion	Suitable for bioencapsulation	High temperature
Interfacial poly condensation	Easy to control	Nonbiocompatible carrier materials, organic solvents
Liposomes	Allow product delivery across cell membranes	Expensive phospholipids, organic solvents
Cyclodextrins	Well established, easy preparation	Cyclodextrins linked to health concerns
Supercritical fluids	Replacement of organic solvents	Still under research

The choice of different microencapsulation technologies depends on many factors:

- The cost of the microcapsules, high-volume microencapsulation.
- The cost of processing into or onto the final product.
- The environmentally friendly aspect of the microencapsulation process.
- The functionality to be incorporated into the microcapsules.
- The compatibility of the microcapsules and the matrix (read: coating, binder) to be incorporated.
- The optimum concentration of the active ingredients in the capsules.
- The ease of application methods without damaging the capsules, thereby avoiding early release of the capsule contents.

- The release mechanism and deliverance profile at a specific time, rate or place of the encapsulated ingredient in function of the application, and the end-use of the textile material.
- The sustained profile required.
- The particle size, density and stability properties or requirements for the active ingredient.

Interesting preparation methods for microcapsules adapted for textile applications include phase separation (coacervation, *in-situ* polymerization, liposomes) and spinning (rotating) disc processes. A detailed discussion on microencapsulation techniques can be found in Chapter 1.

ThermaCell Technologies, Inc. [3] has introduced the new Essential Stones Vax-Cell microspheres, a new fragrance delivery product designed to reduce the risk of leaking oils discoloring the fabrics or textiles. VaxCells are evacuated microspheres of glass which, when mixed in coatings, form a barrier that prevents the exchange of temperature. As a byproduct, the embedded oil is held firmly in place in microscopic spheres of glass which, when fused together, form capillary-like vents that absorb oil. The oils are locked in place and then released over time in order to avoid the threat of leakage. As a result, the oils cannot discolor fabrics or textiles. The microspheres deliver long-lasting scent, for example for use in to air fresheners or in aromatherapy.

Microsphere®F (Matsumoto) [4] are thermoexpandable microspheres (Fig. 7.1), the particles having a wall of copolymers of vinylidene chloride, acrylonitrile. They are mixed with various resins and formed into a layer containing separate pores at low temperatures for a short time through the steps of coating or impregnation. The diameters of the particles range from 10 to 30 µm. Potential applications are thermoinsulation effects and slip-preventive functions.

If the release of the active agent should either be spontaneous or slowly over a very long time, the matrix-based capsules cannot fulfill the requirements of the application. For these cases, a core-shell-encapsulation is preferred. The suitable range of shell materials is very broad, since most gelling systems can be used. The

a)   b)

Figure 7.1 Unexpanded capsules (a) and expanded capsules (b).
Source: Matsumoto.

capsules are still mostly made from gelatin, but more recent advances in polymer technology have widened the scope of applications. Other shell materials used include hydrocolloids such as alginate and agar-agar, as well as gum arabicum, latices, polymethacrylates, polyether sulfones, and waxes.

A common shell material to contain the functional compounds for healing is based on urea-formaldehyde (UF) or melamine-formaldehyde (MF). Both, UF and MF microcapsule shells exhibit good stability against the functional compounds with which they are loaded. Furthermore, these microcapsules exhibit long-term stability, yet are easy to break and allow release of their constituents upon demand. In this case, although some trace levels of formaldehyde are present, the level is less than that found in the industrial guideline for cosmetics, or even oral hygiene products [5]. Microcapsules using MF systems containing oil-based ingredients are also available (e.g., Celessence; a TXT Capsule system [6]). In either UF or MF shells, trace levels of formaldehyde may be present, although these are mostly in nonharmful concentrations.

Interesting properties are obtained with chitosan-based microcapsules (Skintex® [7]; Cognis). Chitosan is a biodegradable, weak cationic polysaccharide which is obtained by extensive deacetylation of chitin, a polysaccharide common in the shells of shrimps. Chitosan exhibits excellent film-forming ability, and is an excellent polymer suitable for shells. The protective chitosan layer protects the contents from warmth, drying out, and cold, and this results in durability against the wear and tear of day-to-day life.

The core material needs to be liquid, which makes both hydrophilic (e.g., aqueous solutions) and also hydrophobic agents (e.g., oils) suitable for the process. For an even finer release tuning, the microcapsules can be processed after their production by coating, polymerization, or crosslinking in order to adjust the properties to the desired needs.

7.4
Commercially Available Microcapsules

Microcapsules can be broadly divided in two groups:
- Those with permanently embedded particles, such as PCMs.
- Those with temporarily embedded particles, such as controlled release particles.

7.4.1
Phase Change Materials (PCMs)

Substances that undergo the process of phase change are known as PCMs. The chemistry of these materials is based on patented technology developed by the National Aeronautical and Space Administration (NASA; space research program) during the 1970s for the US spaceflight sector, and is based on the use of selected paraffins. PCM material confinement by microencapsulation facilitates their incorporation into a wide variety of applications – including fibers, fabrics, coatings, floor coverings, and packaging materials – where their storage properties are put to good use. The thermal energy absorbed or released by the material when its changes state (e.g., from a liquid to a solid) without changing temperature is called "latent heat". Much research has been conducted – and is indeed ongoing – in an aim to exploit this energy. Many textiles, including technical textiles, are currently being developed and tested which contain PCMs and are capable of optimum heat absorption.

7.4.1.1 Mechanism of Action

PCMs have high heats of fusion and so are capable of absorbing and storing large quantities of energy from the environment. The heat is re-released at low temperatures, and this leads to an active thermoregulation and insulation which is adapted to physical activity and to the ambient temperature.

A PCM eliminates fluctuations in temperature, an effect which is useful for maintaining a subject at uniform temperature. The PCM also possesses the ability to change its physical state within a certain temperature range. When the melting temperature in a heating process is reached, the phase change from solid to liquid state occurs. During this physical transformation, the PCM absorbs and stores a relatively large amount of latent heat. When a PCM undergoes a phase change, a large amount of energy is required. In the case of waxes, typical transition temperatures range from 15 to 30 °C. When the wax melts, its molecular bonds are loosened through the absorption of thermal energy, but as the wax hardens thermal energy is released as its molecular bonds are reformed.

One important characteristic of latent heat is that it involves the transfer of much larger packets of energy than does sensible heat transfer. The effect of a PCM can be illustrated by a natural PCM material, namely water. For example, 1 kg of water requires ca. 4.2 kJ of sensible heat energy to reduce its temperature by 1 °C – that is, to reduce the temperature of water from its boiling point to 0 °C requires 420 kJ of energy. Then, to completely freeze the water at 0 °C a further 330 kJ of energy (known as latent heat energy) must be removed. This is why ice is such an effective cooling medium when the process is reversed, to turn ice back into water, when 1 kg of ice requires ca. 335 kJ of energy at 0 °C.

7.4.1.2 Characteristics of PCMs

In addition to ice (water), more than 500 natural and synthetic PCMs are known. These materials differ one from another in terms of their phase change temperature range and their heat-storage capacities.

PCMs can be broadly grouped into two categories: (i) organic compounds (e.g., waxes or polyethylene glycol); and (ii) salt-based products (e.g., Glauber's salt). Each group has both advantages and disadvantages. Some PCMs are mixtures of two chemicals which, when mixed in a particular ratio, form a "eutectic system". The details of some PCMs are summarized in Table 7.2.

A comparison between organic and salt-based PCMs is provided in Table 7.3.

Today, most popular PCMs used in textile applications are based on paraffins (waxes). Compared to other PCMs, paraffins have high heat-storage capacities.

Table 7.2 Available phase change materials (PCMs) and their characteristics.

Material	Melting point [°C]	Heat of fusion [kJ kg^{-1}]	Latent heat [MJ m^{-3}]
$MgCl_2 \cdot 6H_2O$	117	169	242
$Mg(NO_3)_2 \cdot 6H_2O$	89	163	252
$CH_3COONa \cdot 3H_2O$	58	226	287
$MgCl_2.6H_2O/Mg(NO_2)_2 \cdot 6H_2O$	58	132	201
$Na_2SO_4 \cdot 10H_2O$	32	251	335
$Na_2CO_3 \cdot 10H_2O$	32	233	340
Waxes	28 to 4	220 to 245	170 to 195
Polyethylene glycols	28 to −15	146 to 155	165 to 175
$CaCl_2 \cdot 6H_2O$	27	191	298
Water	0	335	335
Range of water/salt eutectics	0 to −64	Wider range	Wide range

Table 7.3 Characteristics of phase change materials (PCMs).

PCM	Advantages	Disadvantages
Organic	Simple to use Noncorrosive No supercooling No nucleating agent	Generally more expensive Lower latent heat/density Often quite broad melting range Can be flammable Lower heat transfer due to low thermal conductivity
Salt-based	Inexpensive Good latent heat/density Good heat transfer due to high thermal conductivity Well-defined phase temperatures Nonflammable	Need careful preparation Need additives to stabilize for long-term use Prone to supercooling Can be corrosive for some metals

Furthermore, paraffins can be mixed in order to realize desired temperature ranges in which a phase change takes place. These products are relatively inexpensive (often byproducts of petroleum refining), hydrophobic, nontoxic, and noncorrosive. In addition, their thermal behavior also remains stable under permanent use.

One disadvantage of paraffin-based PCMs is their low resistance to ignition, but the addition of fire retardants can overcome this problem. Furthermore, paraffins have, in general, a lower density than PCMs produced by TEAP (www.teappcm.com); consequently, paraffins have less thermal capacity per unit volume. Paraffins also have a lower thermal conductivity, both in solid and liquid phases, than inorganic salt-based PCMs, and this may present problems for designers in the heat-transfer process.

A number of different companies have introduced commercialized wax-based PCMs onto the market.

Outlast controls the production of PCM-based microcapsules for incorporation into textiles and other materials, and the company frequently licenses other companies to produce and sell the microcapsules. Commercial examples include the Lurapret® PCMs produced by BASF [8]. These are water-based products that create an optimum temperature balance in textiles. The new treatment offers a high degree of temperature regulation, resulting in a very consistent level of thermal comfort. The material ensures that the wearer feels equally comfortable in conditions of either extreme heat or cold. They function by absorbing excess body heat on physical exertion and returning the stored heat to the body when the temperature falls.

Thermasorb® is a registered trademark of Outlast technologies, and is an innovative thermal management system for advanced materials, used in a broad range of consumer and industrial products. It is sold as Thermosorb® powder additives, with microsized particles containing PCMs or component products for interlinings and casual winter gloves. The interlinings are both lightweight and breathable, and are easily incorporated into end products such as footwear and clothing. They also have automotive applications.

Ciba Specialty Chemicals has introduced Ciba Encapsulence® PC140 [9], a microencapsulated PCM. The confinement of PCMs by microencapsulation facilitates their incorporation into a wide variety of applications from fibers to coatings for automotive, clothing, buildings, and floor coverings where their energy-storage properties can be put to good use.

TEAP and PlusIce both market new salt-based PCMs. TEAP [10] introduced different hydrated salt-based PCMs with temperatures ranging from −32 °C to 89 °C. PlusIce [11] introduced mixtures of nontoxic eutectic solutions having freezing and melting points higher or lower than those of water, and offering thermal energy storage capability between −62 °C and 117 °C. These inorganic PCMs have opened new windows for a wide range of technical applications with insulation properties. Indeed, the only limit of how and where to apply these energy-saving and environmentally friendly solutions is the imagination of design engineers!

7.4.1.3 Optimizing PCM Finishing for Innovative Activity Textiles

The use of PCMS based on waxes in textiles has resulted in the following thermal benefits and interactive functions:

- Thermoregulatory effects, creating a microclimate resulting from either heat absorption or heat emission, keeping the temperature of the surroundings almost constant.
- An active barrier effect in the surrounding substrate, which regulates heat flux through the substrate and adapts heat flux to thermal needs.
- A thermoinsulation effect: a heating effect caused by heat emission of the PCM, and a cooling effect caused by heat absorption of the PCM.

The efficiency and duration of each of these effects are determined by:

- The particle size.
- The thermal capacity of the PCM.
- Phase change temperature.
- Structure and type of the carrier as well as the textile substrate.

The particle size of the PCM must be adapted to the application, as both the microencapsulation process and the size of the particles affect temperatures. The larger the particles, the closer the phase change temperature is to that for the bulk material. In contrast, the smaller the particles, the greater the difference between the melting and freezing temperatures of the PCM. The small size of the capsules provides a large specific surface area for heat transfer, and this results in an extremely high rate at which the PCM reacts to an external temperature change.

The particle sizes used range from 0.5 to 100 µm. For use in textile coatings, wax-based particles of 3 to 10 µm or smaller are common. The particles core constitutes ca. 80–85% of the particle's volume, while the outer shell is approximately 1 µm thick. Larger particles ranging from 10 to 100 µm can be incorporated into composites such as foams to provide higher thermal capacitance, as well as insulation.

The thermal capacity (latent heat) is, of course, a function of the type of PCM (see Table 7.4) for some medium chain-length alkanes (see also Table 7.2). The total thermal capacity of the PCM in certain products depends on the specific thermal capacity and the quantity used.

The amount of PCM must be tailored to the nature and duration of the physical activity in order to ensure a suitable and durable active thermal insulation effect of the PCM (e.g., in activity-related garments). The quantity needed can be estimated by taking the application conditions and textile substrate into considerations, the estimated duration of the application, and the thermal capacity of the specific PCM.

In order to achieve a successful application, the phase change temperature range and the application temperature range must also correspond. In garments, the actual temperature of the phase change should be 30–35 °C, close to the body temperature. By selecting a blend of PCMs with phase change temperatures above and below the skin temperature, it becomes possible to impart the same protection against discomfort caused by overheating.

Table 7.4 Characteristics of wax-based phase change materials (PCMs).

PCM	Crystallization point [°C]	Melting point [°C]	Heat of fusion [kJ kg^{-1}]
Licosane	30.6	36.1	226
Nonadecane	26.4	28.2	222
Octadecane	25.4	22.5	244
Heptadecane	21.5	18.5	218
Hexadecane	16.2	18.2	237

Heat of fusion (kJ kg^{-1}) = amount of energy required to melt 1 kg of material.

The result of a PCM is a tailor-made thermoregulatory effect, though this can only be achieved if the PCM is optimally positioned. For this, certain design principles must be taken into account.

First, the textile substrate construction influences the efficiency of the active thermal regulating and insulating effect of the PCMs. For example, thinner textiles with higher densities (more dense structures) readily support the cooling process. In contrast, the use of thicker and less-dense textile structures leads to a delayed and therefore more efficient heat release of the PCM. Furthermore, the phase change temperature range and the application temperature range should correspond in order to realize the desired thermal benefits.

Selecting a suitable substrate requires the consideration of whether the textile structure is able to carry the necessary quantity of PCM that will provide the required heat transference to and from the microcapsules. Further requirements of the textile substrate in garment applications include sufficient breathing ability, high flexibility, and mechanical stability. In addition, the clothing should be designed so that it supports – and does not hinder – the effect of the PCMs.

7.4.2
Controlled Delivery Microcapsules

7.4.2.1 Range of Microcapsules
Companies involved currently in microencapsulation technology are listed in Table 7.5, the large majority of which are active outside the textile industry.

Much effort has been made to adapt microcapsules to textiles, with microcapsules containing perfumes, cosmetic products (moisturizers, fresheners, toners, etc.), bactericides, acaricides or even a combination of ingredients (e.g., a perfume

Table 7.5 Companies currently active in microencapsulation technology.

3M	Arcade Marketing	Balchem Encapsulates
BASF	Bayer	Bio Dar
Brace	Brookstone Chemicals	Capsulis
CavisCelessence/Devan	Cerexagri	Ciba Specialty Chemicals
Cognis	Coletica	Euracli
Drückfarben	Frisby Technologies	Follmann
Glatt	Hallcrest	Haarman and Reimer
Kappa-Biotech	Karmat	Kobo
Kwizda	Lallemand	Lipo Technologies
Lipotec	LJ Specialities	Makhteshim-Agan
Mane	Microtek	OmniTechnik
Orlandi	Outlast Technologies	Particle Dynamic
Raps	Ronald T. Dodge	Comp.Rotta
Sederma	Sekisui	Sipcam South West Research Inst.
Syngenta	Tagra Biotechnologie	TasteTech
TEAP	Thies Technology	Vertis
Watson Foods		

and bactericide). An overview of microencapsulated products currently available commercially for different purposes is listed in Table 7.6.

Most commercially available microcapsules are sized between one and several hundred micrometers (typically, microcapsules are 20–50 µm in diameter) and a concentration of 20 to 45% (by weight).

7.4.2.2 Release Delivery Mechanisms
The ingredients are released by several mechanisms:

- mechanical stimulus
- chemical stimulus
- thermal stimulus

Table 7.6 Examples of microencapsulated products.

Fragances Different fragances	apple, alpinair, banana, bergamot, blackcurrant, bubblegum, caramel, chocolate, cinnamon, citrus, coconut, eucalyptus, floral, grass, jasmine, lemon, lavender, lilac, lime mint, orange, peppermint, pine forest, toffee, rose, cola, pizza, freesia, leather, teatree, vanilla
...	
Aromatherapiy blends Antihistamine	a blend of fragances which may help relieve the symptoms of hay fever
Comforting	Lavender, Pine & Geranium
Refreshing	Bergamot, Peppermint & Coriander
Relaxing	Lavender, Sandalwood & Ylang Ylang
Rivitalizing	Eucalyptus, Mandarin & Peppermint
Reviving	Lemon, Rosemary & Bergamot
Sensuous	Patchouli, Sandalwood & Ylang Ylang
Soothing	Lavender, Chamomile & Rose
Tonic	Juniper, Bergamot & Lime
Moisturizers and Benefits Activera	*Aloe Barbadensis* extract in light mineral oil
Fragant aloe	as above but incorporating aloe vera fragance oil
Vitamin E *Moisturizer*	Wheat oil in light mineral oil
Bladderwrack/kelp	Seaweed extract in light mineral oil
Dermo-enzymatic deodorant Cosmacol Eli	Moisturizer emollient and enzymatic deodorant efficient alternative of Triclosan safe inhibitor of the body odor
Insect repellents	permethrin
Thermoregulating systems	Phase Change Materials (PCM)

Source: Devan – Celessence International LTD, TXT™ – capsules.

A schematic overview of the different release mechanisms of the embedded ingredients is provided in Figures 7.2 and 7.3. First, the ingredients can be delivered by breakage of the shell under mechanical stress (e.g., friction created by wearing clothes or walking over a carpet) (Fig. 7.2). Depending on the distribution of microcapsules in the material, all of the encapsulated content will be released at once or upon different stress loads until they are exhausted. Generally, microcapsules can control the release rate for the core by governing pressure and temperature. If the temperature rises, the molecule's momentum of core in the inside will increase to break the wall. Even though the pressure and temperature are not changed, the core will be steadily released in time. The radiation character is dependent upon the core class, the wall thickness, and the nature of the high molecular-weight compound from which the wall is constructed.

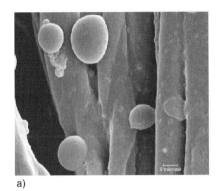 

a)   b)

Figure 7.2 Microcapsules on textile (a) before and (b) after rubbing.

A second possibility is release after degradation of the shell; this can be caused by dissolution as a result of either chemical or enzymatic breakdown of the shell. In this way, the release is delayed until certain environmental conditions occur. When the environmental conditions are set, the time of release can be tailored by varying the thickness and composition of the shell.

The ingredients in chitosan-based microcapsules are released through two mechanisms. First, if the microcapsule is pressed, crushed or rubbed, the core explodes and emerges. Second, the chitosan layer is slowly reduced over time through the action of the body's own enzymes; this activates the ingredients and enables them to move from the fabric into the skin, thereby refining their structure (Fig. 7.3a).

Other mechanisms involve fracture of the shell upon swelling of the core (Fig. 7.3b). In these case the shell must allow the diffusion of solvent into the core. Encapsulated dried products will swell when they come into contact with a suitable solvent. Encapsulated products that decompose can also swell as a result of changes in osmotic pressure within the capsules. Controlling the time of release is rather complex in these cases, as not only the swelling capacity of the core but also the permeability for the solvent and the mechanical properties of the shell determine the

Release mechanism 1

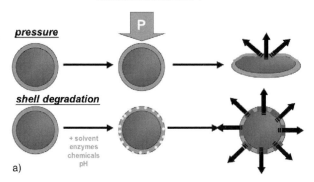

a)

Release mechanism 2

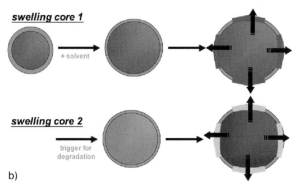

b)

Figure 7.3 (a) Release mechanism 1 of microcapsules.
(b) Release mechanism 2 of microcapsules.

release. The latter effect demands a high-quality uniform coating, because when the shell is stretched it will break at its weakest point

In general, the rate of release of microcapsules depends largely on the structure of the polymer shell, which in turn is influenced by the conditions employed in the preparation. Important shell characteristics determining the release rate are crystallinity, crosslink density, and porosity. As the crystallinity and crosslink density of the wall is increased, the release rate reduces substantially. Another important factor is the outside environment; if it is of the same type as that of the core material, the rate of release will be high. The rate of release can be normally be expressed as a first-order rate process, (i.e., $-dc/dt = kc$), where k is the diffusion constant and c is the concentration gradient [12].

7.5
Application Procedures of Microcapsules on Textile Substrates

Microcapsules can be applied on any fabric (woven, knitted, nonwoven or garments). The substrate may be wool, silk, cotton, flax, or synthetic fibers such as polyamide or polyester, or mixtures. Most of the microcapsules can be applied using conventional finishing techniques without altering the feel and color of colored fabrics or printed patterns. Microcapsules can also be applied during the rinse cycle of a washing machine.

Not only efficacy but also durability should be of prime concern. The major challenge is the durability of the microcapsules in relation to their effectiveness. In some end uses the microcapsules must survive at least 20 wash cycles, conventional ironing and other high-temperature processes such as tumble-drying without affecting the controlled release properties of the microcapsules.

During washing, the chemical action of alkali products, the mechanical action, and the temperature can each alter the microcapsules. Nevertheless, improvements in microcapsule lifetime can be achieved by using selected specific binder or coating systems for each type of fabric. Wash durability must be varied for each application. Euracli [13] claims that on cotton or polyamide, a remainder of 30% of the initial payload of active agent is achievable, even after ten washings. As the result of a project conducted with the research institute TO_2C (linked to the Technical University Gent, Belgium), the Belgian company Devan [14] has optimized the binder and coating systems, and this has resulted in a wash durability of up to more than 30 extensive wash cycles. To date, the company has applied this technology to clothing, socks, curtains, mattress ticking, cushions, and upholstery, as well as to technical textiles.

The best way to obtain most benefit from textiles treated with microcapsules is, of course, to wash them by hand. This minimizes the loss of actives during washing and maximizes their effect on the skin when the item is worn.

Finally, it must be borne in mind that in the case of controlled release capsules the effect is still limited – when all the capsules are empty they cannot be refilled.

7.5.1
Conventional Application Techniques

For particle fixation, the textile material is exposed to a solution, dispersion or emulsion of the textile-reactive particles by conventional methods such as padding, dipping, spraying, soaking, or screen printing. For all of these methods, a binder is required. This may be acrylic, polyurethane, or silicone, its role being to fix the microcapsules onto the fabric and to hold them in place during washing. A catalyst may be present in the medium. Actual fixation takes place during the curing stage, after the drying stage. The curing of crosslinking with textiles takes place under conditions of precise temperature and speed in a stenter facility. The procedure for conventional padding consists of several steps (Fig. 7.4).

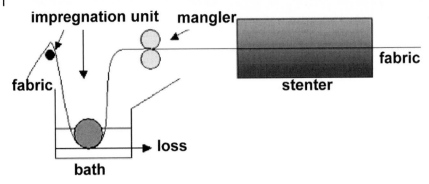

Figure 7.4 Schematic illustration of a padding process.

The use of binders has some drawbacks, however. The amount used should be high enough for a good fixation, but should not disturb the controlled release. A microcapsule covered with a binder may be harder to break, or the binder layer may prevent the release from the fabric after the microcapsule has been broken. Binders may also mask the surface properties of the fibers in the fabric. In order to increase and to ameliorate the use of microcapsules as controlled delivery vehicles for textiles, microcapsules with reactive shells that bond covalently to the fibers have been developed (e.g., reactive groups used in reactive dyestuffs). Such choice is not unlimited because the microcapsules should be able to withstand the conditions necessary for the binding reaction to occur.

In some cases (e.g., chitosan and gelatin capsules), ionic interaction can create ample binding during bath (exhaustion) treatment. Chitosan is a biocompatible, biodegradable, and nontoxic polysaccharide which, as a result of its cationic character, is able to react with polyanions and give rise to polyelectrolyte complexes. For this reason, chitosan has been adopted for use in bath treatment processes. Because of these interesting properties, chitosan has become the subject of numerous scientific reports.

7.5.2
Microcapsules in Coated and Laminated Textiles

Coating and laminating are very exciting technologies with a myriad of possibilities [15]. The aim of coating or laminating a textile substrate with a polymer layer is to significantly influence its external characteristics and physical properties. Textiles coated with, or laminated, bonded with a continuous polymer layer acquire new properties that cannot be achieved by the base fabric alone.

It is desirable from many aspects to examine those coatings in coated and laminated structures which are able to contribute to properties other than strength and barrier. The polymeric components in these structures could be provided with smart functions such as shape-memory and phase-change effects. During the past few years, some interesting developments have taken place with regard to smart materials, where suitable coating and laminating techniques have achieved a value-added function. Among these developments are photoluminescent materials, shape-memory polymers, phase-change materials, and light-protective sheets.

These coated fabrics can be classified roughly into two main groups:

- A group including all of those materials that still exhibit their essentially textile characteristics after being combined with the coating – that is, the textile dominates the quality of the product obtained.
- A group in which the textile functions simply as a supporting material or carrier and serves merely to reinforce and consolidate the polymer layer – there is no clear external indication of its existence. In this case the polymer largely characterizes the end product.

A coated or laminated material consists of two main components: (i) the textile itself; and (ii) the coating or laminate. The choice of both items has important consequences affecting the application procedures to be used and the final properties of the product. Both components must be carefully attuned to one another in combination with the coating or laminating technology.

At the level of the compound (basic materials and additives), new or modified polymers and additives have recently been marketed. Although, in the past, few investigations have been conducted on the use of microcapsules in coating and laminating, these technologies are well suited for incorporation with these microcapsules. Consequently, in today's world efforts are very much speeded up at the research and development stage.

Microcapsules are well suited for incorporation in coating and laminating. The integration of encapsulation technology in coating is relatively new, and offers an entirely unique range of marketing opportunities and product enhancements.

The main advantage over direct application onto the textile surface is the far larger quantity of microcapsules that can be added without altering the properties of the coated textile. This is especially important for PCMs or encapsulated flame-retardants as their quantity is strongly related to their performance.

Different coating techniques are possible, although it is important that the microcapsules remain intact during application, thus avoiding premature release of the active ingredients. Engineering of the coating at the level of technology and polymer is important in order to obtain an optimal effect with respect to product durability and release rate. It is important to develop a release rate that can be tailored, without having to use more sophisticated microcapsules, and in this respect stimuli-sensitive coatings have become increasingly utilized in the recent past.

7.5.2.1 Conventional Coatings

Currently, several specially designed machines and techniques are available for applying a coating to a textile substrate. In general, the particular type chosen is governed mainly by the consistency of the coating medium, which may be in the form of a liquid, paste, melt, powder, granulate, or foam. Another factor which deter-

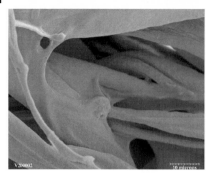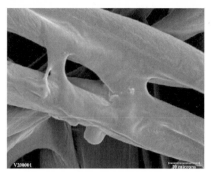

Figure 7.5 Microcapsules in a thin coating layer.

mines the choice of spreading device is, to some extent, the ultimate coating effect desired, the thickness of the coating, and the structure and mechanical behavior of the textile material (e.g., yarn, tapes, nonwoven, woven or knitted fabrics, open or closed structure) employed. The embedding of microcapsules in a coating is illustrated in Figure 7.5.

The choice depends not only on the textile substrate and the polymer formulation, but also on economic and ecological factors, and the suppleness and versatility of the chemical finish in relation to changing substrates and preparations. In this regard, the best system is the one that yields the desired product, manufactured by the most economical and ecological method.

All common coating techniques can be used for PCMs or controlled delivery capsules, on the condition that the system is able to apply a coat mass (dry weight) ranging from 50 to 350 g m^{-2} (Figs. 7.5–7.7). For PCMs, values of 140 to 210 g m^{-2} are preferable. These quantities would be easily applicable in clothing, furnishings, and floor coverings.

Finally, when incorporating microcapsules into coatings it is extremely important that, during the coating process, the microcapsules remain intact. Microcap-

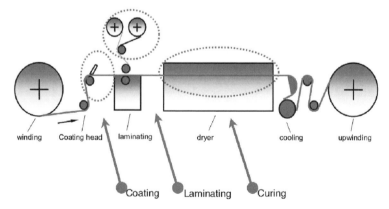

Figure 7.6 Schematic representation of a coating process.

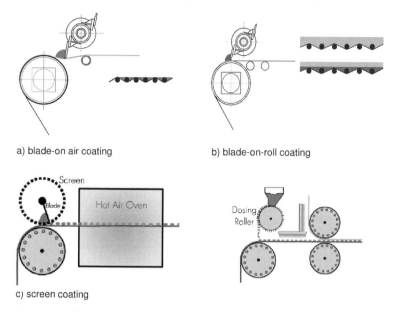

a) blade-on air coating   b) blade-on-roll coating

c) screen coating

Figure 7.7 Different conventional coating techniques.

sules are applied in coatings either by mixing them in the coating formulation or by adding them as powder after application of the coating. Once coated, the treated fabric is then dried and cured (as in a conventional scheme) to achieve an efficient and durable effect.

Depending on the method used, the viscosity of the coating formulation and the compatibility of the coating and the shell material, the microcapsules will either be fully entrapped in the coating or will remain partly at the surface. When fully enclosed, the release will be regulated not only by the strength of the shell but also by the mechanical properties of the coating and the diffusion barrier for the released product formed by the coating. In this way, the engineering of the coating can also regulate (slow down and average) the release rate. In this way, a more tailorable release can be obtained without having to use more sophisticated microcapsules. With this method similar applications to those used on uncoated fabrics can be considered. When the microcapsules are only partially embedded they can suddenly release their content by a mechanical stress, or they may slowly degrade under the influence of the environment into which they are brought into contact. Sudden release applications may include lubricants or reactive products for gradual release (Fig. 7.8).

In applications involving square-edged blades with a sharp profile, the greatest amount of shear is exerted on the coating, and this often results in streaking effects, especially when using high-viscosity coatings or coatings that do not flow readily under shear. These forces often result in a premature release of the active ingredients. A slight radius at the entrance of the blade or a slight draft angle to the flat land portion of the tip can solve this problem, however.

Microcapsules in coatings

- **inclusion**
  - FR, PCM, pigments
  - damping characteristics
- **controlled release**
  - slow release (fragrance, druges, vitamins,..;)
  - trigger controlled
- **sudden release**

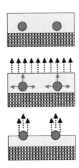

Figure 7.8 Schematic representation of the embedment of capsules in coatings and the associated release mechanisms.

In screen printing or coating (wet-on-wet process) the microcapsules are mixed with the coating. If possible, when using different screens (multilayer coating) it is recommended that the capsule coating must be the last pass under a screen in order to avoid damage to the shell by further screens.

All common coating techniques can be used for PCMs on the condition that the system is able to apply a coat mass (dry weight) ranging from 70 to 340 g m^{-2}, although values of 140 to 210 g m^{-2} are preferable. These quantities would be easily be applicable in clothing, furnishing and floor coverings. The application of PCMs by foam coating (open cell types) is particularly recommended for three reasons:

- Because a greater number of PCMs can be introduced.
- A large amount of PCMs can be used, thereby making available a wider range of regulation temperatures.
- The possibility of anisotropic insertion of the capsules into the foam can reinforce the regulator effect by concentrating PCMs towards the interior of the clothing. This provides excellent moisture management as well as thermal control.

Finally, using microcapsules, foam coating can be prepared without foaming the coating formulation; instead, foaming agents (organic or inorganic such as sodium bicarbonate) can be used. The coating will foam due to the release of carbon dioxide formed by the degradation of the bicarbonate during the thermal curing process.

7.5.2.2 Laminating

In laminating processes microcapsules can be added to the glue formulation under the condition that the microcapsules can withstand the lamination conditions (Fig. 7.9). The glue formulation itself can also be encapsulated to preserve its reactivity until the actual laminating. A straightforward example is separate encapsulation of the components of a two-component glue. The two components can be added in a single step and are mixed after breakage of microcapsules during lamination.

In Lamination

- Apply as a powder or suspension
    - two-component glues
    - non-compatible additives
    - slow release formulations

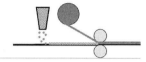

Figure 7.9 Application of microcapsules in lamination.

7.5.2.3 Miscellaneous Applications

Reactive atmospheric plasma technology

Current research into reactive atmospheric plasma techniques for the nanocoating of textile substrates includes a new atmospheric CoatingStar [16] technology for the application of microcapsules [17] in aerosol form. Atmospheric plasma treatments represent an alternative to the (post) treatment of textiles. A schematic representation of the reactive plasma system is illustrated in Figure 7.10.

This new technology combines conventional corona treatment with simultaneous nano-finishing, such that numerous effects can be obtained. The conventional corona treatment modifies and etches the surface of various (hydrophobic) polymers, resulting in:

- improvement of the adhesion (chemical, mechanical)/lifetime of finishes and product take-up;
- a modification of surface tension; and
- hydrophilic enhancement, improving wettability.

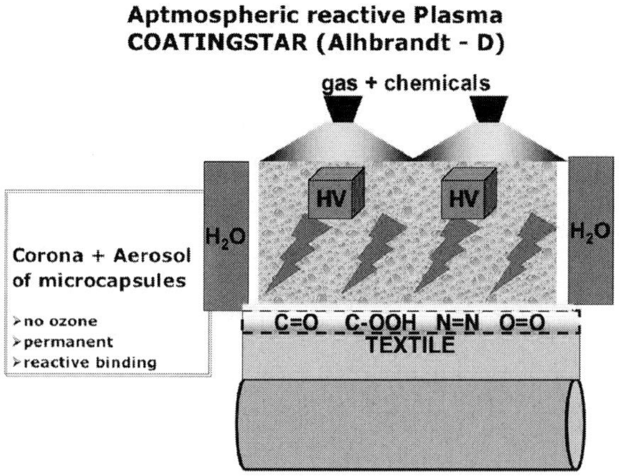

Aptmospheric reactive Plasma
COATINGSTAR (Alhbrandt - D)

gas + chemicals

HV   HV

H_2O H_2O

Corona + Aerosol
of microcapsules

➤ no ozone
➤ permanent
➤ reactive binding

$C=O$ $C-OOH$ $N=N$ $O=O$
TEXTILE

Figure 7.10 Schematic representation of the CoatingStar system.

Customization of the chemicals and gases allows a very wide variation of the effects. One promising possibility is to use an AS Coating Star system, which allows numerous combinations of surface treatments with corona, gases, and aerosols.

In using this new approach, conventional corona treatment is combined with injection in a reactive medium of microcapsules in aerosol form. Chemical nano-treatment or coating occurs when the aerosol condenses on the substrate at low temperature, resulting in a durable effect (most likely due to the formation of covalent bonds). An improved antistatic behavior is also obtained on textile substrates. Application is possible on both natural and synthetic fibers, as well as on thermosensitive hydrophobic substrates such as polypropylene and polyethylene.

It must be noted also that this technology is nonpolluting, since the air-cooling of a conventional corona is replaced by a temperature control system with water. When temperatures exceed 40 °C, the ozone molecules produced immediately disintegrate into oxygen, thereby avoiding ozone emission.

Digital coating or printing

Digital coating or printing opens new opportunities for "smart" applications of microcapsules on textile materials. The process, using valve jet technology, allows the application of microcapsules in combination with a suitable coating or printing system. Microcapsules with diameters <10 μm are simply mixed with water-based, solvent-free inks, print pastes or coatings and applied to the fabrics. Compared to the uniform application of microcapsules obtained with other technologies, the digital technology combined with CAD-CAM systems shows great promise for the "smart" application of microcapsules at well-located, predestined locations. This segmented application results in an efficient use without high additional costs for the use of relatively expensive microcapsules. Such a combination of digital and microencapsulation technologies offers allows designers to create customized products with unique competitive advantages. For example, it might be used to develop creative or educative textile wall-coverings, as flowers or fruit could be printed with their associated fragrances. Rubbing on a particular picture would simultaneously release the smell of the object.

Conventional digital printing heads based on drop-on-demand (bubble jet or piezo-technology) and continuous inkjet systems (due to clogging of the nozzles) are not suitable for this process unless the particle size of the microcapsules is <1 μm.

7.6
Smart End-Uses of Microencapsulated Textile Substrates

From sheets and towels that smell fresh every day to disinfectant wipes and garments with insect repellent, microencapsulation has truly "taken off" in textile finishing. However, microencapsulation technology remains complex and relatively expensive, and is not considered to be a routine finish or a substitute for other commonly used techniques that can be accomplished by conventional means.

Microencapsulation can be very expensive, with costs of up to € 60 kg^{-1} product on a 200-kg scale batch process. Although the price of these capsules is relatively high, adding extra cost to a textile substrate, the added value of the product is often easily recovered, with prices generally several-fold higher than for equivalent non-treated textiles.

The process of microencapsulation has innumerable current and potential applications in textiles, providing extra added value and extra functionality. However, its applications have proliferated greatly, and today the technology is used on textiles to impart smart finishes and desirable properties that are simply not possible using other technologies, or are not cost-effective. In applications where delayed action is essential, this technique must be strongly considered. A typical example would be in medicine, where the slow release of a drug from a textile product would ensure prolonged protection against bacteria. It may also be useful when direct contact of the skin with active ingredients is undesirable, for example with antimicrobials and insecticides.

Although this technology provides unheard of longevity for such treatments through its active-release mechanisms, much additional research must be conducted in order to fully exploit its great benefits.

7.6.1
Smart Use of PCMs

PCM microcapsules can be applied to a wide variety of textile substrates for thermoregulatory and insulation purposes. PCMs have long been used in acrylic fibers in a wet spinning process (Accordis) [18], and are also incorporated into foams or embedded in a coating compound and used in many types of fabrics (nonwoven, knitted or woven). PCMs act as the "engine" behind the thermoregulation in many textile products such as clothing, interlinings, absorbent pads, or technical textiles. In clothing, the use of PCMs can be divided into two general areas of active coolant and passive thermal storage. In coatings or foams, their presence can enhance the effective thermal capacitance of the materials by up to ten-fold.

7.6.1.1   PCMs in Protective Clothing and Attributes
Clothes have always fulfilled a variety of functions, be it fashion, warmth, protection, or support. PCM-based microcapsules add extra value to the clothes and also tolerate several washes. As mentioned above, PCMs act as a thermostat, releasing stored heat and keeping the wearer warm and comfortable outdoors (Fig. 7.11). They are used in all types of clothing, including shirts and tops, shorts, underwear, socks, outerwear, jackets, rainwear, gloves, caps/hats, scarves, pants, and backpacks, although the main merits are in sport and protective clothing.

Other applications of PCM are in cold storage transport for maintaining product temperature under control.

In the patented ComforTemp nonwovens [19] polyurethane-based foams, the incorporated PCMs surround the wearer with dynamic climate control (DCC). When the wearer is active – for example, during skiing, snowboarding or mountaineer-

New Additives & Formulations

Phase Change Materials (PCMs)

- **Revolution in insulation - Thermoregulating**
- **All common coating application techniques can be used on the condition that the system is able to apply a coat weight ranging from 70-340 g/m²**

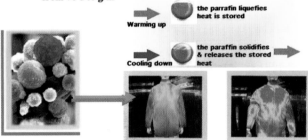

Warming up → the parrafin liquefies heat is stored

Cooling down → the paraffin solidifies & releases the stored heat →

Figure 7.11 Phase change material (PCM) microcapsules.

ing – the foam is able to absorb excess body heat and to wick away moisture, thus preventing the body from overheating. In this way it provides excellent moisture management as well as thermal control.

Another example of a comfortable garment is made from microencapsulated PCMs, packed into a space between two fabric layers in order to produce a garment with passive, regenerative thermal protection. The garment offers improved comfort due to not only enhanced thermal control but also by enabling the transport of humidity through the PCM layer and away from the body.

PCM-coated materials are often used in diving suits for military, recreational, and commercial purposes. The material allows divers to stay for longer periods in frigid waters than ever before. The PCM technology is also used in ski parkas, with the microencapsulated paraffin beads being incorporated into a polyurethane coating. As the skier descends a slope, body heat is generated and absorbed by the PCM. On the lift up the mountain the body is cooling, but heat given off by the PCM keeps the wearer at a comfortable temperature. Other applications in cold climates include hunting clothing and boots, ski and snowboard helmets, fishermen's waders, shooting, working and grain leather gloves, ear-warmers and other protective equipment (helmets, knee and elbow pads).

At higher temperatures PCMs could also provide cooling to those working in a hot environment, though in this case the material would need to be re-cooled after the phase change has occurred.

Finally, PCMs [20] are becoming increasingly popular in improving comfort properties in barrier textiles such as protective or medical textiles against particles and liquids (e.g., blood) that can carry infection. Microencapsulated fibers or coatings for surgeons' gowns contribute to their comfort in use, and the effect is lon-glasting. PCM-finished underwear and nightgowns can also assist in the healing process of patients.

7.6.1.2 Miscellaneous and Technical Applications

Previously, PCM technology has been used for home textiles such as duvets, pillowcases, sleeping bags, curtains, and furniture, while other applications of PCM finished textiles include horse blankets, dog coats/beds, body armor, and helmets.

Microclimate control in hospital bedding using microencapsulated mattresses and blankets may aid healing, as the technology may assist the thermoregulation of patients recovering in intensive care units or undergoing surgery.

The PCM concept has also been applied successfully to coated curtains and furniture, contributing to a thermal comfort improvement. Windows are the main cause of large thermal fluctuations in buildings, as a result of the low thermal insulation properties of glass. During the day, a large amount of sunlight (heat) penetrates the building, but at night – and especially during winter nights – the windows are the main source of heat loss. Blinds or curtains with PCM treatment could solve this problem. For example, a sandwich structure of a two-sided laminate to thin foam treated with PCM could meet these needs, as the PCM incorporated in the foam would regulate heat flux through the different layers by either absorbing or releasing energy. In other words, PCMs offer thermal regulation rather than insulation.

Other applications can be found in the thermal comfort of seating furniture. The heat flux of the body through a seat into the environment is essentially reduced, leading to a rapid temperature increase in the microclimate. A reduction in moisture transfer from the body through the seat, resulting in a substantial increase in microclimate moisture, would also be apparent.

In architectural membrane applications, the PCM must be contained in order to prevent its dissolution while in a liquid state. A newly developed membrane material in which PCM is incorporated into a silicone rubber matrix and applied topically to a basic fabric has been used for architectural membranes, and offers improved thermal performance capabilities previously unattainable in such a fabric [21]. These abilities would lead to a substantial improvement in the thermal management of buildings with membrane enclosures, reducing the air-conditioning and heating demands of the building, and making it more energy efficient. In time, the presently used multi-layer systems might be reduced to an one-layer system, with substantial materials savings. The reduced temperature fluctuations to which the membrane material is exposed might also positively influence its aging behaviour, leading to a longer service life. A unique feature of the newly developed membrane material is the change in light transmission as a result of temperature changes; this might be especially beneficial in architectural applications.

The thermal effects provided by the use of PCMs in membrane structures are durable. In addition, the thermal solution is maintenance free, cost-effective, and does not require an external energy supply.

PCM microcapsules also play an important role in exploiting latent heat in building construction. Due to a large surface area being formed by the numerous microcapsules, heat from the immediate environment is rapidly absorbed by the PCMs. If the room temperature exceeds 22 °C, the paraffin inside the wall begins to melt, absorbing heat from the room's environment. At night, however, when the wall cools the material will harden and become ready again for heat absorption the next day.

Another application is to include microencapsulated polymer mixtures with various glass transition temperatures. The incorporation of such material into a coating leads to a change in vibrational damping properties; this is important for coated textiles used, for example, as transportation belts in industrial machinery.

7.6.2
Controlled Delivery of Active Agents

The future applications of controlled delivery from textiles are virtually endless, and will not be restricted to fragrances. Recently, microcapsules have been applied to many functional and technical textiles. Examples of this include healing agents against hay-fever, antibiotics, aromatic deodorants, and sensitive and ultraviolet-protecting textiles. Sanitary products are most likely to be used on clothing, with their activity being prolonged by controlled release. In agriculture, herbicides, pesticides and fertilizers can each be applied to technical textiles; similar applications are possible for geotextiles. Clearly, endless possibilities exist for controlled delivery systems, and even more complex release mechanisms will undoubtedly be developed in future. Some exciting, new applications are discussed in the following sections.

7.6.2.1 Well-Being: Aromatherapy

Today, the subject of well-being is an area that is receiving much interest, with scent being one of the most important aspects of personal care. With new and unique fragrances being integrated into products, and consumers making purchases based not only on their favorite scents but also on the appeasement of other senses, it is important to stay at the cutting edge of personal care delivery systems such as microencapsulation for controlled release.

Energizing aromatherapy
Microcapsules can also be filled with high-grade ingredients that slowly release a revitalizing aroma. Many ingredients are based on organic plants and fruits such as menthol, orange, ginger, or rosemary. This application is especially suitable for bathrobes and underwear. Even after a number of washes, Skintex® microcapsules remain active.

Relaxing aromatherapy
Fragrance encapsulation can be used simply for visual appeal, for controlled release of a fragrance, or even as protection system for the fragrance, allowing for innovation in a wide range of personal care products. Fragrance-embedded active agents

with defined controlled release curves and an enhanced stability of the embedded chemicals are produced for many textile applications. The microcapsules are specifically tailored to impart a clean and fresh feeling, which then translates to a "feel-good" factor. Different types of fragrance can be provided (see Table 7.7): fragrances designed to be deployed via SPT™ microencapsulation [22] can be developed with far more quality top notes than is usual. This is because the microcapsules effectively provide a reservoir of fresh perfume, which reduces the need to use large quantities of less-volatile bottom notes. Normal-fragrance volatiles are dissipated in only a few minutes (which is why perfume bottles are always airtight, and why fragrances cannot normally be persuaded to stay for any length of time), and lose their initial freshness after this short time. Microencapsulation can prevent premature delivery of the fragrances.

Clothes containing fragrances are able to release a pleasant and relaxing scent, which is ideal for calming both body and soul. Based on microcapsule technology, Skintex® (Cognis) slowly releases natural-based aromas known to relax, such as valerian, amber tree resin oil, or lavender. A protective layer of chitosan protects the capsule; this saves the contents from being dried out by body warmth, and provides it with durability against the wear and tear of day-to-day life. At the same time, the chitosan helps to protect the skin from dehydration and retains a supple and velvety soft feeling.

Cognis claims that the effect of Skintex® persists over several washes. The active components of Skintex® microcapsules are tested dermatologically and fulfill the criteria for certification under the Ecotex standard 100. When combined in sophisticated formulas, these oils form high-grade active ingredients that are anchored into the textile

Finally, the manufacturers of these microcapsules claim that aromatherapy effects also help people to overcome insomnia.

7.6.2.2 Body Odor: Deodorant Effects

When a person sweats, their body chemicals mix with bacteria that produce enzymes which in turn cause the formation of body odor. Intense physical activity, together with the use of synthetic fabrics and foods, increases perspiration which leads to the generation of organic acids. In order to keep this phenomenon under control, cosmetic research has always taken into consideration either microbial inhibition or the reduction of perspiration, but both approaches interfere with skin physiology. *Antimicrobial deodorants* function directly on skin bacterial flow, but have the potential to be toxic not only towards the person wearing them but also for the environment. *Antiperspirants* act directly on reducing the activity of the sweat glands; they are mainly based on modified aluminum salts and are, by nature, potential irritants and sensitizers.

Today, dermoenzymatic deodorants can be used in microcapsules to combat this problem. In the past, many different enzymatic deodorants have been identified which act on lipase activity, preventing the generation of perspiration odor. Cosmacol Eli (Celessence), in contrast to other enzymatic deodorants, is a stable product (it is an ester derivative of lactic acid) which acts as an enzymatic deodorant by disturbing the enzymes involved in the production of body odor. It also has moisturizing and emollient effects, but it does not have antimicrobial properties.

7.6.2.3 Skincare Active Ingredients

Clothes (shirts, underwear, hosiery) equipped with microcapsules containing skincare active ingredients can deliver a valuable skincare regime all day long. Microcapsules filled with natural-based, high-grade care ingredients are released gradually onto the skin, ensuring a supply over many hours. The appearance of extremely dry skin may be noticeably improved. Moisturizing cream, for example, consists of two active ingredients, namely water and oil. As textiles generally cannot be equipped with an unlimited number of microcapsules, a very high percentage of active ingredients in each microcapsule is essential to attain the care effect. In most commercial microcapsules the skincare complex is so highly concentrated that even when the smallest amounts are released, the effect on the skin can be seen and felt.

Aloe vera, known as "Lily of the desert", is an efficient skincare ingredient, and is found today in virtually all cosmetic products. Modern scientific research has shown the Aloe leaf to contain over 75 nutrients and 200 active compounds, including 20 minerals, 18 amino acids and 12 vitamins. The main constituents are amino sugars, amino acids, enzymes, sterols, vitamins, inorganic salt, mono- and polysaccharides, giving the aloe vera gel special properties as a skin care product. It also combines different interesting healing features, in that it has a moisturizing effect and it protects and cares for the skin. Furthermore, it has an anti-inflammatory effect with benefits to the blood circulation, a regenerative effect, and it also promotes suppleness.

SPT™ microencapsulated aloe vera is used on hosiery, underwear (including maternity wear) and other textiles in contact with the skin, such as bedding. It has often been applied to underwear/hosiery as an additional ingredient to aloe vera or kelp to help combat the effects of stretch marks (i.e., for maternity wear). It has also been applied to walking socks as a blister-healing product.

Microcapsules containing *kelp* or *caffeine* as the active ingredient are used in order to assist the control of cellulites and to improve blood circulation. Kelp is a brown alga which grows affixed to rocks along marine shores and is well known for its stimulatory effect of the thyroid gland. Thyroxine, the product of the thyroid gland, not only controls and regulates metabolism, cellular vitalization, toxins and cholesterol elimination, and fat deposition in the body, but also stimulates blood flow. The amino acids belong to the Natural Moisturizing Factor (NMF) and other minerals present in kelp have moisturizing properties, and aid in the development of suppleness of the skin. Kelp has also anticoagulant properties and, with an ability to fluidize blood, improves blood flow. Being free radical scavengers, activates will participate in the reduction of inflammation in the cellulite tissue. Kelp is recommended for products formulated to fight obesity, local fat excess, or cellulite deposition.

Vitamin E is also widely used in many skin creams for medical and cosmetic applications. It is believed to contain powerful antioxidants that can protect skin cells against the effects of free radicals. Medically, it is used to promote skin healing, from scar tissues to sunburn and plastic surgery. Cosmetically, it is included in many moisturizer products to promote a healthy, blemish-free skin epidermis.

Moist oils (essential oils, herbal oils, oils from flower seeds, etc.) also have skincare benefits in that they provide an occlusive layer that lubricates the epidermis, together with a moisturizing effect that helps to prevent excess water loss. Today, the most popular use of essential oils is for aromatherapy. Essential oils are attributed with a range of properties that help to achieve physical and emotional balance. Examples are passion fruit oil, Monoï de Tahiti and squalane (derived from olive oil).

Other skincare products, such as encapsulated glycerol stearate and silk protein moisturizers, have been used for application on bandages and support hosiery. The material maintains comfort and skin quality through extensive medical treatment where textiles are in direct contact with the skin [23].

7.6.2.4 Textiles with a Cooling Effect
High-tech microcapsules are built directly into the fabric and release their caring natural-based ingredients such as Myritol® and menthol. In fact, the fresh Skintex® effect is ideal for fashion which is worn close to the skin. Skintex® chemicals are made from naturally occurring materials; indeed, all of the ingredients used have a well-established reputation for good skincare. A wide range of fragrances can be used, making Skintex® a key competitive differentiator for fashion collections.

7.6.2.5 Drug Delivery for Medical Textiles
Due to the often permeable, breathing structures and absorptive capacity as the result of finishing processes, textile materials are more frequently used in the medical field. Over the years, extensive research has been conducted with regard to drug delivery systems, including drug-loaded hollow fibers, ion-exchange fibers, fibers with bioactive side-chains, and textiles finished with cyclodextrins, azacrown ethers or fullerenes [24]. The most important aspects of a drug delivery system are its biocompatibility ("the ability of a material to perform with an appropriate host response in a specific application") and controllability (avoiding the risk of overdose), as well as colorfastness and washing ability.

Microencapsulation technology represents another promising approach to drug delivery. For drug delivery systems, both hydrophobic or hydrophilic monomers, oligomers, or polymers can be used during the encapsulation, depending on the drug(s) to be encapsulated. As another option, systems can be co-polymerized with soft or rubbery monomers or polymers (e.g., acrylates or methacrylates), to improve wearing properties such as "skin softness" and wrinkle and abrasion resistance. The monomer or polymer systems used contain at least one reactive component able to react with a textile material, even after polymerization.

7.6.2.6 **Polychromic and Thermochromic Microcapsules**

The concept of producing textiles that readily change in color has long been an anathema to textile researchers. The use of polychromic microcapsules represents an interesting commercial outlet. Polychromic dyes undergo reversible changes in color upon the application of an external stimulus. The color change phenomena are classified and named in different groups after the stimulus that causes the change. The two major groups of color-changing dyes used in textile applications exhibit thermochromism [25] and photochromism. Photochromism is a change in color, usually colorless to colored, when exposed to ultraviolet (UV) light (primarily sunlight). Thermochromic dyes change color reversibly with variations in temperature.

Other, emerging color-changing technologies include hydrochromics, which change in response to water, solvatochromics in response to solvents, and piezochromics, which change color in response to pressure.

The most important phenomena – namely photochromic and thermochromic – are discussed in some detail below.

Photochromic dyes are relatively new, having been introduced during the early 1990s. Unfortunately, photochromics are inherently unstable, and actually change their chemical structure when exposed to UV light. Because the dye is so vulnerable in its excited state, stabilization is the prime challenge for the dye manufacturers, since without stabilization the shelf life is very limited. Microencapsulation can overcome this shortcoming, however. When photochromic crystals are exposed to UV light, the dye undergoes a temporary chemical change, but when the UV source is removed the molecules reform their original bonding structure (Fig. 7.12). In the example, the back-reaction is predominantly thermally driven (the assistance of heat can be regarded as an example of thermochromism). In other photochromic dyes the back-reaction occurs photochemically.

Because of their ability to show resistance to thermal fade rates, fulgides have been the class most commonly used for this application. Typical uses are garments, children's toys, and logos on T-shirts.

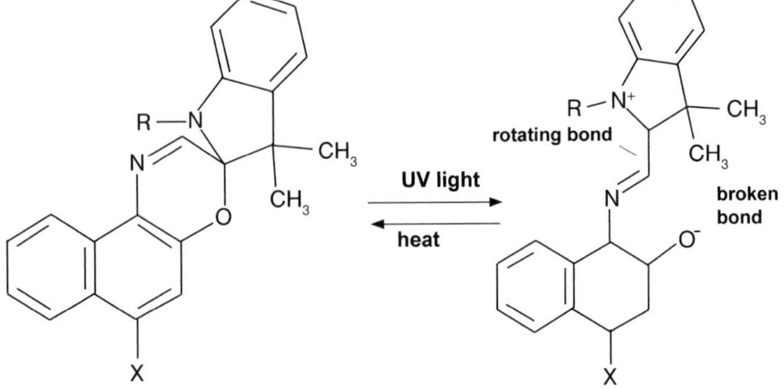

Figure 7.12 Mechanism of photosensitive dyes.

When incorporated into coatings, the photochromic microcapsules can cause fancy effects. Indoors, away from UV light, the design is white on the fabric but changes its color on exposure to UV radiation in daylight. A range is available from yellow through purple and green. In spite of the fugitive nature of the color-formers towards light, microencapsulated formulations are marketed in attention-grabbing product labeling and for anti-counterfeiting measures. Other applications include textiles that evolve new patterns upon exposure to sunlight, or security devices that are observable spectroscopically or with specialized instruments. Thermochromic organic colorants can also produce novel coloration and fancy effects in textiles. Aesthetics share the focus in the textile sector, with the emphasis firmly on novelty items, for example ski-wear or promotional textiles used in display markets. Extensive research investigations are currently under way into thermochromic colorants that are able to meet performance requirements for functional textile applications such as chameleonic smart fabrics, as exemplified by adaptive camouflage materials.

Thermochromic effects can be used on textiles for hidden messages, or heat-sensitive labels. Liquid crystal microcapsules give precise color changes in response to specific temperature changes, while leuco dyes respond over a more general temperature traject. Liquid crystal inks are used to make objects for measuring and indicating temperature variations or as gadgets (e.g., heat-sensitive labels, promotional items).

Analogous to photochromics, thermochromic dyes are also unstable and so have a limited shelf life. Microencapsulation can protect these dyes and safeguard their thermochromic properties, though the protection is not longlasting.

Microencapsulated formulations are available from a number of suppliers, including Solar Active, Chromatic technologies Ltd, Matsui Shikiso Chemical Co., Kiroku Sozai Sozo Kenkyusho, Color Change Corp., Tarzana, and others.

Commercial examples include PPG's photochromic dyes (Photosol) [26]. These chameleon-like dyes respond to natural solar irradiation as well as to artificial sources such as 365-nm "black light". When sunlight or UV radiation is applied, the dye becomes excited and the molecular structure changed, allowing a color to appear. When the stimulus (sunlight/UV) is removed, the dye returns to a state of rest, which is its colorless form.

These proprietary organic materials are available in a variety of activated colors, such as blue, yellow, burgundy, and red. When combined, additional colors such as green, brown and gray can be produced. By further adjusting the ratios or combinations of these photochromic dyes, various additional color shades can be achieved. It is possible to combine both fluorescent and/or organic pigments with reversible photochromics to produce permanent colors indoors and, by the color additive effect, different colors outdoors. This allows thousands of additional color possibilities.

The range of applications is limited only by the imagination. The microcapsules can be used for screen printing and laminate, automotive and industrial coatings. In processing, the dyes survive temperatures of 180 °C or even higher for a few minutes, without damage or degradation.

7.6.2.7 Insect Repellents

The use of microcapsules containing insecticides and acaricides to combat dust mites and insects such as mosquitoes has been investigated in several laboratories. Microcapsules containing essential oils and other plant extracts are available commercially. These have the advantage of providing a longlasting protective effect for significant periods without exposing the user to excessive dosages of hazardous chemicals.

The natural organic ester plant oils (mentholglycol) of SPT™ microcapsules have evolved over the millennia to keep insects away from the flowers and plants, and have been used by humans for thousands of years. The oils act effectively as a nerve agent on the insects. Recent research has shown that it has a very subtle effect on mosquitoes in particular; this is termed "Jamming", where apparently the "black box" which helps the mosquito search for victims is turned off.

The periodic release of these oils (which microencapsulation is ideally suited to deliver) has been shown to significantly reduce the number of bites per minute in test boxes, from 50 to virtually none (Fig. 7.13). Textiles finished with the capsules dramatically reduce the numbers of landings and bites of mosquitoes on the wearer. Applications are many-fold, and include hunting clothes, tents, marquees, swimming suits and even mattress ticking, bed sheets and covers.

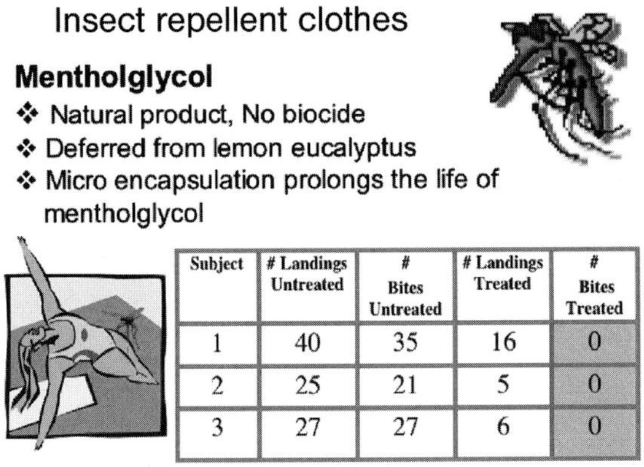

Insect repellent clothes

Mentholglycol
❖ Natural product, No biocide
❖ Deferred from lemon eucalyptus
❖ Micro encapsulation prolongs the life of mentholglycol

Subject	# Landings Untreated	# Bites Untreated	# Landings Treated	# Bites Treated
1	40	35	16	0
2	25	21	5	0
3	27	27	6	0

Figure 7.13 Insect-repellent results with microcapsules containing mentholglycol.

7.6.2.8 Anti-Counterfeiting Measures

In high added value textiles, and in branched and designers goods, there is a great pressure to protect from illegal copying within the marketplace. Microcapsules can be used to help with this problem by offering a covert, yet distinctive, marking system.

NoCopi [27] has patented Covert Ink Solutions. The CopiMark ink is totally invisible under any light – UV or otherwise – until activated with a special chemical pen. The covert ink systems protect the textile products from counterfeiters and diverters through an invisible replication of, for example, tracking numbers or barcode on the product or product packaging.

7.6.2.9 Miscellaneous Applications

Microencapsulated tea tree oil has multifunctional properties, including natural anti-bacterial and anti-fungal effects. Tea tree oil has been shown to be effective at relieving dandruff, and if used on laundry and bedding provides good protection against bed mites which, if eliminated, can be beneficial to asthma sufferers.

Exotic fragrances are already used widely on materials such as ties, bedding, floorcloths [28] T-shirts, scarves, towels, and bedding. For example, a lavender-fragrant pillow induces relaxation and promotes sleep.

Microcapsules with deodorization properties can be used on clothing, socks, curtains, and automotive textiles to eliminate smoke and other malodors, thereby creating pleasant surroundings. Textile coatings can also be infused with fragrances, which may intake unpleasant odors, while imparting pleasant ones, with potential applications for marketing.

7.7
New Developments and Future Perspectives: Smart Coatings

Conventional coatings used today in textile applications are "dumb" in the sense that they have no capability to respond quickly to changes in the service environment – that is, to compensate autonomously for environmental stresses. They also have no ability to indicate to the user any degradation in coating integrity, or to signal any emerging problems such as coating damage by impact, abrasion, or corrosion of the substrate.

Smart materials sense, react and adopt themselves to environmental conditions or to external and environmental stimuli. A few examples of interesting topics for the future are highlighted in the sections below.

7.7.1
Nanoencapsulated Antimicrobial Finish

Recently, Ciba Specialty Chemicals introduced nanocapsules with antimicrobial products. Washing durability is obtained by engineering the cotton surfaces with a covalently bonded encapsulation system, which enables the controlled delivery of antimicrobials. As a consequence of a slow-release mechanism, a small quantity of active material is released step-by-step. In short, the new system captures the best of both worlds: permanently fixed capsules to achieve durability, and mobile antimicrobials to inhibit bacterial growth on the entire surface (Fig. 7.14).

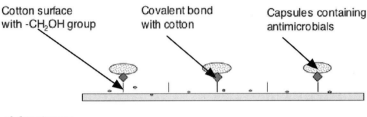

Cotton surface with -CH_2OH group   Covalent bond with cotton   Capsules containing antimicrobials

•**Advantages**:
- Durable
- Washfast
- Controlled release
- Compatible with other topical finishes

• **Disadvantages:**
- May slightly influence repellency finishes
- Needs crosslinker

Figure 7.14 Nano-encapsulated antimicrobial.

7.7.2
Incorporation of Microcapsules in Stimuli-Sensitive or Adaptive Coatings [29]

"Plastic memory" is well known in the preparation of polymers used as "shrink wrap" films. These polymer films are oriented by stretching at elevated temperatures and then being set into a stretched shape by cooling. Reheating causes shrinkage, enveloping the items to be wrapped. Other responsive materials are shape-memory alloys. The shape-memory effect in polymers (e.g., heat-shrinkable films) is not a specific bulk property, but rather results from the polymer's structure and morphology in combination with a processing and programming technology. Unlike their metallic counterparts, shape-memory plastics exhibit a reduction in elastic modulus with increasing temperature. Around the glass temperature, the changes in elastic modulus are rather sharp. Above the glass transition temperature (T_g) these materials are soft and compliant, but as the temperature sinks below T_g their hardness increases rapidly increases until they become rigid and eventually brittle. In addition to the elastic modulus differential, shape-memory polymers have a time response, which is temperature-dependent.

The technical challenge for many textile chemists is to develop analogous structured coating systems that can sense external stimuli and environmental variations (e.g., temperature, pH, electrolyte, stress, light, strain, or chemical potential), process the information, and then respond in a predetermined manner to restore the undisturbed state (Fig. 7.15).

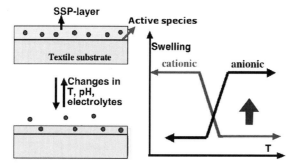

Figure 7.15 Schematic illustration of smart adaptive coatings.

They react to outside conditions in a selective manner. The magnitude of their stimulus-driven response can be quite large, for example increasing 10- to 100-fold in volume when warmed or cooled slightly.

Whereas numerous one-way systems are already used, the development of two-way responsive materials (i.e., capable of reversible switching between different properties) remains in its infancy.

Examples of stimuli-sensitive or adaptive coatings are numerous. Many stimuli-sensitive coatings (e.g., poly N-isopropylacrylamide; PNIPAAm) are temperature-sensitive in that a phase transition occurs at a certain temperature, called the lower critical solution temperature (LSCT). At low temperature the polyamide forms a highly swollen gel, absorbing large quantities of water at equilibrium. Above the transition temperature, the gel collapses, expelling water along with active solute (e.g., microcapsules) carried by it. Other stimuli-sensitive coatings (e.g., polycarboxylic polymer; PAA) formed from ionic polymers with weak acid groups (such as carboxyls) are sensitive to both pH and electrolytes. At lower pH, the carboxyls lose their charge and reduced repulsion allows the gel to collapse. Adding salt to the solution reduces the effective distance over which repulsion acts, collapsing the gel just as effectively as does low pH. A new generation of promising hydrogels are based on chitosan (sensitive to pH, electrolyte, and temperature) and polyethyleneglycol (PEG) [30]. PEGs are very different to other hydrogels because they are rubbers that can, nevertheless, be highly crystalline. While PAA and PNIPAA are amorphous and glassy in the dry state, PEG-based gels are rubbers, able to withstand the stress of aqueous swelling without significant risk of surface cracking.

The intelligent combination of adaptive coatings and microencapsulation could result in the manufacture of numerous coating systems that embody smart concepts. Our institute [31] and others [32] have investigated the possibility of using these adaptive coatings in controlled-release systems for modifying the discharge of physically active agents. With some ingenuity, these items can be combined with novel engineering to produce truly smart coatings. Upon the application of an external stimulus (e.g., sweat or an increase in temperature) such a responsive polymer can return to its previously memorized shape or reduced thickness and release – in a controlled manner – the incorporated microcapsules.

Possible applications to be studied are the controlled delivery and release of functional substances (e.g., drugs, nutrients, perfumes, herbicides) for use in active sportswear, medical and hygienic textiles, such as babies' diapers, healing underwear, bandages, plasters, incontinence pads or sanitary towels and socks and agro-textiles (herbicides, pesticides). These approaches, which are currently under investigation, combine the mechanical properties of textiles and the environmental responsiveness of adaptive coatings. An added advantage is the greatly lowered response time associated with the high surface areas common to adaptive coatings.

7.7.3
Other Smart Opportunities

In the near future these combined concepts of adaptive coatings and microcapsules may be extended to other smart coated textiles with chemomechanical response (cf. litmus paper changing color when the pH changes). These smart coatings could be used to alarm and signal the presence of harmful chemical or biological threats. Other possibilities include the development of coated textiles which release neutralizing agents when chemical or biological agents are detected.

One innovative application is the use of microcapsules for self-repairing or self-healing coatings. Man-made materials lack the ability for self-repair, whereas biologically produced materials sense and repair defects through cellular activities. Living cells act as the microreactors during repair. In an effort to mimic the self-repair functions of living systems, structural polymeric materials have been developed in the paint industry that possess the ability to self-heal. Self-healing is accomplished by incorporating a microencapsulated healing agent and a catalytic chemical trigger within a polymer matrix. When the coating cracks, microcapsules along the crack will break and the monomers are released. Upon contact with the crosslinker in the coating the monomers begin to polymerize, filling the crack and stopping its progression.

In paintings, the healing effect can be the result of the formation of a calcium carbonate ($CaCO_3$) "healing" film via the chemical reaction of released calcium hydroxide ($CaOH$) with carbon dioxide (CO_2) in the air when the coating is damaged.

Analogous concepts may be applicable to textile coatings to repair damage and protect the underlying textile substrate. In the near future, self-healing coatings for textiles will become a hot issue in the textile industry. The challenge for the researchers is to find appropriate microcapsules to imbue the coating with specific characteristics, such that the coatings become self-healing, chemical-resistant, contain passive sensors (such as an indicator dye), and have the ability to alert personnel or the wearer (e.g., the surgeon, soldier, fire-fighter) to potential coating deficiencies. Moreover, these coatings could be able to modify their physical characteristics on command, increasing the reliability and service life of coated materials or polymer composites used in a wide variety of applications ranging from protective textile to other high-performance technical textiles.

Interesting smart coatings are based on microcapsules containing a marker dye that can be released when the coating is damaged by cutting, by abrasions, and by

impact, or is otherwise stressed. In this way, the "smart" coating can alert maintenance personnel to the need for repair. The marker dye may be the sole constituent
of these microcapsules, or it may be mixed with coating healants.

Other possibilities include smart coatings that would provide radar invisibility by
using microcapsules that have radar-absorbing capabilities, or the proper camouflage in varying optical settings.

Finally, for a technology such as microencapsulation that has been recognized for
more than 50 years, controlled release is not even close to becoming passé, as over
the years a wide variety of alternative slow-release systems have been developed.

Ingredient supplier Penreco [33] has recently developed a unique patented system for the thickening and gelling of hydrocarbon materials. The gels act as film-
formers and delivery vehicles for personal care products, with commercial products including Versagel® and Synergel®. The new Versagel® technology can be
used for the controlled release of fragrance, causing a perfumed product to have a
generous smell for its lifetime, and not simply overpowering the initial top notes.
The fragrance is chemically linked to the gel network, and will have a slower and
more even release through the time of exposure as compared to a product that does
not contain any polymeric gels. Synergel® gels also provide controlled release of active ingredients. The Synergel® products which are designed for aerosol applications are marketed under the Synergel® A designation. All of these gels possess excellent thermal and UV stability.

Other promising systems are almost too numerous to mention, but include cyclodextrins, aza-crown ethers [34], ion-exchange fibers, and drug-loaded hollow
fibers. One example already introduced commercially is that of Nouwell E (CHT-
D), a cyclodextrin–vitamin E complex with a wellness function. The active ingredient of vitamin E (α-tocopherol) is encapsulated in the cyclodextrin units, which results in an excellent protection against UV-light and high temperature. The complex is anchored to the textile substrate using a self-crosslinking polyurethane compound. When the complex comes into contact with the skin, the vitamin E is gently
released and transferred to the body through the skin (Fig. 7.16).

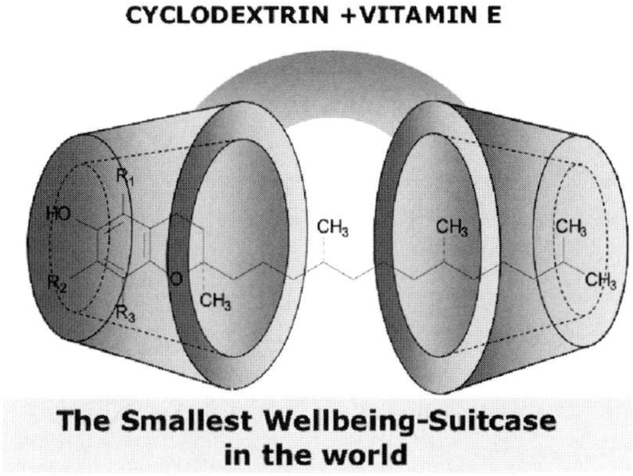

CYCLODEXTRIN +VITAMIN E

**The Smallest Wellbeing-Suitcase
in the world**

Figure 7.16 The cyclodextrin–vitamin E complex.

Abbreviations

DCC	dynamic climate control
LSCT	lower critical solution temperature
MF	melamine-formaldehyde
NMF	natural moisturizing factor
PAA	polycarboxylic polymer
PCM	phase change material
PEG	polyethyleneglycol
PNIPAAm	poly N-isopropylacrylamide
T_g	glass transition temperature
UF	urea-formaldehyde

References

1. G. Nelson, *Rev. Prog. Color.* **2001**, *31*, 57.
2. www.microlithe.com/microcap.
3. ThermaCell Technologies Inc. (NAS-DAQ:VCLL), USA.
4. www.mtmtys.co.jp/.
5. Int. dyer June **2000**, 26.
6. www.celessence.com/.
7. www.cognis.com.
8. http://corporate.basf.com.de.
9. www.cibasc.com/.
10. www.teappcm.com.
11. www.epsitd.co.uk.
12. L.J.H. Nayes, E.H. Wissler and D.P. Colvin, A model for encapsulated Phase Change Material in a Conductive Media (available on the internet).
13. Euracli (www.euracli.fr).
14. Sponsored by IWT – Flemish Government, **2003**.
15. M. Van Parys "Textile Coating", Dr., 1st ed., UNITEX, Gent **2005**.
16. CoatingStar AS – Ahlbrandt – Germany, www.ahlbrandt.de/.
17. Research project concerning atmosferic plasma technologies at TO₂C- Technical University of Gent, info@unitex.be, www.unitex.be.
18. R. Cox, *Chem. Fibre Int.* **1998**, *48*, 475.
19. www.comfortemp.com.
20. B. Pause, *Tech. Text. Int.* **1999**, *8* (7), 23.
21. B. Pause, Membranes with thermo-regulating properties for applications in fabric structures, 2nd Colloquium Coating and Surface Functionalisation of Technical Textiles, Denkendorf-D, **2005**.
22. www.expresstextile.com/20050515/processworld01.shtml.
23. S.A. Dim, French Patent 2 780 073 A1 (**1999**).
24. M.R. Ten Breteler, V.A. Nierstrasz, M.M.C.G. Warmoeskerken, Textile slow-release systems with medical applications, AUTEX Research Journal, *Vol. 2*, No. 4, December **2002**, pp. 175–189.
25. A. Towns, *Journal of Society of Dyers and Colorists* **1999**, *115*, 196.
26. www.ppg.com.
27. www.nocopi.com.
28. Hako-Werke Gmbh (DP 1 9545 242 (**1997**).
29. The European Coatings Conference: "Smart Coatings III", June **2004**, Berlin, Germany; lectures of Prof. Dr. André Laschewsky, Fraunhofer Institute for Applied Polymer Research, Germany, Prof. Dr. Andreas Lendlein, mnemoScience GmbH, Germany.
30. N. Graham, Swell Gels, *Chemistry in Britain*, April **2001**, 42–44.
31. TO₂C Research Laboratoria – Technical University Gent – Belgium, e-mail: info@unitex.be.
32. National textile Center research – USA, www.ntcresearch.org/.
33. www.penreco.com/.
34. D. Knittel, E. Schollmeyer, Chitosan und seine Derivate für die Textilveredlung, Teil 1: Ausgangssituation, *Textilveredlung* **1998**, *33*, Nr. 3/4, 67–71.

8
Encapsulations Through the Sol-Gel Technique and their Applications in Functional Coatings

Isabel van Driessche and Serge Hoste

8.1
An Outlook on Coatings and Paints

Materials chemistry is a field of high priority internationally, in terms of both fundamental and applied science. Research contributions to the advancement of this field come from a wide range of scientists who develop the synthetic processes, perform the fundamental characterizations, and employ the materials in a wide range of applications. During recent years, sol-gel chemistry has become a major topic of research in materials science and synthesis.

The conventional synthesis for multicomponent materials involves a solid-state reaction in which appropriate precursors (usually oxides or carbonates) are mixed together. These precursors are often ball-milled to enhance mixing and to reduce their particle sizes, so that maximum contact surface can be obtained between the particles [1]. The reaction occurs initially at the points of contact between the components [2]. With increasing reaction time, the reaction slows down because of the longer interdiffusion distances through the end product. The mixture is fired at relatively high temperature to allow interdiffusion of the cations. Repeated cycles of milling and calcination are usually carried out to improve homogeneity.

Recently, the sol-gel method has become very attractive for materials synthesis because it permits direct fabrication of multicomponent materials in different configurations (monoliths, coatings, and fibers) without powder intermediates [3] or without the use of expensive vacuum technologies [4,5]. The diversity with which materials can be obtained, has made the sol-gel method an important synthesis route in several domains of research, including optics, electronics, biomaterials, and semi- and superconductors.

The term "sol-gel" denotes a process by which largely inorganic materials are synthesized. The term is an abbreviation for "solution-gelling" which denotes its principle: a solution or sol (which is a dispersion of colloidal particles) starting from precursors which are dissolved in a liquid phase, is transformed to the solid

Functional Coatings. Edited by Swapan Kumar Ghosh
Copyright © 2006 WILEY-VCH Verlag GmbH & Co. KGaA, Weinheim
ISBN 3-527-31296-X

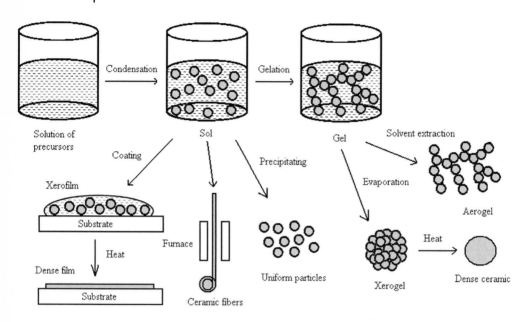

Figure 8.1 Functional materials based on sol-gel chemistry.

state through a sequence of chemical reactions which involve polymerization at ambient temperatures [6] (Fig. 8.1). A "gel" is an interconnected polymeric network formed by assembly of the sol. The gelation proceeds through stages by which the rigidity of the product is increased. The material produced before the final process is a porous glass-like solid which is termed a "xerogel". Although xerogel implies a dry material, pore water is still present.

If the term "sol-gel" is used in a strict manner – that is, it must include processing in which a sol undergoes a transition to a gel characterized by an infinite three-dimensional network structure spreading uniformly throughout the liquid medium – only two sol-gel processing techniques in which a true gel is prepared, are recognized:

- the formation of a network by destabilization of dispersed colloidal particles in a liquid, resulting in particulate gels
- hydrolysis and polycondensation reactions of metal alkoxides, resulting in alkoxide gels [2]

However, the term "sol-gel" is now often assigned to any wet chemical process which can produce inorganic materials by a variety of solution routes.

Chemists continue to explore means of synthesizing xerogels with tailored properties. One property which will be described in detail in this chapter is the internal pore structure. This pore structure can be controlled in a manner that makes it possible to trap species such as complexing agents and biomolecules such as enzymes. This type of process is denoted with the term "encapsulation" or "immobilization". Results of this new synthetic method are formulations and materials that are ideal for applications such as chemical and biochemical sensors, optical lenses, and controlled-release drugs.

Recent progress in the development of materials from structural to functional, to "smart" or even "intelligent", has been remarkable over the past decade. Admittedly, the present "intelligence" of materials does not yet rival that of bacteria, but a trend towards a focus on the reaction of materials to external triggers has been set and this may yield a measure of remarkable interplay between the mineral and organic worlds in the near future. "Smart" materials possess the ability to recognize changes in their environment by undergoing structural variations within their framework, and can therefore be tuned to take suitable action against environmental changes. This "smartness" derives not so much from the phenomenology, because such effects are after all quite natural, but from their focus and the driving forces to develop those materials in the first place. Depending upon the applications and types of environmental stimuli, there has been an extensive investigation of several types of smart materials, such as shape-memory metals, alloys [7], polymers [8] and even ceramics [9], electrorheological and magnetorheological fluids [10] and ormosils (organically modified silicates) [11]. The different stimuli to which many of these materials are envisaged to react cover a broad range of physical, chemical and biological properties such as electromagnetic waves (light) and electrical potential, mechanical phenomena (heat, movement), chemical forces (pH, solvent polarity) and the presence of biomolecules.

Not only do these materials and their responses require the presence of a finely tuned chemical structure and composition, but their technological application also necessitates a useful shape, method, format or mode into which they can be formulated. This content versus format relationship is one that pervades many modern chemical developments, and it is here that the use of sol-gel system provides its highest impact. As will be see in the following sections, both shape and composition can be heavily influenced through the judicious use of sol-gel techniques in order to create or improve the reactivity of these materials.

8.2
Rationale for the Use of Sol-Gel Chemistries

There are a number of reasons for the particular value of, and high interest in, sol-gel synthesis routes [1,12–15]:

- Since liquid precursors are used, it is possible to fabricate materials in a range of *complex shapes* such as coatings or fibers, which are of technological importance. The use of liquid media in combination with high-speed deposition techniques such as spin-coating or dip-coating renders these methods ideally suited for continuous coverage of large areas of suitably pretreated substrates.

- The sol-gel method has the potential for cation mixing at the molecular level. The use of gels finds its rationale in their ability to immobilize large concentrations of different species, together with molecular or atomic resolution. It is assumed that this level can be maintained during the subsequent conversion to the solid compound. In that way, the *homogeneity* is obviously much better than can be achieved with the conventional solid-state reaction.

- As the sol-gel method relies on "soft chemistry" ("chimie douce"), a greater control is possible over *stoichiometry*. It is also easier to incorporate dopants in a homogeneous manner.

- Precursors such as metal alkoxides or metal salts are available at a high level of purity. The *purity* of sol-gel products depends not only on the purity of the relevant starting materials but also on the degree of care taken throughout the whole process. Compared to the conventional solid-state reaction, the high level of purity can be more easily maintained in the sol-gel synthesis as no mechanical mixing is required.

- The temperatures required for thermal decomposition are low. Furthermore, *shorter thermal treatments* are necessary because of the higher homogeneity at the start of the process.

- The sol-gel process also has the economic value that it is characterized by *low financial costs*.

- One merit of special attention in this chapter is the fact that sol-gel synthesis offers a convenient method for hosting chemical reactions, a process which is not possible using other synthesis techniques. Typically, in sol-gel encapsulation, silica nanoparticles surround the captive molecules during gel formation. In principle, the sol-gel process can be considered as a phase separation by sol-reactions, sol-gelation and finally, removal of the solvent resulting in a ceramic material. Depending on the preparation, dense oxide particles or polymeric clusters will be obtained [16].

- Many types of chemical species can be impregnated or encapsulated within the pores of a xerogel by adding them to the sol. Organics [17–19], organometallics [20], proteins [21] and enzymes are a few such examples. By appropriate chemical modification of the precursors, control may be achieved over the rate of hydrolysis and condensation (see below) and over colloid particle size and thus pore size, porosity and pore wall chemistry of the final material. An important example that has been emphasized in the recent literature is the doping of xerogels with biological compounds. Since the chemical conditions under which sol-gel materials are produced are mild, it is found that biomolecules retain their activity when encapsulated in silica prepared by sol-gel chemistry. The sol-gel processes are catalyzed by acids and bases, but extreme pH conditions may easily be avoided. In this way, the sol-gel process allows the xerogel matrix to form under conditions compatible with the biomolecules and – upon encapsulation – protects them from an otherwise hostile environment because of reasons such as high or low pH values. The biomolecule can thus retain its normal stability and reactivity. Examples in the literature include using xerogels doped with hemoglobin, myoglobin and glucose oxidase for the detection of oxygen, carbon monoxide and glucose, respectively [22].

These potential advantages have given rise to the increased application of sol-gel chemistry for the development of different materials. The most important commercial sol-gel products are films and coatings. A great variety of coating films have been prepared for:

- optical (optical waveguide, optical absorption and coloring, reflection and anti-reflection coatings, fluorescent films, electrochromic films, films for patterning)
- electronic (ferroelectric films, piezoelectric films, films showing electronic or ionic conduction, superconducting films, magnetic films)
- chemical (photoelectrochemical catalyst films, films as catalyst support, films for chemical protection of substrates)
- mechanical applications [23,24]

8.3
Historical Evaluation of Sol-Gel Encapsulation

The development of different sol-gel encapsulation methods and their applications form one of the chapters in sol-gel science in general. Their origin is buried in ancient history as the use of gelatinous substances for coating and embellishment of artifacts were developed, largely on an empirical and artistic basis. A rediscovery of sol-gel science during the second half of the twentieth century [25] has led to an explosion of applications and spin-offs during the past twenty years. Today, the potential rewards of sol-gel chemistry are so great that a deeper insight in its chemical and physical roots is absolutely necessary in order to allow predictions, to hasten the targeted development of new sol-gel materials, and to improve their formulation, efficiency and specificity. That is the framework into which sol-gel encapsulation was developed.

The field of sol-gel encapsulation is relatively new. A search of the Web of Science for "sol gel encapsulation" in title, keywords or abstracts yielded 73 papers. Their distribution over the past two decades is sketched in Figure 8.2, wherein the data point clearly to 1993 as being the year in which the topic "caught on".

year	publications	year	publications
1990	0	1998	8
1991	0	1999	7
1992	0	2000	4
1993	1	2001	8
1994	2	2002	7
1995	1	2003	8
1996	4	2004	11
1997	1	2005	9

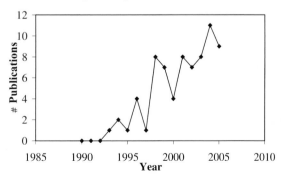

Figure 8.2 Evolution of the number of papers referenced in the Web of Science related to sol-gel encapsulation over the past 15 years.

Two articles appear to have sparked off research in different fields – one article by Klein on the sol-gel encapsulation for optical materials [26], and another heavily cited article by Dave in the field of biosensors [22]. As far as the biochemical application of sol-gel encapsulation is concerned, it was shown as early as 1955 [27] that several enzymes could be entrapped into silicic acid-derived glasses with partial retention of their biological activity, and a true sol-gel encapsulation method was subsequently used to do this for the first time in 1990 [28]. All this has set the scene for a steady increase from 1993 onwards, the result being the production of more than twenty papers cited in the Web of Science in 2005. A Google-scholar search on "sol-gel encapsulation" yielded 1970 hits at the time of writing of this chapter. Although the sol-gel research area has its own website (http://www.sol-gel.com/index.asp) and journal (*The Journal of Sol-Gel Science and Technology*), the latter scored only 24 papers overall when a search was restricted to "encapsulation". Every sol-gel scientist is obviously indebted to the work of Brinker and Scherer [29], who firmly established the field of sol-gel chemistry and physics in their monumental monograph. In summary, the ideas related to encapsulation, its effects and uses have developed from the field of sol-gel science around 1993, and the flow of scientific research papers does not yet show any sign of saturation.

8.4
Sol-Gel Chemistries

Sol-gel materials encompass a wide range of inorganic and organic/inorganic composite materials which share a common preparation strategy. They are prepared via sol-gel processing involving the generation of colloidal suspensions ("sols") which are subsequently converted to viscous gels and then to solid materials [29]. In general, the different stages involved in a sol-gel process include [15]:

- hydrolysis: the formation of hydroxide species
- condensation: the formation of oxide species
- gelation: the formation of a "spanning cluster", yielding a network which entraps the remaining solution, with high viscosity. The gelation step includes hydrolysis and condensation
- ageing: a range of processes, including the formation of further crosslinks, associated shrinkage of the gel as covalent links replace nonbonded contacts, Ostwald ripening and structural evolution with changes in pore sizes and pore wall strengths
- drying: the loss of water, alcohol and other volatile components, first as syneresis (expulsion of the liquid as the gel shrinks), then as evaporation of liquid from within the pore structure with associated development of capillary stress. This may also include supercritical drying, in which capillary stress is avoided by the use of supercritical fluids in conditions where there are no liquid/vapor interfaces

- densification: thermal treatment leading to collapse of the open structure and formation of a dense material

The sol-gel processing of materials can refer to a multitude of reaction processes which employ a wide variety of chemical precursors to prepare many different products [30]. According to Kakihana, there are essentially three different kinds of sol-gel routes: (i) colloidal sol-gel route; (ii) inorganic sol-gel routes; and (iii) polymeric sol-gel routes [2]. The primary goal in all of these sol-gel processes is the preparation of a homogeneous precursor solution from which a homogeneous solid compound can be obtained. This classification of sol-gel routes is shown in Table 8.1.

The precursors for synthesizing these colloids consist of a metal or metalloid element surrounded by various reactive ligands. Metal alkoxides are most popular because they react readily with water. The most widely used metal alkoxides are the alkoxysilanes, such as tetramethoxysilane (TMOS) and tetraethoxysilane (TEOS). However, other alkoxides such as aluminates, titanates, and borates are also commonly used in the sol-gel process, often mixed with TEOS.

Following the classification described in Table 8.1, other precursor materials, such as metal nitrates or metal acetates combined with complexing and polymerizing agents such as citric acid, ethanolamines, tartaric acid and ethyleneglycol, are frequently used in more recently reported studies.

Table 8.1 Classification of sol-gel routes.

	Supporting structure	Bonding fashion	Source
Colloidal sol-gel route	Colloid	*Particles* interconnected by van der Waals or hydrogen bonding	Mostly from oxide or hydroxide sols
Inorganic sol-gel route	Metaloxane polymer	*Inorganic polymers* interconnected by van der Waals or hydrogen bonding	Hydrolysis and condensation of metal alkoxides
	Metal-complex	*Associates* weakly interconnected by van der Waals or hydrogen bonding	Concentrated metal complex solution
Polymeric sol-gel route	*In-situ* polymerizable complex	*Organic polymers* interconnected by coordinate, van der Waals or hydrogen bonding	Polymerization between α-hydroxy carboxylic acid and polyhydroxyalcohol in the presence of metal complexes
	Polymer-complex solution	*Organic polymers* interconnected by coordinate, van der Waals or hydrogen bonding	Coordinating polymer

8.4.1
Colloidal Sol-Gel Route

Colloids are suspensions of particles of linear dimensions between 1 nm and 1 μm. The formation of uniform suspensions of colloidal particles can be understood by calculation of the sedimentation rates, assuming that the particles are spherical so that Stokes' law may be applied [15].

The sol is converted to a gel by controlling the electrostatic or steric interactions between the colloidal particles [2,13,31].

During the 1940s, Derjaguin, Verwey, Landau and Overbeek (DVLO) developed a theory which dealt with the stability of colloidal systems. DVLO theory suggests that the stability of a colloidal system is determined by the sum of the van der Waals attractive and electrical double-layer repulsive forces that exist between particles as they approach each other due to the Brownian motion they are undergoing. This theory proposes that an energy barrier resulting from the repulsive force prevents two particles approaching one another and adhering together (Fig. 8.3). However, if the particles collide with sufficient energy to overcome that barrier, the attractive force will pull them into contact where they adhere strongly and irreversibly together. Therefore, if the particles have a sufficiently high repulsion, the dispersion will resist flocculation and the colloidal system will be stable. However, if a repulsion mechanism does not exist then flocculation or coagulation will eventually take place.

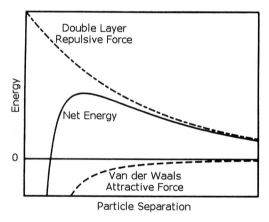

Figure 8.3 Schematic diagram of the variation of free energy with particle separation according to DVLO theory. The net energy is given by the sum of the double layer repulsion and the van der Waals attractive forces that the particles experience as they approach one another.

Therefore, to maintain the stability of the colloidal system, the repulsive forces must be dominant. How can colloidal stability be achieved? There are two fundamental mechanisms that affect dispersion stability (Fig. 8.4):

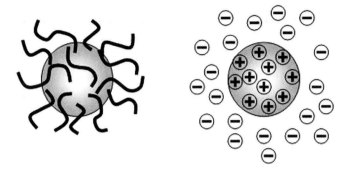

Steric stabilization   Electrostatic stabilization

Figure 8.4 Types of colloidal stabilization.

- Steric repulsion: this involves polymers added to the system adsorbing onto the particle surface and preventing the particle surfaces from coming into close contact. If enough polymer is adsorbed, the thickness of the coating is sufficient to keep particles separated by steric repulsions between the polymer layers, and at those separations the van der Waals forces are too weak to cause the particles to adhere.
- Electrostatic or charge stabilization: this is the effect on particle interaction due to the distribution of charged species in the system.

Each mechanism has its benefits for particular systems. Steric stabilization is simple, requiring simply the addition of a suitable polymer. However, it can be difficult subsequently to flocculate the system if this is required, the polymer may be expensive, and in some cases the polymer used is undesirable – for example, when a ceramic slip is cast and sintered, the polymer must be "burnt out". This causes shrinkage and can lead to defects.

Electrostatic or charge stabilization has the benefits of stabilizing or flocculating a system by simply altering the concentration of ions in the system. This is a reversible process, and is potentially inexpensive.

It has long been recognized that the zeta potential is a very good index of the magnitude of the interaction between colloidal particles, and this parameter is often measured to assess the stability of colloidal systems [32–34].

The liquid layer surrounding the particle exists as two parts: (i) an inner region (Stern layer) where the ions are strongly bound; and (ii) an outer (diffuse) region where they are less firmly associated. Within the diffuse layer there is a notional boundary inside which the ions and particles form a stable entity. When a particle moves (e.g., due to gravity), ions within the boundary move with it. The ions beyond the boundary do not travel with the particle. The potential at this boundary (surface of hydrodynamic shear) is the zeta potential (Fig. 8.5).

The magnitude of the zeta potential provides an indication of the potential stability of the colloidal system. If all of the particles in suspension have a large nega-

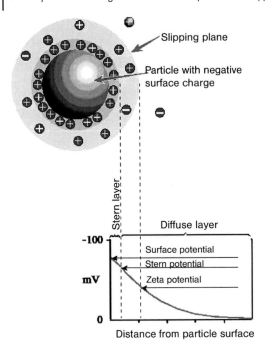

Figure 8.5 Schematic representation of the zeta potential.

tive or positive zeta potential, then they will tend to repel each other and there is no tendency for them to come together. However, if the particles have low zeta potential values then there is no force to prevent them from coming together and flocculating. The general dividing line between stable and unstable suspensions is generally taken at either +30 or −30 mV. Particles with zeta potentials more positive than +30 mV or more negative than −30 mV are normally considered stable.

Zeta potential can be affected by changes in pH, conductivity (concentration and/or type of salt) or changes in the concentration of an additive (e.g., ionic surfactant, polymer).

Because the interparticle interactions in those sols are dominated by physical forces such as van der Waals forces, electrostatic forces and Brownian motion, the colloidal sol-gel method is termed a "physical" gel route [2]. The next two sections are restricted to two chemical gel routes: (i) inorganic polymerization; and (ii) organic polymerization.

8.4.2
Inorganic Sol-Gel Route

Inorganic polymeric gels are made basically in two ways: either from metal alkoxides in organic solvents, or from metal salts dissolved in aqueous or organic solvents and stabilized by chelating ligands. In both cases, the chemistry is dominated by the high electropositive character of the metal cations.

8.4.2.1 Alkoxide Gel Method

Metal alkoxides have the general formula $M(OR)_z$ where M is a metal ion, R is an alkyl group and z is the valence state of the metal. On addition of water, the metal alkoxides readily hydrolyze as follows [35]:

$$M(OR)_z + x\, H_2O \rightarrow M(OH)_x(OR)_{z-x} + x\, ROH \tag{1}$$

Hydrolysis is followed by condensation to form -M-O-M- bonds via either dehydration or dealcoholation:

$$\text{-M-OH} + \text{HO-M-} \rightarrow \text{-M-O-M-} + H_2O \quad \text{(dehydration)} \tag{2}$$

$$\text{-M-OH} + \text{RO-M-} \rightarrow \text{-M-O-M-} + ROH \quad \text{(dealcoholation)} \tag{3}$$

In case of the preparation of multi-metal-oxides, both self-condensation (i.e., formation of -M-O-M- bonds) or cross-condensation (i.e., formation of -M-O-M'- bonds) may occur [1]. Using these condensation reactions, an inorganic polymeric oxide network is built up progressively.

For coordinatively saturated metals the hydrolysis and condensation of metal alkoxides $M(OR)_z$ corresponds to the nucleophilic substitution of alkoxy ligands which can be described by a S_N2 mechanism as follows:

$$
\begin{array}{c}
\text{H} \\
\backslash \\
\text{O}^{\delta-} + M^{\delta+}\text{-O}^{\delta-}\text{-R} \\
/ \\
\text{X}
\end{array}
\longrightarrow
\begin{array}{c}
\text{H}^{\delta+} \\
\backslash \\
\text{O-M-O}^{\delta-}\text{-R} \\
/ \\
\text{X}
\end{array}
\longrightarrow
\begin{array}{c}
\text{H}^{\delta+} \\
/ \\
\text{XO-M-O} \\
\backslash \\
\text{R}
\end{array}
\longrightarrow \text{XO-M} + ROH \tag{4}
$$

where X stands for hydrogen (hydrolysis), a metal atom (condensation), or even an organic or inorganic ligand (complexation).

The reaction starts with the nucleophilic addition of negatively charged $HO^{\delta-}$ groups on to the positively charged metal atom, $M^{\delta+}$, leading to an increase of the coordination number of the metal atom in the transition state. The positively charged proton is then transferred towards an alkoxy group and the positively charged protonated ROH ligand is finally removed.

The chemical reactivity of metal alkoxides towards hydrolysis and condensation depends mainly on the positive charge of the metal atom $\delta(M)$ and its ability to increase its coordination number N. As a general rule, the electronegativity of metal atoms decreases and their size increases when going from the top right of the periodic table to the bottom left. The corresponding alkoxides become progressively more reactive towards hydrolysis and condensation. Silicon alkoxides are rather stable, while cerium alkoxides are very sensitive to moisture. Alkoxides of electropositive metals must be handled with care under a dry atmosphere, otherwise precipitation occurs as water is present.

The oxidation state z of metal ions is usually smaller than their coordination number N in the oxide compounds. Coordination expansion is therefore a general tendency of the sol-gel chemistry of metal alkoxides $M(OR)_z$. Positively charged metal atoms $M^{\delta+}$ tend to increase their coordination number by using their vacant orbitals to accept electrons from nucleophilic ligands. Coordination expansion usually occurs via nucleophilic addition, leading to the formation of oligomers, solvates, or heterometallic alkoxides.

The hydrolysis, condensation, and polymerization reactions are governed by several factors, including the molar ratio of water to alkoxides, choice of solvents, temperature and pH (or concentration of acid or base catalysts). Depending on these factors, either a linear polymeric gel or a more crosslinked polymeric gel can be formed. A homogeneous network is built up as the polymerization reaction continues.

Since silica is a commonly used material in sol-gel chemistry, the next section will concentrate on the reaction mechanisms involved in that system. Here, we will describe several mechanistic principles, related to the general mechanisms described above, which allow some prediction of the effects of varying the nature and amounts of the silicon precursor, co-solvent and water, and of variables such as pH and temperature. For convenience, the stages of sol-gel processing will be considered separately, although it must be realized that this is an artificial scenario since in real systems several of the steps may occur concurrently.

Reaction mechanisms involved in silica sol-gels

Generally speaking, the hydrolysis reaction, through the addition of water, replaces alkoxide groups (OR) with hydroxyl groups (OH). Subsequent condensation reactions involving the silanol groups (Si–OH) produce siloxane bonds (Si–O–Si) plus the byproducts water or alcohol. Under most conditions, condensation commences before hydrolysis is complete. However, conditions such as, pH, H_2O/Si molar ratio (R), and catalyst can force completion of hydrolysis before condensation begins. Additionally, because water and alkoxides are immiscible, a mutual solvent such as an alcohol is utilized. With the presence of this homogenizing agent, alcohol, hydrolysis is facilitated due to the miscibility of the alkoxide and water. As the number of siloxane bonds increases, the individual molecules are bridged and jointly aggregate in the sol. When the sol particles aggregate, or inter-knit into a network, a gel is formed. Upon drying, trapped volatiles (water, alcohol, etc.) are driven off and the network shrinks as further condensation can occur. It should be emphasized, however, that the addition of solvents and certain reaction conditions may promote esterification and depolymerization reactions according to the reverse of hydrolysis and condensation equations.

In the following sections, specific factors that influence the hydrolysis and condensation reactions of the sol-gel process will be discussed. It has been established that certain reaction parameters are more important than others. Hereinafter, we will focus primarily on the following influences: pH, nature and concentration of catalyst, and H_2O/Metal molar ratio (R).

During hydrolysis, the following reaction parameters are of most importance.

- *Influence of pH*

  Normally, the polymerization process is divided into three pH domains: < pH 2, pH 2–7, and > pH 7. However, regardless of pH, hydrolysis occurs by the nucleophilic attack of the oxygen contained in water on the silicon atom as evidenced

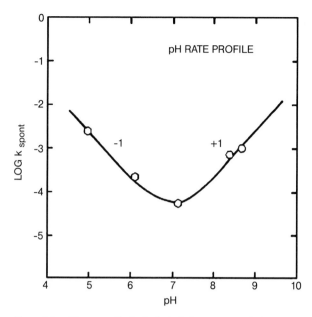

Figure 8.6   pH rate profile for hydrolysis in aqueous solution.

by the reaction of isotopically labeled water with TEOS that produces only unlabelled alcohol in both acid- and base-catalyzed systems:

$$-\overset{|}{\underset{|}{Si}} - OR + H^{18}OH \rightleftharpoons -\overset{|}{\underset{|}{Si}} -^{18} OH + ROH \tag{5}$$

It can be seen in Figure 8.6 how pH affects the hydrolysis rate.

- *Nature and concentration of catalyst*
  Although hydrolysis can occur without the addition of an external catalyst, it is most rapid and complete when they are employed. Mineral acids (HCl) and ammonia are most generally used, though other catalysts such as acetic acid, KOH, amines, KF, and HF are also used. Additionally, it has been observed that the rate and extent of the hydrolysis reaction is most influenced by the strength and concentration of the acid- or base catalyst.

  Under acidic conditions, it is likely that an alkoxide group is protonated in a rapid first step. Electron density is withdrawn from the silicon atom, making it more electrophilic and thus more susceptible to attack from water. This results in the formation of a penta-coordinate transition state with significant S_N2-type character. The transition state decays by displacement of an alcohol and inversion of the silicon tetrahedron (Fig. 8.7).

  Base-catalyzed hydrolysis of silicon alkoxides proceeds much more slowly than acid-catalyzed hydrolysis at an equivalent catalyst concentration. Basic alkoxide oxygens tend to repel the nucleophile, –OH. However, once an initial hydrolysis has occurred, the following reactions proceed stepwise, with each subsequent

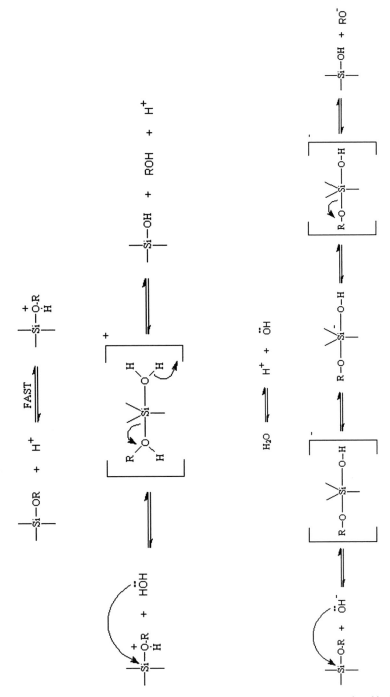

Figure 8.7 Acid-catalyzed hydrolysis.

Figure 8.8 Base-catalyzed hydrolysis.

alkoxide group more easily removed from the monomer then the previous one. Therefore, more highly hydrolyzed silicones are more prone to attack. Additionally, hydrolysis of the forming polymer is more sterically hindered than the hydrolysis of a monomer. Although hydrolysis in alkaline environments is slow, it still tends to be complete and irreversible.

Thus, under basic conditions, it is likely that water dissociates to produce hydroxyl anions in a rapid first step. The hydroxyl anion then attacks the silicon atom. Again, an S_N2-type mechanism has been proposed in which the –OH displaces –OR with inversion of the silicon tetrahedron (Fig. 8.8).

- *H_2O/Si molar ratio (R)*
  The hydrolysis reaction has been performed with R values ranging from less than 1 to over 50, depending on the desired polysilicate product. An increased value of R is expected to promote the hydrolysis reaction. The most obvious effect of the increased value of R is an acceleration of the hydrolysis reaction. Additionally, higher values of R caused more complete hydrolysis of monomers before significant condensation occurs. Differing extents of monomer hydrolysis should affect the relative rates of the alcohol- or water-producing condensation reactions. Generally, with understoichiometric additions of water (R << 2), the alcohol-producing condensation mechanism is favored, whereas, the water-forming condensation reaction is favored when R > 2.28.

  Although increased values of R generally promote hydrolysis, when R is increased while maintaining a constant solvent:silicate ratio, the silicate concentration is reduced. This in turn reduces the hydrolysis and condensation rates, resulting in longer gel times. This effect is evident according to Figure 8.9, which shows gel times for acid-catalyzed TEOS systems as a function of R and the initial alcohol:TEOS molar ratio.

  Finally, since water is the byproduct of the condensation reaction, large values of R promote siloxane bond hydrolysis.

During condensation, the same parameters can be controlled.

- *Influence of pH*
  Polymerization to form siloxane bonds occurs by either an alcohol-producing or a water-producing condensation reaction. A typical sequence of condensation products is monomer, dimer, linear trimer, cyclic trimer, cyclic tetramer, and higher order rings. This sequence of condensation requires both depolymerization (ring opening) and the availability of monomers which are in solution equilibrium with the oligomeric species and/or are generated by depolymerization.

- *Nature and concentration of catalyst*
  As with hydrolysis, condensation can proceed without a catalyst, although their use in organosiloxanes is very helpful. Furthermore, the same type of catalysts are employed: generally, those compounds which exhibit acidic or basic characteristics.

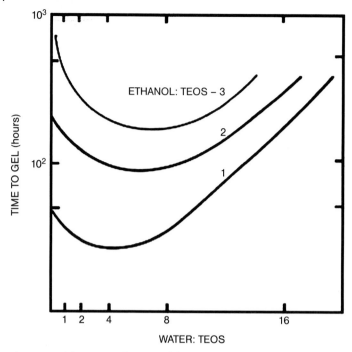

Figure 8.9 Gel times as a function of the water:TEOS ratio (R).

It has been shown that condensation reactions are acid- and base-specific. Again, catalysts which dictate a specific pH can, and do, drive the type of silica particle produced, as seen in the previous discussion on pH.

It is generally believed that the acid-catalyzed condensation mechanism involves a protonated silanol species. Protonation of the silanol makes the silicon more electrophilic and thus susceptible to nucleophilic attack. The most basic silanol species (silanols contained in monomers or weakly branched oligomers) are the most likely to be protonated. Therefore, condensation reactions may occur preferentially between neutral species and protonated silanols situated on monomers, end groups of chains, etc.

The most widely accepted mechanism for the base-catalyzed condensation reaction involves the attack of a nucleophilic deprotonated silanol on a neutral silicic acid.

Furthermore, it is generally believed that the base-catalyzed condensation mechanism involves penta- or hexa-coordinated silicon intermediates or transition states, similar to that of a S_N2 type mechanism.

As with hydrolysis, the relative rates of reaction of different species depend on steric effects and the charge on the transition state. Thus, for acid hydrolysis with a positively charged transition state stabilized by electron-donating groups, $(RO)_3SiOH$ condenses faster than $(RO)_2Si(OH)_2$, which condenses faster than $(RO)Si(OH)_3$, etc.

This means that for acid-catalyzed reactions, the first step of the hydrolysis is the fastest, and the product of this first step also undergoes the fastest condensation (Fig. 8.10). Hence, an open network structure results initially, followed by further hydrolysis and cross-condensation reactions. In contrast, in base-catalyzed conditions the negatively charged transition state becomes more stable as more hydroxyl groups replace the electron-donating alkoxy groups. Thus, successive hydrolysis steps occur increasingly rapidly, and the fully hydrolyzed species undergoes the fastest condensation reactions. As a consequence, in base-catalyzed reactions highly crosslinked large sol particles are initially obtained which eventually link to form gels with large pores between the interconnected particles. Hence, the choice of acid or base catalysis has a substantial influence on the nature of the gel which is formed, and will have this effect on the encapsulation properties.

- Acid-catalyzed
    - yield primarily linear or randomly branched polymer

- Base-catalyzed
    - yield highly branched clusters

Figure 8.10 Gel structure for acid- and base-catalyzed reactions.

The situation is more complicated in the preparation of multi-component gels comprising more than one type of metal ion. The degree of chemical homogeneity of the resulting gel is strongly affected not only by the level of mixing of different alkoxides in the precursor solution, but also by reactivity of each alkoxide species towards water. In a mixture of alkoxides with different hydrolysis rates, the hydrolysis and condensation regimes should be separated to form a homogeneous gel. Decoupling the two reaction regimes allows hetero-condensation between the different hydrolyzed alkoxides and reduces the degree of cation segregation as a result of homo-condensation reactions. Homogeneity can be obtained by chemically modifying the alkoxide with the highest hydrolysis rate or by partial hydrolysis of the metal alkoxide with the lowest hydrolysis rate prior to mixing with other alkoxides. Several other approaches are used to separate the hydrolysis and condensation regimes and hence improve the gel homogeneity. These include, lowering the concentration of water for hydrolysis, partial hydrolysis of the slowest-reacting species, acid or base catalysis, formation of hetero-metallic alkoxides, and reducing the hydrolysis rate of highly reactive species using chelating agents [13].

8.4.2.2 Metal–Chelate Sol-Gel Method

Another way to trap the randomness of the solution state is to use precursor solutions that can be converted to an amorphous glassy state when the solvent is removed. The basic idea behind the metal–chelate gel formation method is to reduce the concentration of free-metal ions (mostly hydrated in aqueous solutions) in the precursor solution by the formation of soluble chelate complexes [2]. From this point of view, strong chelating agents such as citric acid and ethylene-diamine-tetra-acetic acid (EDTA) are preferably used in the synthesis of multi-metal oxides which greatly expands a range of experimental conditions such as pH of the solution, temperature, and metal concentrations where gelation can occur upon evaporation of the solvent. On concentrating a solution of metal–chelate complexes, the viscosity rises. This high viscosity prevents the precipitation of metal–chelate complexes and a clear gel forms upon further dehydration. On cooling, the viscous dehydrated gel undergoes a rubber to glass transition and converts to a brittle polymer glass [13].

Inorganic precursors are much cheaper and easier to handle than metal alkoxides. Water, however, behaves both as a solvent of ionic compounds and as a chemical reagent for the hydrolysis of molecular precursors. Hydrolysis and condensation reactions are therefore more difficult to control. Several monomeric or oligomeric solute species are often formed simultaneously when a metal salt is dissolved in water. These species are in dynamic equilibrium and it is not easy to know which one is going to nucleate the solid phase. Moreover, anions are known to play an important role even when they do not seem to be chemically involved in the hydrolysis and condensation of metal cations. As for the chemical modification of metal alkoxides by organic ligands, anions are able to form complexes with cationic precursors, to provide a chemical control of the hydrolysis and condensation reactions. The aqueous chemistry of inorganic salts is quite complicated owing to the occurrence of hydrolysis and condensation reactions which leads to formation of numerous molecular species [36]. Their analysis is based on the so-called "partial charge model" that provides a useful guideline for a rational explanation of these reactions. This model applies not only to the sol-gel synthesis of metal oxides but also to other wet chemical methods from aqueous solutions such as (co-)precipitation methods or thermohydrolysis [37–40].

The foundations for understanding the aqueous chemistry of metal ions were laid by Bjerrum [41], Werner [42] and Pfeiffer [43] in 1907. More detailed development of the ideas only occurred much later, by Sillen [44] in 1959 and quite recently in the partial charge model of Livage et al. [45]. Here, the key to understanding hydrolytic equilibria in aqueous solutions of metal salts lies in the calculation of the partial charges of the atoms.

When two atoms of different electronegativities form a bond, an electron transfer occurs. The more electronegative atom acquires a negative charge and its electronegativity decreases. The less electronegative atom acquires a positive charge and its electronegativity increases. The electronegativities of each atom converge towards an average value χ, indicating that equilibrium is attained and electron transfer is completed. This is known as the electronegativity equalization principle stated by Sanderson [46]. The Partial Charge Model (PCM) is based on this principle.

When two atoms combine chemically, both the electronegativity χ_i of a given atom i and its partial charge δ_i will vary. These two parameters are related, and a linear relationship is usually assumed as follows:

$$\chi_i = \chi_i^o + \eta_i^o \delta_i \tag{6}$$

where η_i^o is the hardness of atom i. Hardness is related to the softness $\sigma_i^o = 1/\eta_i^o$ which provides a measure of the polarizability of the electronic cloud around atom i. Softness increases with the size of the electronic cloud (i.e., with the radius r of atom i). Therefore, hardness varies as $1/r$. According to the Allred–Rochow scale, electronegativity is based on the electrostatic interaction between electrons and nucleus and is proportional to Z_{eff}/r^2, where Z_{eff} is the effective charge on the nucleus and r is the covalent radius. Hardness may then be approximated as:

$$\eta_i^o = k\sqrt{\chi_i^o} \tag{7}$$

where k is a constant that depends on the electronegativity scale. Using the Allred–Rochow scale, it can be shown that k = 1.36.

At equilibrium, in the chemical combination, the equalization principle gives $\chi_i = \chi$ or

$$\chi_i^o + \eta_i^o \, \delta_i = \chi \tag{8}$$

and the partial charge:

$$\delta_i = (\chi - \chi_i^o) / \eta_i^o \tag{9}$$

If softness (the reciprocal of hardness) is used, Eq. (9) can also be written as: $\delta i = \sigma_i^o (\chi - \chi_i^o)$ where

$$\sigma_i^o = \left(k \sqrt{\chi_i^o} \right)^{-1} \tag{10}$$

Charge conservation in the chemical combination imposes that the total charge z of a given chemical species is equal to the sum of the partial charges of all individual atoms $z = \Sigma_i \delta_i$. This, together with Eqs. (7) and (9) leads to the following expressions for the mean electronegativity:

$$\chi = \frac{\Sigma_i \sqrt{\chi_i^o} + k z}{\Sigma_i \left(1 / \sqrt{\chi_i^o} \right)} \tag{11}$$

The PCM provides an easy way to obtain the charge distribution on each atom and the mean electronegativity of chemical species. From Eq. (10), χ may be calculated for any atomic combination, from the composition, the total charge z of the chemical combination and the electronegativity of each atom.

The aqueous chemistry can thus be described as a function of three parameters: the charge z of the cation; the pH of the solution; and the mean electronegativity of the chemical species (metal or ligands). Three diagrams can be proposed to describe each chemical process in the aqueous chemistry:

- By plotting z as a function of pH, a *pH-charge diagram* is obtained, accounting for the hydrolysis of cations. This indicates which ligands are involved in the coordination sphere of the metal cation: H_2O, OH^- or O^{2-}. Several monomeric precursors have usually to be taken into account. However, some elements cannot be protonated or deprotonated in aqueous solutions. They give strong acids or bases, and condensation is not possible in these cases.
- By plotting z against χ , an *electronegativity-charge diagram* can be built, accounting for condensation of cations. This allows one to predict which precursors are able to give condensed species. However, condensation cannot occur unless some charge conditions are fulfilled ($\delta_{OH} < 0$ and $\delta_M \le 0,3$). Olation of low-valent cations leads to polycations and the precipitation of hydroxides or oxy-hydroxides. Oxolation of high-valent cations leads to polyanions and the precipitation of hydrous oxides.
- By plotting χ as a function of pH, a *pH-electronegativity diagram* is found, accounting for complexation of cations. This indicates in which pH range complexation may occur, and it can be used for the chemical modification of aqueous precursors with anions. Monovalent anions X^- are either complexing, or not. In the first case, they lead to the precipitation of basic salts, while in the second case they simply behave as counterions. Multivalent anions X^{n-} are more interesting, as these exhibit several protonated forms and can be complexing in a given range of pH, but not complexing in another range. In this case, complexation occurs during the first steps of the hydrolysis and condensation process. Anions are then removed before the precipitation of solid phases. Such multivalent anions can be used to control the size, shape and morphology of colloidal particles.

When these three diagrams are used together, they provide a very useful and powerful tool to forecast the chemical behavior of an element in aqueous solution.

The PCM is a simple and convenient method for the broad prediction of the behaviour of elements in solution. Electronegativity – that is, the electronic chemical potential – seems to be an adequate and useful concept in the understanding of the direction of electron transfers between atoms. Therefore, it is a useful concept in predicting the evolution of a chemical system, since the principle of electronegativity equalization can broadly characterize an equilibrium situation.

Because of its many approximations – and mostly because it contains no concentration or structural considerations – the model is unable to provide quantitative information, and may in no circumstances serve as a substitute for thermodynamic data. When structural issues such as the presence of multiple bonds are likely to affect the reactivity of a species, the model should be used with extreme care. It must be noted that the model does not allow for any distinction between isomers, and therefore it must not be used to identify any property related to the structure of the combination. In addition, the analysis of hydrolysis and condensation reactions with the PCM would require modifying the hardness value for each cation.

This is also true for the study of complexation, which may only be treated from an extremely qualitative standpoint.

In spite of these disadvantages, the PCM does rationalize the chemistry of cations in solution with regard to hydrolysis and condensation processes.

8.4.3
Polymeric Sol-Gel Method

The third sol-gel route involves the formation of an organic polymeric network. This can be carried out in two ways. The first method is an "in-situ" polymerization route where the gel network is made by polymerization of organic monomers. If this is carried out in the presence of metal ions, it can be expected that all metal ions are distributed homogeneously into the organic polymeric network. When the process is a combination of metal–chelate complex formation and an in-situ polymerization, the process is termed the "polymerized complex method" (Fig. 8.11).

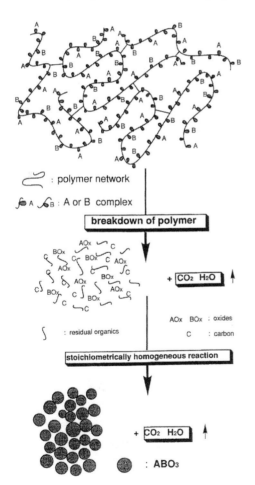

Figure 8.11 Polymerized complex or polymer precursor method [134].

The second method involves the preparation of a viscous solution system containing metal ions, polymers and a suitable solvent and the process will be simply called the "polymer precursor method".

8.4.3.1 Polymeric Complex Method

One of the characteristics of this method [36] is that it includes a combined process of metal-complex formation and in-situ polymerization of organics. The basic idea is to reduce the individualities of different metal ions, and this can be achieved by encircling stable metal–chelate complexes steadily by a growing polymeric network. The principle of the polymer complex method is thus to obtain a polymeric resin comprising randomly branched polymers, throughout which the cations are uniformly distributed. The immobilization of metal–chelate complexes in such a rigid organic polymeric network can reduce segregations of particular metals during the decomposition process of the polymer at high temperatures.

The Pechini method is a representative example of this approach as it is the most widely used polymer complex method. The original patent [47] describes the formation of a polymeric network by polyesterification of an α-hydroxy-carboxylic acid (e.g., citric acid) and a polyhydroxy-alcohol (e.g., ethylene glycol). It is an effective method for the synthesis of multicomponent oxides by dissolving metal nitrates in water, adding an α-hydroxy-carboxylic acid to chelate the metal cations, and adding ethylene glycol. Upon heating at moderate temperature (100 °C to 1500 °C), a polyesterification reaction occurs that results in an ester which still contains two alco-

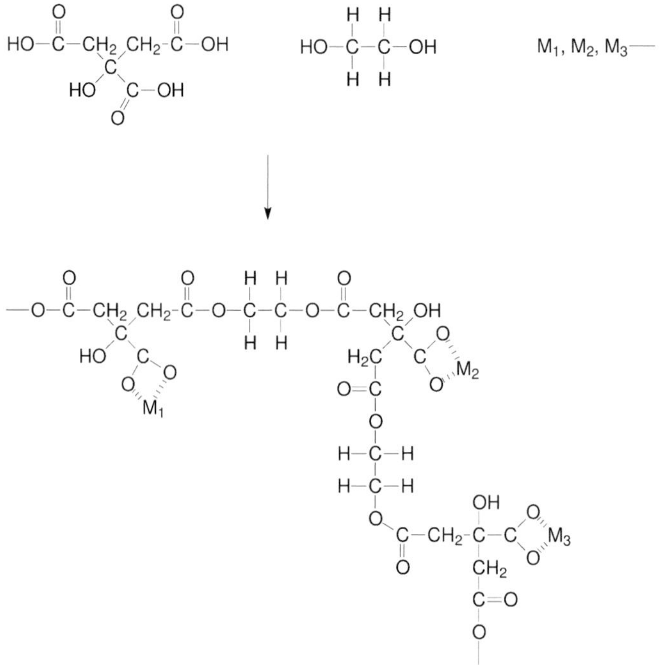

Figure 8.12 The sol-gel Pechini process.

holic hydroxyl groups and two carboxylic acid groups; hence, further esterification reactions can take place. This results in the formation of a stable three-dimensional gel network in which the individual metal-complexes are immobilized, preserving the initial stoichiometric ratio of metal ions. Calcination of the polymeric resin under relatively mild conditions ($>300\ °C$) causes a breakdown of the polymer, and further appropriate heat treatments lead to formation of a homogeneous crystalline oxide.

The advantage of the Pechini method is that the viscosity of the solution and the polymer molecular weight of the solution can be tailored by varying the citric acid:ethylene glycol ratio, and also the synthesis temperature. This allows thickness to be controlled in the preparation of films. However, the large shrinkage during heat treatment can render the fabrication of fibers or films rather difficult.

8.4.3.2 Polymer Precursor Method

The first step is to prepare an aqueous precursor solution containing metal salts, followed by addition of a water-soluble polymer such as polyvinyl alcohol (PVA), polyacrylic acid (PAA), polyethylene imine (PEI), all of which are organic polymers coordinating to cations. The basic idea behind this sol-gel route is to lower the mobility of free metal ions in a polymer solution by increasing interaction between metals and polymers [36]. Removal of the excess water forces the polymer species into closer proximity, which increases the probability of crosslinking between them. The metal ions act as crosslinking centers between polymers when the polymer possesses the ability to form metal–polymer complexes. Random crosslinking across polymer chains would entrap water in growing three-dimensional networks, which can convert the system into a stable gel. The immobilization of cations in highly crosslinked polymers is thought to imply inhibition of segregation of metal ions throughout the resulting gel. The polymer is chemically modified during the gel formation, and modifies the rheological characteristics of the aqueous solution.

8.5
Applications

8.5.1
Scope

A reviewer of modern advances in sol-gel encapsulation is faced with two options. The subject can be organized along the lines of application of sol-gel encapsulation, or it can be organized based on the chemical composition. Both approaches have their advantages and disadvantages. The application-oriented subdivision will clearly follow today's trend for the strategic development of materials and place emphasis on their implementation in various processes. On the one hand, the materials-oriented approach is probably more fundamental and may draw generic conclusions and, hopefully, allow extrapolation or prediction by developing models, theorems, mechanisms related to the formation and behavior of the encapsulated molecules and the functional description of the encapsulator. However, the vast majority of data available are more concerned with the specifics of the application than with the generalities of the phenomena, and it is for this reason that the authors have also included the first option in this part of the chapter under the heading "Smart materials". This view is reflected in a review published in 1998 by Uhlman et al. [24], who described a state of the art and future prospects based on a questionnaire received from almost 50 leaders in the sol-gel community. The authors acknowledged the potential of the field, but argued that its achievement was to depend upon greatly increased involvement of the sol-gel community with applications and with carrying out a much larger fraction of the sol-gel research in the context of applications.

A more fundamental view is reflected in the second approach, however, where sol-gel encapsulation science is divided into sections related to molecular confinement studies, the use of sol-gel encapsulation as a synthetic intermediate, and the structural studies. The majority of the articles and ideas reviewed were published in 2000 and later. Earlier studies were only included if they represented a major breakthrough or bore a special significance to the field. Finally, although the following review describes applications grouped into certain sections, much of the material may bear relevance to other sections, and it is clear that the emergent science of sol-gel will only benefit from such cross-fertilization.

8.5.2
Smart Materials

8.5.2.1   Smart Materials as Biomedical Assays and Biosensors
The major advantages of sol-gel-derived silicate materials for the immobilization of proteins are that:

- they can be made to be optically transparent, making them ideal for the development of chemical and biochemical sensors that rely on changes in absorbance of fluorescence
- they are open to a wide variety of chemical modifications based on the inclusion of various polymer additives, redox modifiers and organically modified silanes, resulting in electrically conducting materials suitable for electrochemical sensors
- they have tuneable pore size distributions, which allow small molecules and ions to diffuse into the matrix while large biomolecules remain trapped in the pores allowing size-dependent bioanalysis. The latter advantage is now routinely applied in size-exclusion gel-permeation chromatography and capillary electrochromatography [48]

The entrapment of proteins via sol-gel processes poses a set of requirements to the type of gel to be used:

1. The method must be amenable to an aqueous environment, since this is required to maintain the biological function of the biomolecules.
2. The polymerization of the gel must be compatible with the range of pH (4–10) and ionic strength (0.01 M to 1.0 M) required by the biomolecule.
3. The processes must occur close to room temperature in order to maintain the proteins in their native conformation.
4. The gel must have a pore structure sufficiently small to prevent leaching of the biomolecule, but simultaneously sufficiently large to allow rapid diffusion of the analyte.
5. The gel backbone should be chemically modifiable and tuneable to maximize the activity of the entrapped molecule.
6. The gel should be transparent or electrically conductive in its final form, depending on the type of detection envisaged.
7. Fabrication of the gel should be straightforward and allow formulation in the shape of bulk, wires, films, columns, arrays, etc.

In practice, gels are usually obtained by mixing TEOS or TMOS in water, with or without the addition of methanol. This sol is then added to a buffer containing the biomolecule and left standing to age. Gelation times clearly range from minutes to hours, depending on the temperature, composition, and pH. Gels can be used in the form of monoliths, but in view of their use as a optically monitored sensor, a thin film is often preferred. This can easily be obtained by coating transparent substrates such as microscope slides or spectrophotometric cuvettes. In order to obtain good wettability of these surfaces, a measure of MeOH is usually added (in addition to the alcohol that is formed during hydrolysis). This addition is however limited by the increased rate of denaturation of the biologically active species that occurs at high doses.

The initial gels are soft, have a high water content (50–80%), and exhibit large pores (up to 200 nm diameter). Aging of the network occurs over a relatively long time (days to weeks) and promotes further condensation and strengthening of the network. This is concurrent with shrinkage (10 to 30 vol.%) and pore size reduction. After the final drying most of the water and alcohol is lost, leaving open pores in the range of 2 to 20 nm and an overall shrinkage of the material by up to 85% of its original volume.

Bioanalytical assay methods have made extensive use of sol-gel encapsulation. One application consists of providing the biomedical and biochemical community with ready-made reactant kits where the different components are already present in a gelled matrix consisting of acid-hydrolyzed TEOS with added bioreactants. This offers the advantage of simplicity to the laboratory technician or scientist. The gellified contents of the reactant medium can even be pre-affixed to the wall of the test vessel, such as a spectrophotometric or fluorimetric cuvette [49]. This was performed by laying a cuvette on its side during the gelation, the result being transparent, uniform films. Such a system clearly requires good permeability of the gel, sufficiently low detection limits, independence from interfering agents, a good linearity of response, and a long shelf life. These points claim to have been achieved for the fluorimetric detection of uric acid by Amplex Red, and for an enzyme-based biosensor immobilized on glass slides for the fluorimetric detection of superoxide radicals in phospholipid vesicles [50]. The sensitivity of these methods is 20 nM, which approaches the sensitivity of much more involved and costly instrumental measurements. These ready-to-use kits were even shown to be reusable at least four times without loss of activity, obviously with intermittent washing.

8.5.2.2 Smart Materials for Controlled Release of Drugs

Porous sol-gel-made silica particles have been investigated as encapsulation matrices for the controlled release of substances in food and pharmaceutical applications. As is the case for the vast majority in this "Smart materials' section, polysiloxane polymers are the preferred medium. The primary synthetic step for the reliable formation of the gels is the pre-doped hydrolysis of TEOS-containing precursors (see Section 8.0.0 on sol-gel chemistries).

Drug delivery systems have been developed for doxorubicin [51] (a drug used to treat leukemia but which has cardiac risk factors), and for flavors such as decanoic acid [52]. In both cases, the encapsulation efficiency as given by load, recovery speed and recovered quantity was assessed. In addition, the stability (shelf life) of such gel matrices in terms of stability of the pharmaceutically active ingredient was increased compared to their free form. Similar applications for coatings for the controlled release of biocides [53] and other pharmaceuticals such as vitamins [54] have also been made. Clearly, gel porosity, pore size distribution, temperature, pH all influence the release, and these parameters must be optimized.

8.5.2.3 Smart Optical Materials

Lisa Klein summarized her early studies of optical applications [55] for sol-gel based silica-glass production which rivals the traditional glass melting technology in terms of optical quality. Whilst no new optical phenomena have been produced with this newer technology, it does offer new ways to assemble materials. Early reports noted that particles trapped in nanosized pores may be described as quantum dots and exhibit quantum confinement phenomena [56]; this led to the observation of higher order optical nonlinearity in encapsulated gold nanoclusters [57] and zinc tetrasulphophthalocyanine molecules [58]. Interestingly, this is one of the few fields where systems other than those based on silica have been used. Particular technological requirements of bulk glasses have resulted in the elaboration of sol-gel syntheses of two- and three-oxide end products based on precursors such as $B(OCH_3)_3$, $Ge(OC_2H_5)_3$, H_3PO_4, $La(NO_3)_3$, $Al(OC_4H_9)_3$, $LiNO_3$ and $PbCOOCH_3$, and a host of other metals in the case of optical coatings. Geometries ranging from monoliths over coatings to fibers are available. Monoliths in particular replicate their molds quite accurately and miniaturize all features upon drying, without distortion. Transparent physical hybrids between silica and a series or organics (PVA, PMMA, PEO) and ormosils have also been produced [59]. These preliminary findings have clearly enabled others to develop optical detection methods based on encapsulates in bulk and thin-film sol-gel matrices [60]. In addition, Sb-doped silica sol-gels have been shown to have light-dependent optical properties, such that under dark conditions they are blue-green, but change their color reversibly to yellow under ultraviolet light. Organic additives may result in an altered photonic response, such as the light-induced electron transfer between flavin mononucleotide (FMN) and nicotinamide adenine dinucleotide (NAD) and methylene blue in TMOS-derived gels [61], and this may open the way to photochromic memories and tuneable optical properties.

8.5.2.4 Sol-Gel Encapsulation for Nonbiological Sensing

Sol-gel-encapsulated molecules may be used for a variety of sensing actions outside the biological context. Two recent examples are optical pH sensing and optical temperature measurement. The use of immobilized ruthenium (II) polypyridyl complexes as chemical transducers for optical pH sensing [62] has been advocated as a preferred, noninvasive analysis of foodstuffs by colorimetric sensor patches. In addition, a solid pH indicator was developed by encapsulating methyl orange into a silica network [63] with full retention of its well-known color changes. Reversible changes in the optical absorption maxima in relation to intramolecular charge transfer transitions in encapsulated pyridinium N-phenoxide betaine dye was used to monitor temperatures between 30 °C and 80 °C. Such a thermochromic system could be used to avoid interference from environmental factors when temperature measurements are to be made at nonaccessible sites. Examples might be underwater environments, underground geochemical sites, or the internal parts building structures [64].

8.5.2.5 Sol-Gel Encapsulation for Coatings

There is a current need for alternative coatings that can provide corrosion resistance to metals and alloy surfaces, without the environmental hazards posed by conventional coatings. Thus, sol-gel-based systems have been developed for this purpose, largely based on the ease of deposition of uniform, crack-free coatings with thicknesses ranging from less than 100 nm to several micrometers over large areas under atmospheric conditions. Coatings based on the hydrolyzates of 2-cyanoethyltriethoxysilane, ethyltrimethoxysilane, *n*-butyltrimethoxysilane and benzyltrimethoxysilane exhibited attractive mechanical properties as derived from peel tests, resistance to acids, contact angles, and potentiodynamic tests [65]. Anticorrosion coatings based on the silane-system with the addition of nonchromate inhibitors such as $NaVO_3$, Na_2MoO_4 and $Ce(NO_3)_3$ also showed an improvement in the potentiodynamic curves, together with self-healing capacity [66], in a defect by slow release of the inhibitor from the nanoreservoir in the sol-gel coating. It is clear that spin-coating, spray-coating and dip-coating represent attractive deposition techniques for encapsulated gels, and these have therefore also been applied to the deposition of transparent SnO_2 films, with high electrical conductivity and transparency [67], as well as on textiles [68].

8.5.2.6 Nanomaterial and Catalyst

Studies of molecules inside the cavities of a three-dimensional network with pore diameters of a few nanometers will clearly impact upon today's interest in nanotechnology. Thus, the dispersion of catalytic particles, biosensitive molecules and dispersed precursor phases for a host of other applications has attracted attention. A true encapsulation of catalytically active nanoparticles was recently achieved using an elegant technique based on the coating of micelles [69] stabilized by organic stabilizers (CTAB, cetyltrimethylammonium bromide) in dry toluene that were charged with aqueous solutions containing $(NH4)_2PtCL_4$ and $Ag(NO_3)$. Upon reduction with hydrazine monohydrate, metallic nanoparticles resulted that could be coated with a nanometric layer of silica following the addition of TEOS. These nanocomposites showed catalytic activity in toluene hydrogenation that was identical to that for metal-impregnated silica at 573 K, but retained 60% of their activity at 1123 K; by comparison, the impregnated material lost 95% of its original catalytic activity. Increased stability of encapsulated particles was also established for many biologically active materials [70], though most likely for reasons other than those operating at the high temperatures encountered here. The next step in the development of encapsulated nanomaterials is to use ordered mesoporous structures in which the chemical and mechanochemical environment of the encapsulate experiences a much narrower degree of freedom. Such ordered mesoporous materials have been developed recently for silica- [71] and alumina-based [72] gels, and future encapsulates will undoubtedly permit much closer investigation into substrate–adsorbate interaction. In fact, "bimodal mesoporous silica spheres" have already been developed which show multiple but discrete ranges in porosity [73]. This holds particular promise in areas such as combinatorial synthesis, selective catalysts, and multifunctional response sensors.

8.5.2.7 Smart Materials for Environmental Technologies

Phenol is an important contaminant in environmental aspects of ground and surface water pollution. Its monitoring has become significant in recent years and has been performed using expensive or time-consuming spectrophotometric [74] as well as chromatographic [75] methods. Good sensitivities (detection limit 10^{-7} M) and linear responses between 1.2×10^{-7} M and 2.6×10^{-4} M were obtained with vapor impregnation of a titania gel matrix onto a glassy carbon electrode into which a solution containing the enzyme tyrosinase was absorbed [76]. This electrode was exposed to Ti-isopropoxide vapor at 25 °C for 6 h, resulting in a porous titania membrane being formed on the surface of the electrode by slow hydrolysis with water which, presumably, was present in the absorbate. Electrical detection was performed using cyclic voltametric and amperometric measurements, and yielded a fast response (<5 s) resulting from the porous structure of the titania gel. The pristine biosensor retained 94% of its activity after storage for 80 days at 4 °C, and intermittently used sensors were somewhat less active. The various advantages of biogels for amperometric biosensing have recently been discussed, along with common designs of sol-gel-derived bioelectrodes [77].

In addition to the detection of environmentally harmful species, patented sol-gel encapsulation chemistries may also play a vital role as sequestrants of waste containing high salt loadings [78]. Customarily, Portland cement has been used for such confinement, but high salt contents preclude a sufficiently extensive hydration of the cement, and this results in poor mechanical quality. Maintaining structural integrity, chemical durability and leach resistance has been the focal point of research involving the use of "polyceram" hybrid organic/inorganic materials as waste containers. The composites obtained by in-situ hydrolysis of acidified TEOS in the presence of HF and polybutadiene can be added to waste sludge, gelled and dried in a vacuum furnace to a compact mass with sufficiently high compressive strength and reasonably low leachability. However, additional gel infiltrations were required to improve the retentive quality, which gives reason to believe that such an encapsulation scheme may require further research to obtain dense materials, before its application. In addition, the preparative steps are rather involved compared to the Portland cement method. Additionally, the resistance to high temperatures may be insufficient in view of the rather large proportion of organics remaining after encapsulation.

8.5.2.8 Bacteria and Cell Therapy

The first group to encapsulate live bacteria [79] was led by Jacques Livage, who has spurred a long tradition of physico-chemical sol-gels in France. An *Escherichia coli* B (lac+) strain was cultivated in silica sol-gel media obtained by two different routes. An aqueous route was based on the neutralization of sodium silicate with HCl to pH 7 to which, under buffered conditions, a centrifuged bacterial culture was added before gellation occurred at room temperature. A comparatively simpler, but more aggressive, route consisted of encapsulating bacteria in a partially prehydrolyzed mixture of TMOS/HCl in acid conditions and neutralizing the whole with suitable buffers. This led to bacterial cell concentrations of 2.4×10^8 cells mL^{-1}, which were then subjected to enzymatic tests (β-galactosidase activity) in order to assess their survival. Cell growth could be followed using simple counting of the developing colonies. Although both encapsulation methods resulted in viable colonies, the methanol produced in the condensation reaction of TMOS could possibly have exceeded the tolerance threshold of the bacteria. Another much-rumored finding was that encapsulation seemed to offer an efficient protection against aging when compared to free bacteria suspensions.

These findings have now been expanded upon in studies performed on encapsulated genetically engineered *Moxarella* sp. cells for the surface expression of the pesticide-hydrolyzing enzyme organophosphorus hydrolase [80]. Again, the immobilization of these cells (using a sodium silicate process as described above) led to a 5% reduction in activity compared to a 30% activity reduction in the case of free cells in buffer; this indicated that immobilization leads to stabilization, a key parameter in biosensor development. Finally, encapsulations of functional cells by sol-gel silica are currently being studied for cell therapy in vivo [81] and after cyropreservation of the encapsulated material in liquid nitrogen.

8.5.2.9 Perovskites and Ceramic Superconductors

The development of sol-gel techniques for the chemical solution deposition of per-ovskite thin films dates back to the mid-1980s [82], for example in the development of lead zirconate titanate (PZT) thin films. The use of sol-gel chemistries finds its rationale in their ability to immobilize large concentrations of different ionic species together with molecular or atomic resolution in the medium under deposition, thereby ensuring maximal homogeneity in composition and distribution in the final layer. Moreover, the use of liquid media renders the sol-gel technique ideally suited for continuous coverage of large areas and complex shapes. The selection of the starting materials is dictated by solubility and reactivity considerations of the different precursor species. The most frequently used sol-gel approaches are based on alkoxide precursors with alcohols [83], chelate processes that use modifying ligands such as acetic acid, citrates and amines [84], and metalo-organic decomposition (MOD) routes [85]. When properly carried out, sol-gel processes offer excellent control and reproducibility. Thin films with excellent properties have been prepared for a number of perovskite materials including PZT, with high dielectric constant and ferroelectric hysteresis, $BaTiO_3$ and $BaZrO_3$ [86] exhibiting high dielectric constants, and ZrW_2O_8 showing negative thermal expansion [87]. Control of nucleation at the interface with the substrate has also been used for the preparation of epitaxial and highly oriented superconducting thin films, such as NBCO and YBCO [88].

8.5.3
Sol-Gel Encapsulation Science

8.5.3.1 Molecular Confinement Studies

Many studies have appeared in which sol-gel methods were applied to biomolecules for their "encapsulation", "entrapment", "confinement", "incarceration", or "bio-immobilization". In this context, it is remarkable that none of those terms appeared in the sol-gel "bible" [29], though this can easily be explained by the change in emphasis from the sol-gel medium during the 1990s to the more recent focus on the phases dispersed in it. Only a relatively small number of investigations have focused on the structural consequences of encapsulation on the dispersed species; however, an excellent review by Jin et al. [89] has discussed the more fundamental aspects of this supposed immobilization.

All authors concur in the statement that their research findings show that biomolecules immobilized in a sol-gel-derived matrix can, to a large extent, retain their functional characteristics such as ligand binding, oxidation/reduction, fermentation and enzyme activity. The porosity of sol-gel glasses allows small analyte molecules to diffuse into the matrix, while the large protein macromolecules remain physically trapped in the pores. The transparency of the matrix enables one to use optical spectroscopic methods to characterize the reactions that occur in the pores [60]. Furthermore, sol-gel materials are ideal candidates as hosts for biomolecular dopants because they are synthesized at low temperatures and under fairly mild reaction conditions. The bio-immobilization of enzymes, antibodies and other proteins has resulted in their use for the selective extraction, delivery, separation, conversion, and detection of agents. The main difficulties encountered are related to leaching and desorption of biomolecules [90], the initial and long-term denaturation of the biomolecules, and the difficulty in controlling the orientation of the biomolecule [91].

The chemical structural changes occurring during gellation were studied using mainly optical methods (UV-visible absorbance , fluorescence spectroscopy, resonance Raman spectroscopy, flash photolysis, circular dichroism) and, to a lesser extent, dielectric relaxation measurements, magnetic measurements, electron paramagnetic resonance (EPR) and nuclear magnetic resonance (NMR). The geometric structures of the gel and pores were determined using transmission electron microscopy (TEM), scanning electron microscopy (SEM) and nitrogen adsorption, while the composition was studied using X-ray diffraction (XRD), thermal analysis, infra-red spectrometry and gas chromatography.

Preliminary studies have been performed to assess the distribution, the conformational stability and the dynamic mobility, the accessibility, the functionality and the response times of the entrapped molecules. Depending on their relative concentration and reactivity, the biomolecules may disperse homogeneously on a molecular scale within the pores, they can be adsorbed on specific sites between the silica–solvent interface, or they can partition into nanocomposite materials or even segregate at high concentration [92]. Furthermore, the presence of hydrophilic, hydrophobic, ionic- or hydrogen-bonding groups within the proteins and/or silica matrix can result in templating effects of the silica around the protein [93]. Overall, many studies have indicated that the confinement of proteins in sol-gel matrices does not simply result in a solution-like environment, and that interaction with the matrix may show a distribution of dynamic motions, stabilities and binding constants that changes as the material ages.

Although some cases of denaturation are known (particularly for large molecules that are unable to unfold completely inside the pores), the biological activity – and therefore also the conformational stability – of many are retained after encapsulation. In fact, a resistance to denaturation (higher temperatures or higher concentrations of denaturant required) has been noted after encapsulation [94]. Not only was enhanced thermal protein stability noted [95], but it was suggested that the rather crowded environment in the encapsulated state mimics the effects of confinement and crowding in a living cell better than does a free protein in a buffered aqueous medium. In order to quantify denaturation, evolution of the mean ellipticity of a protein molecule was taken as a measure of its geometric deformation, and followed by monitoring optical rotary dispersion spectra at different temperatures. It was concluded that although the onset of denaturation (60–70 °C) did not change noticeably, its extent was greatly diminished upon encapsulation and that in some cases the denaturation was even partly reversible [95].

The accessibility has been modeled phenomenologically [89] in terms of four types of microenvironments present in encapsulated form: accessible functional molecules present in large pores; nonaccessible ones; and denatured forms in accessible and nonaccessible environments. A further numerical or analytical modeling based on this picture may lead to a theoretical understanding of the kinetics encountered. In a survey of more than ten kinetic studies a general trend was discovered wherein reaction rate constants (k_{cat}) for the binding of an analyte to an encapsulated protein were similar or lower than non-entrapped species. In addition, the binding constants (K_M) were systematically higher, yielding a k_{cat}:K_M ratio that was (only!) zero to three orders of magnitude smaller in encapsulated form. These data were in quite good accordance with a model where the encapsulated molecule is rather more difficult to access and where the substrate–enzyme complex is rather more stable in the encapsulated environment. Such studies have a direct bearing on response times and regeneration or recovery of the sensors. In addition to purely kinetic and stability considerations, a reduction of diffusion constants due to increased viscosity of entrapped solvents and electrostatic interactions with the pore walls may lead to very low response times (10 min and more). This finding clearly demonstrates that ultra-thin films (of the order of a few hundred nanometers) with high doses of active molecules in porous matrices may be required to obtain response times of seconds or less.

A BET study of tetrasulfophthalocyanine as a model for a biologically active oxidant, in TMOS gels [96] revealed specific areas between 400 m^2 g^{-1} to 700 m^2 g^{-1} and a specific vol.% of mesopores between 10 and 80%, together with some hysteresis effects in adsorption–desorption isotherms. In addition, ^{29}Si-NMR, diffuse reflectance and UV-visible spectroscopy revealed that the phthalocyanines occurred in aggregated form, and that isolation by covalent grafting may be more promising for these relatively small species.

8.5.3.2 Synthetic Intermediates

The development of society has often been related to the nature of the materials that man was using or developing (stone, iron, bronze, biomaterials) and, from the beginning, the development of many materials was based on high-temperature processing. During the twentieth century, XRD and quantum chemistry spurred the development of materials science and solid-state chemistry, which was still based on the mixing of solid powders and their treatment at high temperatures. Thus, the solid-state synthesis of many materials depends on an experimental approach termed "shake-and-bake", thereby illustrating the often painstaking cycles of comminution and thermal treatment in order to obtain sufficient homogeneity and chemical specificity in the fired materials.

Living organisms have developed a totally different strategy from those used by engineers [97]. Silicate glasses, for example, are made by melting silica sand above 1000 °C, whereas diatoms and radiolarians build sophisticated silica structures at room temperature directly from the small amounts of silica dissolved in seawater [98]. This is an example of "chimie douce", a term introduced by French scientists during the 1970s [99]. The transformation of a set of precursors in solution into a three-dimensional sol-gel network via inorganic polymerization reactions including hydrolysis and condensation are of the same order. Such sol-gel-based chimie douce saves energy, allowing the formation of oxides at temperatures much lower than those used in classical synthetic routes based on solid precursors. It is a convenient method for the powderless processing of glasses and ceramics directly from solutions, and has therefore a special appeal in the deposition of such oxides as films via techniques such as spin-coating, ink-jet spraying or dip-coating. Chimie douce is much more than a new process, however. Since it is relatively undemanding as to the specificity of its precursors and as mild starting conditions suffice, organic chemists have found here a new way to produce hybrid materials by mixing inorganic and organic precursors on a molecular level [100]. One of the great advantages of sol-gel processing observed during the past few years has undoubtedly been the synthesis of hybrid organic-inorganic materials. This field was initiated during the 1980s, and pioneered by Schmidt and co-workers at the Frauenhofer Institute [101]. Such hybrids have been termed ORMOSILS (organically modified silicates), ORMOCERS (organically modified ceramics), CERAMERS (ceramic polymers), or POLYCERAM (polymeric ceramics) [102].

These nanocomposites fill the gap between glasses and polymers, and some of them have already been commercialized. Thus, the synthesis of materials by chimie douce and sol-gel encapsulation routes has also become a meeting point of many synthetic chemists who use these new materials as reaction intermediates.

The large majority of electroceramic materials are complex oxides in which high phase purity, high crystallinity and small grain size with narrow size distribution are often required. Thus, ball milling in a wet state was often the preferred way to prepare multicomponent oxide electroceramics in the past. However, due to its practical and energetic cost, the risk of contamination, stoichiometry loss and amorphization [103], this method is to be applied with diligence and a concept of soft mechanochemical processing is now developing [104] in which the enhancement of chemical reactivity noted in milled electroceramic precursors is retained while minimizing the introduced damage (e.g., aggregation during milling). In this way, prolonged heating at high temperature is minimized and the much sought after restriction in grain growth can be obtained. Quick nucleation and limited growth are essential in electroceramics, and the best way to achieve this is by using sol-gel precursor materials. These allow a high density of nucleation sites by using high concentrations of reactants; they also reduce diffusion distances and promote growth, and they achieve a kinetic stabilization of the reaction front in controlling the separation and homogeneity of the constituents. Thus, high-temperature ceramics, negative thermal expansion ceramics and their nanocomposites, piezoceramics and highly refractive $BaZrO_3$ have all been synthesized using sol-gel techniques (see Section 8.5.2.9).

Another synthetic area which has experienced the advantageous introduction of sol-gel encapsulation is that of combinatorial chemistry. The recently developed combinatorial approach to many complex synthetic efforts may benefit from sol-gel encapsulation in so-called "one-pot" reactions. Different reagents such as multiple catalysts may be encapsulated ("site-isolated") in a sol-gel matrix, thereby offering the possibility of performing two or more successive or simultaneous reactions in a single vessel. One-pot synthesis may result in more sustainable synthetic routes as long as catalyst selectivity can be maintained at a level sufficient to allow good control over the succession of the different reactions [105].

8.5.3.3 **Simulations**

Simulation models, unlike experimental systems, can be quickly and easily characterized by many computational methods. In such a study, the systematic optimization of synthesis conditions is much less time-consuming than in the real world. Simulations can also reveal why and how particular structures appear and therefore, the design cycle can be greatly accelerated by incorporating the predictive capacity of realistic computer modeling. At present, the simulation of encapsulated species has not been performed, though simulation of the gels themselves is well underway. Very early stages of the polycondensation in silicates can probably best be modeled by atomistic computations, and appropriate potentials have been developed with convincing results [106]. Aging can be seen as an extension of the gelation step in which the gel network is reinforced through further polymerization and by syneresis (the expulsion of solvent due to gel matrix shrinkage). Simulation of aging requires methods that can access long time scales. Special techniques were implemented by moving the silica system repeatedly onto saddle points of the potential energy hypersurface and allowing them to relax, thereby efficiently sampling many potential minima. These methods were successfully applied to amorphous silica [107].

Large-scale molecular dynamics simulations [108] of the polymerization of silicic acid in aqueous solution using specially adapted potentials have yielded realistic activation energies, and extrapolations based on simulated reaction rates agree within one order of magnitude with observations. In view of the complexities involved, a simulation of host–guest relationships inside a sol-gel-encapsulated medium is still awaited, but theoretical adsorption–desorption studies have already been initiated [109].

8.6
Conclusions and Outlook on Future Fundamental Developments

Based on the present review, a few future trends or fields of development with great potential may be discerned:

- As more than 90% of all sol-gel chemistries rely on the hydrolysis of organosilanes, the development of new backbone chemistries is urgent. Admittedly, the (tetravalent) silica system is versatile and well-known today, but its acidity is quite different from that which may be envisaged from more basic networks based on Al_2O_3 and MgO. Furthermore, titania is the only other metal-based gel network that has been (minimally) investigated, though examples based on Ce, Zr, Y, and binary or ternary mixtures await development. Finally, encapsulation in sol-gel systems existing of organic fractions such as organic acids and their esterification products, which act as precursors for multimetal ceramics, have been investigated to only minimal extent.
- Better control over morphological aspects also remains a matter of concern. The formation of ordered gels and the use of templating additives during gel

formation may be important to allow a better understanding of dynamic mobility of encapsulates, solute movement, shrinkage, and syneresis.

- A development of numerical or analytical modeling of sol-gel encapsulation and of the host–guest relationships inside a sol-gel-encapsulated medium is urgently needed. This will lead to improved understanding of the kinetics encountered.
- Apart from controlling the pores and their distribution, much research remains to be conducted to asses the possibilities of covalent grafting of spacer molecules or ionic bonding of exchange-controlling species inside the cavities. These could allow fairly generic gels (e.g., silica-based ones) to be tuned a much higher degree of specificity. In reality, this amounts to the encapsulation of single molecules with other molecules [110] and represents a fascinating chemical nanotechnology which would make the best possible use of self-assembling phenomena.

Acknowledgments

The authors are very grateful to their former co-worker, Dr. Greet Penneman, for her efforts and scientific contribution in the framework of her PhD thesis. They also thank their co-workers and students Dr. Els Bruneel, Dr. David Van de Vyver, Dr. Bart Schoofs, Dr. Klaartje De Buysser, Dr. Tran Thi Thuy, Veerle Cloet, Pieter Vermeir and Nigel van de Velde for fruitful discussions.

Abbreviations

CERAMER	ceramic polymer
EPR	electron paramagnetic resonance
FMN	flavin mononucleotide
MOD	metalo-organic decomposition
NMR	nuclear magnetic resonance
ORMOCER	organically modified ceramic
ORMOSIL	organically modified silicate
PAA	polyacrylic acid
PCM	Partial Charge Model
PEI	polyethylene imine
POLYCERAM	polymeric ceramic
PVA	polyvinyl alcohol
PZT	lead zirconate titanate
SEM	scanning electron microscopy
TEM	transmission electron microscopy
TEOS	tetraethoxysilane
TMOS	tetramethoxysilane
XRD	X-ray diffraction

References

1. D. Segal, *J. Mater. Chem.* **1997**, 7(8), 1297–1305.
2. M. Kakihana, *J. Sol-Gel Sci. Technol.* **1996**, 6, 7–55.
3. J. George, *Preparation of thin films*, ed. Dekker, **1992**.
4. H. Kinder et al, *IEEE Trans. Appl. Supercond.* **1995**, 5 (2), 1575.
5. R. Wördenweber, *Superconductor Science and Technology* **1999**, 12, R86–R102.
6. S. Rabaste, *Microcavités optique élaborées par voie sol-gel: applications aux ions terre rare d'Eu³⁺ et aux nanocristaux semiconducteurs de CdSe*, Ph.D. Thesis, Lyon, **1993**.
7. K. Otsuka, X. Ren, *Intermetallics* **1999**, 7, 511–528.
8. See for example, J.D. Chiodo, D. Harrison, E.H. Billet, *Proc. Inst. Mech. Eng.* **2001**, 215 Part B, 733–741; T. Tanaka, *Sci. Am.* **1981**, 244, 124; A.S. Hoffman, *Polymer gels*, D. DeRossi, et al. (Eds.), Plenum Press, New York, **1991**.
9. See for example, A.P. Jardine, *Smart Materials Structures* **1994**, 3, 140–148; P. Pandit, S.M. Gupta, V.K. Wahawan, *Solid State Commun.* **2004**, 131, 665–670; J.D. Chiodo, N. Jones, E.H. Billet, D.J. Harrison, *Materials and Design* **2002**, 23, 471–478.
10. T. Hao, *Adv. Mater.* **2001**, 13, 1847; N. Yao, A.M. Jamieson, *Macromolecules* **1997**, 30, 5822.
11. M.S. Rao, B.C. Dave, *Materials Res. Soc. Symp.* **2002**, 726, Q1.3.1–Q1.3.6.
12. H. Dislich, Sol-Gel Technology for Thin Films, Fibers, Preforms, Electronics, and Specialty Shapes. Klein, L. (Ed.), Noyes Publications, New Jersey, **1988**.
13. Y. Narendar, G.L. Messing, *Catalysis Today* **1997**, 35, 247–268.
14. H. Schmidt, *Structure and Bonding* **1992**, 77, 119–151.
15. J.D. Wright, N.A.J.M. Sommerdijk, *Sol-Gel Materials: Chemistry and Applications*. Gordon and Breach Science Publishers, Amsterdam, **2001**.
16. L.L. Hench, J.K. West, *Chem. Rev.* **1990**, 90, 33–72.
17. O. Lev, B. Kuyavskaya, Y. Sacharov, C. Rottman, A. Kuselman, D. Avnir, M. Ottolenghi, *Chem. Mater.* **1994**, 6, 1605–1614.
18. O. Lev , M. Tsionsky, L. Rabinovich, V. Glezer, S. Sampath, I. Pankratov, J. Gun, *Anal. Chem.* **1995**, 67, 22–30A.
19. R. Zusman, C. Rottman, M. Ottolenghi, D. Avnir, *J. Non-crystalline Solids* **1990**, 122, 107–109.
20. D. Avnir, S. Braun, O. Lev, M. Ottolenghi, *Chem. Mater.* **1994**, 6, 1605–1614.
21. S. Braun, S. Shtelzer, S. Rappoport, D. Avnir, M. Ottolenghi, *J. Non-crystalline Solids* **1992**, 147/148, 739–743.
22. B.C. Dave, B. Dunn, J.S. Valentine, J.I. Zink, *J. Anal. Chem.* **1994**, 66, A1120–A1127.
23. S. Sakka, T. Yoko, *Structure and Bonding* **1992**, 89–118.
24. D.R. Uhlmann, G. Teowee, *J. Sol-Gel Sci. Technol.* **1998**, 13, 153–162.
25. See for example, H. Schroeder, *Phys. Thin. Films* **1969**, 5, 87–141; B.E. Yoldas, *J. Mater. Sci.* **1977**, 12, 1203–1208; M. Yamane, A. Shinji, T. Sakaino, *J. Mater. Sci.* **1978**, 13, 865–870.
26. L. Klein, *Annu. Rev. Mater. Sci.* **1993**, 23, 437–452.
27. F.H. Dickey, *J. Phys. Chem.* **1955**, 58, 417.
28. S. Braun, S. Rappoport, R. Zusmann, D. Avnir, M. Ottolenghi, *Mater. Lett.* **1990**, 10, 1–5.
29. C.J. Brinker, G.W. Scherer, *Sol-gel Science*, Academic Press Inc., **1990**.
30. B.D. Fabes, B.J.J. Zelinski, D.R. Uhlmann, 'Sol-gel-derived ceramic coatings, in: J.B. Wachtman, R.A. Haber (Eds.), *Ceramic Coatings and Films*. Noyes Publications, New Jersey, **1993**.
31. M. Henry; *Sciences du Vivant*, 1990,1(2),11–63.
32. G. Schmid, *Clusters and Colloids: from Theory to Applications*. VCH, Weinheim, **1994**.
33. S.S. Zumdahl, *Chemical Principles*. Houghton Mifflin Company, USA, **1995**.
34. Malvern, private communication, equipment manual.
35. J. Livage, Sol-gel synthesis of solids, in: *Encyclopedia of Inorganic Chemistry*. John Wiley Publications, **2000**, pp. 3836–3851.
36. M. Kakihana, M. Yoshimura, *Bull. Chem. Soc. Japan* **1999**, 72, 1427–1443.
37. M. Henry, J.P. Jolivet, J. Livage, *Structure and Bonding* **1992**, 77, 153–206.

38. J.P. Jolivet, *Metal Oxide Chemistry and Synthesis: From Solution to Solid State.* John Wiley & Sons Ltd., Chichester, **2000**.

39. J. Livage, M. Chatry, M. Henry, F. Taulelle, *Chem. Proc. Adv. Mater.* **1992**, *271*, 201–212.

40. J. Livage, M. Henry, Sol-Gel Synthesis of Multifunctional Mesoporous Materials – II. Inorganic Precursors, in: C.A.C. Sequeira, M.J. Hudson (Eds.), *Multifunctional Mesoporous Inorganic Solids.* Kluwer Academic Publishers, Dordrecht, **1993**.

41. N. Bjerrum, *Zeitschrift Physilogische Chemische* **1907**, *59*, 336.

42. A. Werner, *Ber.* **1907**, *40*, 272.

43. P. Pfeiffer, *Ber.* **1907**, *40*, 4036.

44. L.G. Sillen, *Q. Rev.* **1959**, *13*, 146.

45. J. Livage, M. Henry, C. Sanchez, *Prog. Solid State Chem.* **1988**, *18*, 259–341.

46. R.T. Sanderson, *Science* **1951**, *114*, 670–672.

47. M.P. Pechini, *U.S. Patent*, no. 3.330.697, **1999**.

48. W. Li, D.P. Fries, A. Malik, *J. Chromatogr. A* **2004**, *1044*, 23–52.

49. D. Martinez-Perez, M.L. Ferrer, C.R. Mateo, *Anal. Biochem.* **2003**, *322*(2), 238–242.

50. I. Pastor, R. Esquembre, V. Micol, R. Mallavia, C.R. Mateo, *Anal. Biochem.* **2004**, *334*(2), 335–343.

51. M. Prokopowicz, J. Lukasiak, A. Pryjazny, *J. Biomater. Sci. Polymers* **2004**, *15*(3), 343–356.

52. S.R. Veith, M. Perren, S.E. Pratsinis, *J. Colloid Interface Sci.* **2005**, *283*, 495–502.

53. H. Böttcher, C. Jagota, J. Trepte, K.-H. Kallies, H. Haufe, *J. Controlled Release* **1999**, *60*, 57–65.

54. M. Ishida, H. Sakai, S. Sugihara, S. Aoshima, S. Yakohama, M. Abe, *Chem. Pharmacol. Bull.* **2003**, *51*(11), 1348–1349.

55. L. Klein (Ed.), *Sol-Gel Optics: Processing and Applications.* Kluwer, The Netherlands, **1993**.

56. M. Nogami, K. Nagasaka, K. Kotani, *J. Non-Crystalline Solids* **1990**, *126*, 87–92; L. Spanhel, M.A. Anderson, *J. Am. Ceramics Soc.* **1990**, *112*, 2278–2284.

57. S.T. Selvan, M. Nogami, A. Nakamura, Y. Hamanaka, *J. Non-Crystalline Solids* **1999**, *255*, 254–258.

58. H. Zhan, W. Chen, J. Chen, M. Wang, *Mater. Lett.* **2003**, *57*, 1483–1488.

59. L. Klein, C.L. Beaudry, A.B. Wijck, M. Mandanas, *Ceramic Trans.* **1998**, *81*, 273–280.

60. B.C. Dave, J.M. Miller, B. Dunn, J.S. Valentine, J.I. Zink, *J. Sol-Gel Sci. Technol.* **1997**, *8*, 629–634.

61. B.C. Dave, J.E.A. Ottosson, *J. Sol-Gel Sci. Technol.* **2004**, *31*, 303–307.

62. C. Malins, H.G. Glever, T.E. Keyes, J.G. Vos, W.J. Dressick, B.D. MacCraith, *Sensors and Actuators* **2000**, *B67*, 89–95.

63. F. Zaggout, N.M. El-Ashgar, S.M. Zourab, I.M. El-Nahlal, H. Motaweh, *Mater. Lett.* **2005**, *59*, 2928–2931.

64. M.C. Burt, B.C. Dave, *Sensors and Actuators* **2005**, *B107*, 552–556.

65. B.C. Dave, W. Hu, Y. Devaraj, *J. Sol-Gel Sci. Technol.* **2004**, *32*, 143–147.

66. M.L. Zheludkevich, I.M. Salvado, M.G.S. Ferreira, *J. Mater. Chem.* **2005**, *15*(48), 5099–5111.

67. A.P. Rizzato, C.V. Santilli, S.H. Pulcinelli, P. Hammer, V. Briois, *J. Eur. Ceramics Soc.* **2005**, *25*(12), 2045–2049.

68. B. Mahltig, H. Haufe, H. Böttcher, *J. Mater. Chem.* **2005**, *15*(41), 4385–4398.

69. K.M.K. Yu, D. Thompsett, S.C. Tsang, *Chem. Commun.* **2003**, 1522–1523.

70. G. Fiandaca, E. Vitrano, A. Cupane, *Biopolymers* **2004**, *74*(1-2), 55–59.

71. J. Shi, Z. Hua, L. Zhang, *J. Mater. Chem.* **2004**, *14*, 795–806.

72. K. Niesz, P. Yang, G.H. Somorjai, *Chem. Commun.* **2005**, 1986–1987.

73. Y. Wang, F. Caruso, *Chem. Commun.* **2004**, 1528–1529.

74. J. Poerschmann, Z. Zhang, F.D. Kopinke, T. Pawliszyn, *Anal. Chem.* **1997**, *69*, 597–600.

75. C.D. Chriswell, R.C. Chang, J.S. Fritz, *Anal. Chem.* **1975**, *47*, 1325–1329.

76. Y. Jiuhong, L Songqin, J. Huanggxian, *Biosensors and Bioelectronics* **2003**, *19*(5), 509–514.

77. J. Wang, *Anal. Chim. Acta* **1999**, *399*, 21–27.

78. NN, *Innovative Technology Report, DOE,* OST reference #2036, http://OST.em.doe.gov, under "Publications".

79. A. Coiffier, T. Coradin, C. Roux, O.M.M. Bouvet, J. Livage, *J. Mater. Chem.* **2001**, *11*, 2039–2044.

80. D. Yu, J. Volponi, S. Shhabra, C.J. Brinker, A. Mulchandani, A.K. Singh, *Biosensors and Bioelectronics* **2005**, *20*(7), 1433–1437.

81. G. Carturan, R. Dal Toso, S. Boninsegna, R. Dal Monte, *J. Mater. Chem.* **2004**, *14*, 2087–2098.

82. J. Fukushima, K. Kodaira, T. Matsushita, *J. Mater. Sci.* **1984**, *19*, 595–598.

83. P. Coffman, S.K. Dey, *J. Sol-Gel Sci. Technol.* **1994**, *1*, 251–257.

84. See for example, T. Mouganie, M.A. Moram, J. Summer, B.A. Glowacki, B. Schoofs, I. Van Driessche, S. Hoste, *J. Sol-Gel Sci. Technol.* **2005**, *36*(1), 87–94; G. Penneman, I. Van Driessche, E. Bruneel, S. Hoste, *Key Eng. Mater.* **2004**, *264-268*, 501–504.

85. Y. Ito, *Int. Ferro* **1997**, *14*, 123–129; O. Castano, et al., *Superconductors Sci. Technol.* **2003**, *16*, 45–53.

86. T. Hayashi, *Jpn. J. Appl. Physics* **1993**, *32*, 4092–4094.

87. P. Lommens, C. De Meyer, E. Bruneel, K. De Buysser, I. Van Driessche, S. Hoste, *J. Eur. Ceramics Soc.* **2005**, *25*, 3605–3610.

88. I. Van Driessche, G. Penneman, E. Bruneel, S. Hoste, *Pure Appl. Chem.* **2002**, *74*(11), 2101–2109; I. Van Driessche, G. Penneman, J.S. Abell, S. Hoste, *Mater. Sci. Forum* **2003**, *426-4*, 3517–3522.

89. W. Jin, J.D. Brennan, *Anal. Chim. Acta* **2002**, *461*, 1–36.

90. E.T. Vandenberg, R.S. Brown, U.J. Krull, in: I.E. Veliky (Ed.), *Immobilized Biosystems in Theory and Practical Applications.* Elsevier, Holland, **1983**, p. 129.

91. B. Lu, M.R. Smith, R. O'Kennedy, *Analyst* **1996**, *121*, 29R–32R.

92. P. Audebert, C. Deaille, C. Sanchez, *Chem. Mater.* **1993**, *5*, 911–913.

93. B. Dunn, J.M. Miller, B.C. Dave, J.S. Valentine, J.I. Zink, *Acta Mater.* **1998**, *46*, 737–741.

94. See for example, K. Flora, J.D. Brennan, G.A. Baker, M.A. Doody, F.V. Bright, *Bio-phys. J.* **1998**, *75*, 1084–1096; J.M. Miller, B. Dunn, J.S. Valentine, J.I. Zink, *J. Non-Crystalline Solids* **1996**, *202*, 279–289; Q. Chen, G.L. Kennausis, A. Heller, *J. Am. Chem. Soc.* **1998**, *120*, 4582–4585.

95. D.K. Eggers, J.S. Valentine, *Protein Sci.* **2001**, *10*, 250–261.

96. A.B. Sorokin, P. Buisson, A.C. Pierre, *Microporous and Mesoporous Materials* **2001**, *46*(1), 87–98.

97. J. Livage, *New J. Chem.* **2001**, *25*, 1.

98. S. Mann (Ed.), *Biomineralisation and Biomimetic Materials Chemistry.* VCH Publishers, New York, **1996**.

99. J. Livage, *Le Monde*, October 26th, **1977**.

100. C. Sanchez, F. Ribot, *New J. Chem.* **1994**, *18*, 1007–1047.

101. H. Schmidt, *Mater. Res. Soc. Symp. Proc.* **1984**, *32*, 327–355.

102. J. Livage, *Curr. Opin. Solid State Mater. Sci.* **1997**, *2*, 132–138.

103. E. Bruneel, K. Verbist, S. Van Tendeloo, L. Fiermans, S. Hoste, *Superconductor Sci. Technol.* **1996**, *4*(9), 357–364.

104. M. Senna, *J. Eur. Ceramics Soc.* **2005**, *25*(12), 1977–1984.

105. S.J. Broadwater, S.L. Roth, K.E. Price, M. Kobaslija, D.T. McQuade, *Organic Biomol. Chem.* **2005**, *3*, 2899–2906.

106. See for example, B.P. Feuston, S.H. Garofalini, *J. Chem. Physics* **1990**, *94*, 5351–5356; S.H. Garofalini, G. Martin, *J. Chem. Physics* **1994**, *98*, 1311–1316; K. Yamahara, K. Okazaki, *Fluid Phase Equil.* **1998**, *144*, 449–459.

107. N. Mousseau, G.T. Barkema, S.W. De Leeuw, *J. Chem. Physics* **2000**, *112*(2), 960–964.

108. N.Z. Rao, L.D. Gelb, *J. Phys. Chem. B* **2004**, *108*(33), 12418–12428.

109. R. Salazar, L.D. Gelb, *Mol. Physics* **2004**, *102*(9-10), 1015–1030.

110. L.C. Palmer, J. Rebek, *Organic Biol. Chem.* **2004**, *2*, 3051–3059.

9
Electrolytic Co-Deposition of Polymer-Encapsulated (Microencapsulated) Particles

Zhu Liqun

9.1
Introduction

Microencapsulation is a process in which liquid or solid is encapsulated by film-forming materials to produce particles with diameters ranging from several micrometers to several millimeters. The process is characterized by the properties of the core material of the microcapsules; these properties are well maintained by the core material being separated from the environment by the wall material. Subsequently, and under certain conditions, the core material is released when the wall material is broken. The combination of microencapsulation with traditional coating technology offers a completely new approach to surface protective treatment. The preparation techniques involved, together with details of the morphology of microcapsules and their properties are discussed in Chapter 1 of this book.

The development of the preparation processes of liquid microcapsules (see Chapter 5) provides a new opportunity for the application of composite coatings in the protective treatment of metal surfaces. Some liquid substances which possess specific properties can be encapsulated (Fig. 9.1) and used to form composites ei-

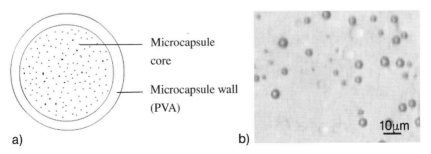

a)   b)

Figure 9.1 Scheme of a liquid microencapsule (a) and morphology of the prepared liquid microencapsules (b).

Functional Coatings. Edited by Swapan Kumar Ghosh
Copyright © 2006 WILEY-VCH Verlag GmbH & Co. KGaA, Weinheim
ISBN 3-527-31296-X

ther by mixing the microcapsules with organic coatings, or by electrolytic co-deposition with metals. The incorporation of microcapsules in the composite coatings helps to improve the surface properties.

The electrolytic co-deposition of metals with microcapsules provides excellent wear resistance, corrosion resistance, and self-repairing characteristics, these being due to release of the core materials from the liquid microcapsules.

It is known that the corrosion of metals always begins at the metal surface. Such surfaces are usually treated with paints or by electroplating to prevent them from becoming corroded. When composite coatings containing microcapsules are used on metal articles, the core materials are released during application due to the effect of external environmental factors such as abrasion, and changes in temperature and stress. The released core materials not only affect the plating surface but also fill up and repair any defects on the composite coating surface; in this way, the service life of the metal is prolonged, and the properties of the articles improved.

The American 3M Company has developed microcapsules containing polycondensates of cellulose, methacrylic acid and polyurethane as wall materials, with corrosion-inhibitors such as benzimidazole and polybenzotriazole as the core materials. Powder coatings have been prepared by mixing the above-mentioned microcapsules with a film-forming matrix such as epoxy resin and solvent-based paints. When microcapsules containing powder coatings are applied to a metal surface and heated, a uniform protective coating layer is obtained. The core, which contains corrosion-inhibitors of the microcapsules in the coating, can then be released and allowed to spread over the coating surface when the latter has been damaged locally by an impact or by stress. At the same time, the concave sites and cracks in the coating are filled up and the damaged sites repaired. If dyes such as phthalic esters are incorporated into the microcapsule-containing coating the damaged regions can be revealed visually and thus become accessible to repair processes. Coatings which contain microcapsules exhibit favorable protection for metal substrates and possess excellent self-repairing properties.

The sol-gel process (see Chapter 8) is a versatile solution process for forming oxide coatings on metal surfaces using metal organic or inorganic compounds as precursors. The process begins with a solidified film-forming reaction through sol-gel transformation, and this is followed by heating treatment at a certain temperature. The zirconium oxide sol-gel film formed on the metal surface tends to crack due to the fact that there is a difference in expansion coefficients between the sol-gel film and the metal substrate (Fig. 9.2a). Thus, the film cannot protect the metal surface effectively from corrosion. Using aqueous phase separation, the present authors [1] have prepared microcapsules with organic silicone as the core material and polyvinyl alcohol (PVA) as the wall material. This was then added to the zirconium oxide sol-gel and applied to the metal surface in order to obtain a zirconium oxide sol-gel composite coating which may prevent cracks occurring on the surface (Fig. 9.2b).

Figure 9.3 shows high-temperature oxidative weight increment curves of stainless steel with or without a zirconium oxide sol-gel coating and a zirconium oxide sol-gel composite coating containing microcapsules. It can be seen from Figure 9.2

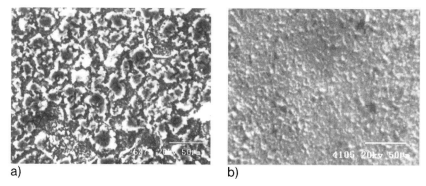

a)   b)

Figure 9.2 Morphology of the zirconium oxide sol-gel coating
(a) and the zirconium oxide sol-gel composite coating containing
microcapsules [1] (b).

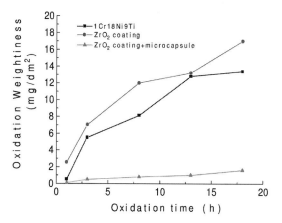

Figure 9.3 High-temperature oxidative weight increment plots of
stainless steel with or without the said coatings [1].

that many cracks are present on pure zirconium oxide sol-gel coated on a stainless steel surface. The oxidative weights of the uncovered stainless steel and the zirconium oxide sol-gel-coated specimen increase rapidly. For the oxidation-treated specimen at the same high temperature which was coated by zirconium oxide sol-gel composite coating containing microcapsules, the decrease in oxidative weight increment is clear. The reason for this is that the wall materials of the microcapsules embedded in the coating have been broken during the high-temperature oxidation, whereupon the organosilicone cores were released to fill the cracks in the zirconium oxide sol-gel coating which were formed by the expansion coefficient difference between the said coating and the substrate. The composite coatings prevent traditional sol-gel films from cracking, and also provide better oxidation resistance for the steel substrate than pure zirconium oxide sol-gel-coated surfaces. In addition, the results of the immersion test in 5 wt.% NaCl solution show that the

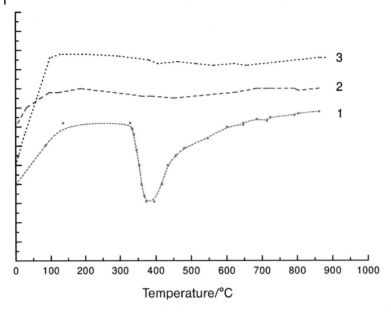

Figure 9.4 DTA plots of silicone sol-gel films with or without incorporated microcapsules [1]. Curve 1: traditional silicone sol-gel coating. Curve 2: silicone sol-gel composite coating containing organosilicone and ZrO$_2$ nanoparticles microcapsules. Curve 3: silicone sol-gel composite coating containing organosilicone and SiC ceramic fiber microcapsules.

composite coatings containing microcapsules also exhibit excellent pitting corrosion- resistance.

Figure 9.4 illustrates differential thermal analysis (DTA) plots of the silicone sol-gel film, with or without incorporated microcapsules. Two types of microcapsules have been prepared using an *in-situ* polymerization method; the composite coatings were obtained by adding them into the sol solution. While the first type of microcapsules contain SiC ceramic fibers as the core material and PVA as the wall material, the second type have ZrO$_2$ particles as the core and PVA as the wall. Both types of microcapsules were mixed with silicone sol-gel solution with organosilicone to obtain the composite coatings. The data in Figure 9.4 show that a pure silicone sol-gel coating produces a distinct endothermic peak at 400 °C, whereas neither distinct endothermic nor exothermic reactions were observed for the sol-gel composite coating containing organosilicone and encapsulated SiC ceramic fiber or ZrO$_2$ nanoparticles until the temperature reached 860 °C. The absence of peaks in the DTA curves for microcapsule-containing coatings indicates that the thermal stability of these coatings is excellent.

This result shows that the thermal stability of the silicone sol-gel coating is improved substantially by the incorporation of these microcapsules. The reason for this is that the microcapsules embedded in the composite coating play a stabilizing role on the thermal expansion-induced volume contraction during heating of the silicone sol-gel coating. This reduces microcracking on the composite coating and improves the coating performance, and especially the corrosion-resistant and oxidation-resistant properties.

Composite coatings can be prepared by mixing microcapsules in a paint formulation, or by electrodeposition [1,2]. Composite electrodeposition (or co-deposition) is a widely used surface treatment process which includes the following steps:

- The addition of solid or liquid microparticles into the normal plating solution, followed by stirring to disperse or suspend the microparticles homogeneously in the solution.
- The co-deposition of microparticles with metal ions in an electrolytic bath under the influence of electric field to form a composite plating coating containing those microparticles.

The technique of electrolytic co-deposition provides the possibility of designing metal surfaces with functional properties. Depending on the type of microcapsule co-deposited electrolytically, several properties such as abrasion, friction, and corrosion-resistance can be achieved. The types of solid particles that can be incorporated into the composite plating coating are almost unlimited; typical examples are SiC, Al_2O_3, BC, ZrO_2, TiC, La_2O_3, PTFE (polytetrafluoroethylene), metal powders, and nanoparticles. Very few research investigations into the electrolytic co-deposition of metals and microcapsules containing liquid active agents have been published. We have reported previously the successful co-deposition of liquid microcapsules with nickel and copper, where excellent properties are exhibited by the composite coatings [3–5]. In this chapter, we will focus our discussion on the electrolytic co-deposition of liquid-containing microcapsules.

The electrolytic co-deposition process of liquid microcapsules clearly includes more complex steps. Today, most research studies concentrate on the co-deposition of nickel, copper, iron-phosphorus, tin-based composite coatings containing microcapsules that include cores such as lubricating oil, corrosion inhibitors, fluorescent agents, welding agents and other special agents [3–12]. The main aim is to improve the wear resistance, corrosion resistance, welding ability and other properties of the co-deposited coatings.

For example, lubricating oil (hereafter termed "lube oil")-encapsulated microcapsules are incorporated into the nickel- or acidic copper-plating solution to perform composite electrodeposition; the result is composite plating coatings with very little frictional force and a small frictional coefficient. The addition of benzotriazole to a copper-plating bath may substantially enhance the corrosion resistance and oxidative discoloring capabilities of the copper-plating coating.

Microcapsules prepared with organosilicone resin as the core and PVA as the wall have been added to an iron-phosphorus plating bath to form the iron-phosphorus-based composite coating by co-deposition onto metal parts. The liquid core substance was released during friction testing of the composite coating containing the microcapsules, which formed a silicone resin film between the sliding frictional surfaces. The composite coating was endowed with a lower frictional coefficient by the said film due to its low shear strength; hence, the composite coating exhibits better friction and wear resistance than that of the usual iron-phosphorus plating layer. Such composite coatings show even better friction and wear resistance after thermal treatment at 150 °C. In addition, the iron-phosphorus-based composite coating containing microcapsules exhibits favorable corrosion resistance in solutions of 5% NaCl and 10% H_2SO_4. In addition, the greater the number of microcapsules embedded in the coating, the better the corrosion resistance obtained.

Chen Yu and co-workers prepared microcapsules containing a fluorescent agent as the core material by solvent evaporation from duplicate phase solution [9]. The composite electrodeposited coatings produced by incorporating microcapsules into the acidic $CuSO_4$ plating bath or nickel-plating bath showed fluorescent characteristics because some of the microcapsules were only half-embedded in the composite coating. Encapsulation of the fluorescent agent prevents direct contact with metal ions in the electrolytic bath and avoids contamination of the bath, as well as providing good properties for the co-deposited coating. Furthermore, it is relatively easy to control parameters such as the concentration of fluorescent agent and pH while preparing the microcapsules. In addition, the wall materials can prevent the fluorescent agents from depigmentation, which in turn prolongs the service life of the fluorescent coating to a remarkable degree.

Due to its good welding ability, tin is widely used in automotive and electronic applications, as well as in printed circuit electroplating. Recent studies related to the development of alternative co-deposited coatings by substituting the lead-tin plating were noticeable due to a need for the lead-free welding plating process. Among these coatings, tin-plating coatings have recently become attractive once more. Masayuki Itagaki et al. prepared composite coatings containing microcapsules and with a welding agent as the core material, by using electrolytic co-deposition [11,12]. The release of welding agents from microcapsules during the welding process enhanced the welding ability of the composite coating. The same authors also prepared microcapsules containing lube oil as the core material; these capsules are added to a Watt's plating bath and co-deposited with Ni to form composite coatings. Different parameters such as distribution of the microcapsules inside the coating, the electrode shape, and the dispersion method used for the microcapsule content in the composite coatings were investigated. The stability of microcapsules inside the coating was explored by re-dissolving the composite coating, and then collecting and examining the number and content of the microcapsules.

9.2
The Preparation of Microcapsules

9.2.1
Methods

Although many methods are available for the preparation of microcapsules (see Chapter 1), they may be broadly subdivided into chemical or physical techniques.

9.2.1.1 Chemical Methods

Interfacial polymerization is a widely used chemical method characterized by the fact that the microcapsules prepared in this way possess high strength and stability; thus, they are not easy to break. In this process, two reactive monomers are dissolved in water and an organic solvent respectively, while the core materials are dissolved in a solvent of the dispersed phase. An oil-in water or water-in-oil emulsion is formed by mixing the two immiscible solutions, with the addition of an emulsifier. The two reactive monomers in the two phases migrate to the interface of the emulsion droplets and react to form a polymer at the interface by polycondensation; this acts as a wall material to encapsulate the core materials in the emulsion droplets to form microcapsules. The major disadvantages of this method are its complexity and the slowness of the reaction. In addition, many unwanted side products are produced.

9.2.1.2 Physical Methods

Physical methods include spray-drying, centrifugal extrusion, solvent evaporation, and phase separation. The complex procedures hinder the widespread use of some of these methods.

Phase separation from an aqueous or organic solvent is straightforward. In this process, the oily cores are dispersed in the aqueous solution of macromolecular wall material. After adding a flocculant to the solution, the quantity of water is greatly reduced as much of it combines with the flocculants. This causes the macromolecular wall material to agglomerate from the aqueous solution and to encapsulate the oily cores, thereby forming microcapsules as its concentration is too high to dissolve completely. No special equipment is needed for this process, and it is used conventionally to prepare microcapsules with oily liquid cores. The main advantages of this method are the simple procedure, the wide selection of wall materials and core materials available, and the ability to control the microcapsule size by changing the emulsification conditions.

Selection of the wall material, core material and other additives has a substantial effect on the preparation of microcapsules with high strength, and with sufficient toughness and desirable size.

Many types of core material are available, but they are broadly divisible into two types according to application, namely aqueous liquid cores and oily liquid c ores, the latter being used widely in medicine, food and other industrial applications.

Selection of the wall material is determined by the core material used and the required size, strength and release performance of the microcapsules. The conventional wall materials include carbohydrates and proteins. Typical carbohydrate wall materials include sodium alginate, agar, gum arabic and gelatin, all of which are natural glues with many advantages, including a simple film-forming procedure and good biocompatibility; they are also "environmentally friendly". However, the strength of these natural wall materials is insufficient to resist breakage, and synthetic wall materials such as PVA – which exhibit higher strength and toughness – are used to prepare smaller microcapsules. One problem here is that these materials are not biodegradeable, and this may lead to environmental contamination.

Using a single water-phase separation process, the present authors have prepared microcapsules of spherical shape and in the size range of 3 to 8 μm; moreover, these microcapsules exhibited high strength, good toughness and good stability in the plating bath.

The fluorescent microcapsules produced by Chen Yu et al., using a solvent evaporation method from duplicate-phase solution, can also be used for electrolytic co-deposition. The microcapsules prepared by Masayuki Itagaki et al. by *in-situ* polymerization are valuable when preparing composite tin platings in which the core materials consist mainly of abietic acid dissolved in organic welding solvents; these microcapsules have a density which is approximately equal to that of the tinning plating electrolyte. Moreover, the microcapsules which contain PVA as the wall and lube oil as the core are also produced by phase separation from the aqueous solution, and exhibit good co-depositing capability.

In short, the preparation method for a microcapsule is selected according to the process conditions of the composite electrodeposition and characteristics of the composite coating.

9.2.2
Selection of the Core and Shell Materials

Selection of the core and shell (wall) materials is very important during the preparation of liquid microcapsules. Selection of the core materials is based on characteristics of the plating coating and the object of improving the surface properties of the coating, such as antirust oil, lube oil, corrosion inhibitor, hydrophobic agent, welding agent, fluorescent agent, or silicone resin [3–8]. In addition, the core material must be easily encapsulated by the wall material, such that the microcapsules prepared are no larger than 10 μm in size, as large microcapsules are not easy to co-deposit with metal ions in the plating bath.

The most common wall materials include PVA, gelatin, soluble starch, and gum arabic.

PVA is soluble in hot water, forming a colloidal solution which may be phase-separated by the addition of inorganic salt or a hydrophilic solvent. PVA is itself a surfactant, and may prevent the aggregation of microcapsules. PVA has good film-forming properties, and a 2 wt.% aqueous solution can produce an elastic film with good toughness and rub resistance. Although PVA aqueous solution may coagulate in the nickel plating bath, it disperses well in a copper plating bath.

Gelatin is a natural macromolecule which maybe biodegraded. Microcapsules prepared from 2 wt.% gelatin aqueous solution possess a good elastic wall and do not coagulate with nickel ion. Microcapsules with gelatin as the wall are well dispersed in a plating bath and useful in a nickel plating solution.

Soluble starch may act not only as the wall to encapsulate the core alone, but it may also form a film after mixing with gum arabic. Microcapsules prepared from a mixture of PVA, gelatin and soluble starch possess a very hard wall after being solidified by formaldehyde solution, but such microcapsules are too large to be co-deposited.

The concentration of the wall material greatly affects the preparation, stability and particle size of microcapsules. The stability and particle size of microcapsules prepared from 1 wt.%, 2 wt.% and 3 wt.% gelatin aqueous solutions respectively are listed in Table 9.1. The composite nickel plating coatings resulting from these three forms of microcapsule possess different morphologies. For example, the plating coating resulting from 1 wt.% gelatin solution was smooth and contained relatively fewer microcapsules, that from 2 wt.% gelatin solution was substantially even and contained relatively more microcapsules, but that from 3 wt.% gelatin solution was rough and contained few microcapsules. Thus, microcapsules prepared from 2 wt.% gelatin solution are preferred for electrodeposition with nickel.

Table 9.1 Stability and particle size of microcapsules prepared from different concentration of gelatin solutions.

Concentration	Particle size	Stability
1 wt.%	10% are 2–8 µm, others are larger	Poor, tend to crack
2 wt.%	70–80% are 2–8 µm	Stable, firm, hard to crack
3 wt.%	90% are 10–30 µm	Stable, firm, hard to crack

9.2.3
Preparation Procedure of Liquid Microcapsules

After determining the appropriate wall and core materials, other factors must be considered such as the particle sizes of the prepared microcapsules and their stabilities in plating baths; thus, a suitable process is essential when preparing specific microcapsules.

During the preparation, a variety of surfactants (e.g., sodium lauryl sulfonate, ammonium alkyl trimethyl chloride) was tested to study their effect on aggregation of the microcapsules prepared. Results showed that Span 80 and Tween 80, when used as stabilizers, were effective in preventing microcapsules with PVA or gelatin as the wall material from breaking or aggregating. Furthermore, microcapsules prepared by *in-situ* polymerization with organosilicone, SiC ceramic fibers or ZrO_2 nanoparticles as the core material respectively, and PVA as the wall material, exhibited high stability in sol-gel composite coatings.

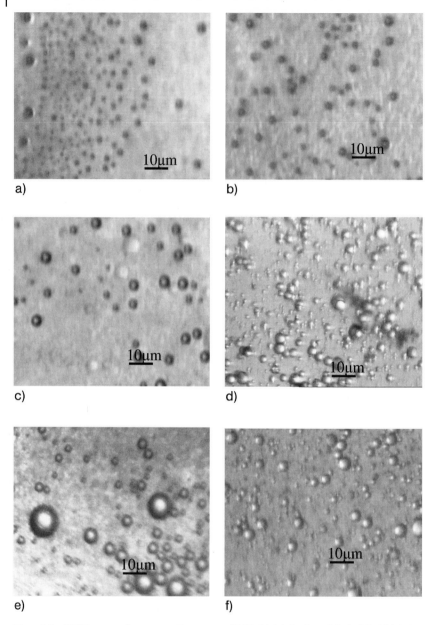

Figure 9.5 SEM images of copper coating containing microcapsules prepared from: (a) hydrophobic agent (PVA); (b) inhibitor (PVA); (c) lubricating oil (gelatin); (d) lubricating oil (PVA); (e) silicone resin (PVA); (f) silicone resin (gelatin).

In addition to the nature of surfactants, other factors such as the ratio of core to wall material components, the surfactant quantity, the feeding order and preparation time also affect the stability and particle size of microcapsules.

9.2.3.1 Procedure

A preparation procedure which is suitable for liquid microcapsules useful in electrolytic co-deposition is briefly described as follows. To a beaker containing 2 wt.% gelatin aqueous solution is added dropwise the core material (e.g., lube oil or hydrophobic agent) while stirring (1200 rotations min^{-1}) at room temperature. When the addition is complete, the mixture is stirred for another 5 min. Addition is then made of 0.1 mL nonionic surfactant, and stirring continued for 1 min. The dehydrating agent (anhydrous alcohol) is then added, after which the stirring rate is slowed and then stopped to obtain the expected liquid microcapsules.

The addition of anhydrous alcohol may result in two phases: one phase will contain more gelatin than the other, and this leads to an enhanced wall macromolecular interaction and the formation of insoluble polymer aggregates that encapsulate the homogenously dispersed core material.

The morphology of microcapsules with liquid corrosion inhibitor, hydrophobic agent or lube oil as the core is illustrated in Figure 9.5. The majority of the liquid microcapsules are seen to be uniform, spherical particles with diameters ranging from 4 to 8 μm. The distribution of microcapsule particle size is shown in Figure 9.6. Due to the homogeneity and smaller particle size (<10 μm) of the prepared microcapsules, they are easy to co-deposit with metal ions in the plating bath to form composite electroplated coatings.

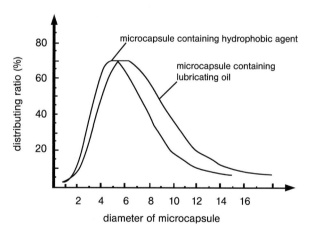

Figure 9.6 Particle size distributions of liquid microcapsules.

9.3

Description and Different Parameters for Electrolytic Co-Deposition Techniques

9.3.1
Electrolytic Co-Deposition

The prepared microcapsules were added into normal nickel-plating or copper-plating baths to determine how they would co-deposit with the metal ions, the effects of different types of metal ion, and the process parameters required for co-deposition.

Electrolytic co-deposition of the prepared microcapsules with metal ions in an acidic copper-plating bath, a nickel sulfamate-plating bath, a ferrophosphorus bath and an acidic tin-plating bath were successfully carried out under different technical conditions. Details of the bath composition and process conditions for the composite copper- and nickel-plating containing liquid microcapsules are summarized in Table 9.2. In general, prepared microcapsules with an appropriate core were added to the plating bath and co-deposited electrolytically onto metal surfaces with different metal ions such as Ni^{2+}, Cu^{2+} and Sn^{2+}; this resulted in the corresponding composite copper- , nickel-, and tin-plating coatings.

Table 9.2 Composition and process conditions of the composite copper- and nickel- plating baths containing liquid microcapsules.

Copper-plating	Conditions	Nickel-plating	Conditions
$CuSO_4 \cdot 5H_2O$	150 g L^{-1}	$Ni(SO_3NH_2)_2 \cdot 4H_2O$	800 g L^{-1}
H_2SO_4	40 g L^{-1}	Boric acid	45 g L^{-1}
Emulsion containing microcapsules	As appropriate	Emulsion containing microcapsules	As appropriate
Temperature (°C)	25–35	Temperature (°C)	55–60
Electric current (mA cm^{-2})	30	Electric current (mA cm^{-2})	150

9.3.2
Morphological Studies of the Composite Plating Coatings

In order to verify whether the microcapsules were co-deposited with different metal ions, scanning electron microscopy (SEM) studies of the composite coatings were carried out; typical micrographs are shown in Figure 9.7. The surface morphology for the composite copper-plating coatings containing four types of microcapsule (encapsulated silicone resin, lube oil, corrosion inhibitors, and hydrophobic agent) and nickel-plating coating containing antirust lube oil were studied in detail. It can be seen from Figure 9.7 that the microcapsules with various core materials are each capable of electrolytic co-deposition, and form the desired compos-

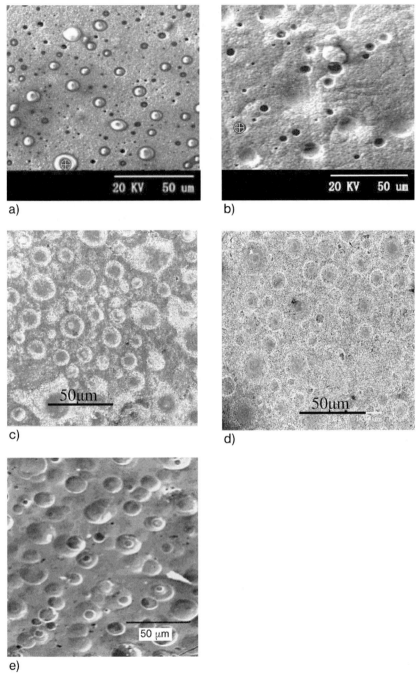

a)

b)

c)

d)

e)

Figure 9.7 SEM images of copper coating containing microcapsules with: (a) silicone resin (PVA); (b) lubricating oil (PVA);

(c) inhibitor (PVA); (d) hydrophobic agent (PVA); (e) lubricating oil (glutin).

ite plating coatings. Furthermore, morphological studies of the composite coatings also revealed substantial morphological differences among the copper-plating coatings containing various microcapsules. In terms of detail, the micrographs showed there to be distinct protrusion (for silicone microcapsules), distinct concavity (for lube oil microcapsules), and shallow traces left by microcapsules.

Figure 9.8 shows cross-sectional SEM images of normal copper-plating coatings, the composite copper-plating coatings, and nickel-plating coatings containing microcapsules. The figure shows that coarse spherical crystals appeared on the normal acidic copper-plating coating, while the addition of microcapsules to the copper-plating bath reduced the crystal size. In the composite coatings, copper or nickel crystals formed through the reduction of Cu^{2+} or Ni^{2+} on the electrode tightened the space for the microcapsules, such that the shape of the microcapsules was deformed from a sphere to an ellipse (Fig. 9.8b) or other shape (Fig. 9.8c). X-ray photoelectron spectroscopy (XPS) analysis of the ellipsoids in the composites confirmed their chemical composition to be the same as that of the wall and core of the microcapsules, which showed that the microcapsules had been co-deposited in the plating coatings.

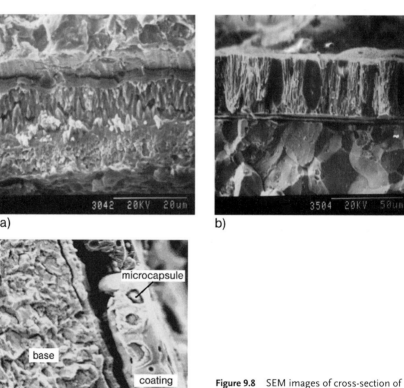

a)

b)

c)

Figure 9.8 SEM images of cross-section of coating without containing microcapsules and with microcapsules. a: normal copper-plating coatings; b: copper-plating coatings with microcapsules; c: nickel-plating coatings with microcapsules [1].

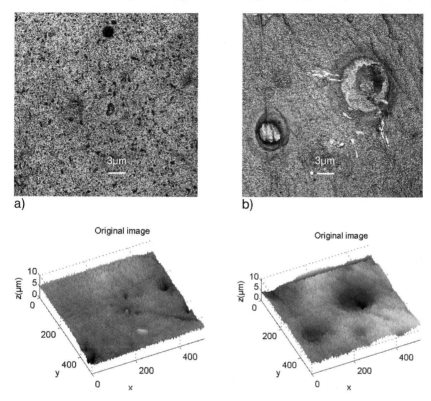

Figure 9.9 Surface morphology of copper coating without micro-capsules (a) and microcapsules with inhibitor (b).

The morphology of the copper-plating coatings containing microcapsules with corrosion inhibitors as core were observed with Nikon E600 confocal laser scanning microscope. These images also confirmed the presence of microcapsules in the copper composite coatings (Fig. 9.9a and b).

Furthermore, it can be seen from the XPS curves (Fig. 9.10) of normal copper-plating coating and composite copper-plating coating containing microcapsules that the peak position of copper (Cu) and oxygen (O) in the said composite coatings were shifted to higher binding energy values compared to the standard XPS peaks of Cu, O (smooth curves). This was due to the presence of microcapsules in the composite coatings, and showed that the microcapsules were co-deposited with Cu^{2+}.

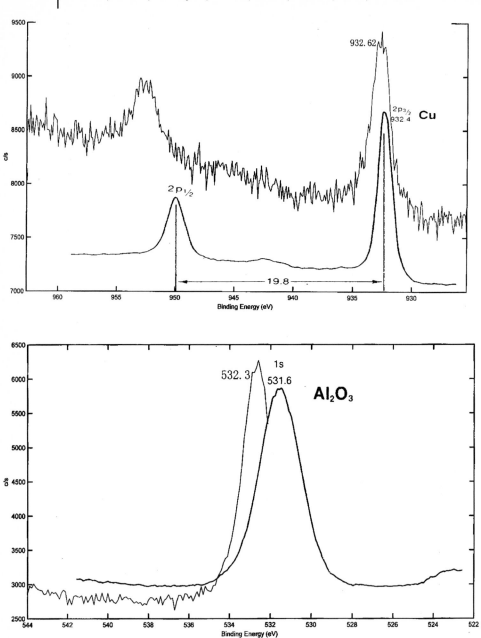

Figure 9.10 X-ray photoelectron spectroscopy of elemental copper and oxygen in copper coating containing microcapsules.

9.3.3

Properties of the Copper-Plating Bath Containing Microcapsules

Microcapsules prepared with silicone resin or lube oil as the core and PVA as the wall were incorporated into the copper-plating bath (for composition, see Table 9.2), and electrolytic co-deposition was carried out at current densities (D_k) of 10, 20 and 30 mA cm^{-2}, respectively.

The times at which microcapsules began to deposit on the electrode surface during the composite copper-plating are detailed in Table 9.3. The amount of deposited microcapsules was found to be directly related to D_k and to the plating time. Microcapsules with a lube oil core were deposited later than the encapsulated silicone resin under the same D_k. This showed that a higher D_k and a longer time would be required in order for the lube oil microcapsules to be co-deposited. It was also noted that an increase in D_k and in plating time had a positive influence on the electrodeposition of the above-mentioned microcapsules. However, a higher D_k leads to coarse composite plating coating at the edges of the substrates. For example, with a D_k of 30 mA cm^{-2}, similar co-deposition behaviors were observed for the above two different types of microcapsules, and these were dispersed homogeneously in the composite coatings if the microcapsules had a diameter of 3 to 10 μm.

Table 9.4 presents weight increments in the co-deposition of microcapsules under different D_k-values. For comparison, the normal acidic copper-plating and composite copper-plating containing microcapsules (with silicone resin or lube oil as core and PVA as wall) were carried out at D_k-values of 5, 10, 20, 30, and 40 mA cm^{-2}, respectively, for 35 min. It can be seen from these data that, during normal copper-plating and composite copper deposition, when D_k was <20 mA cm^{-2} the current efficiency of the normal acidic copper plating was relatively high, and the weight increment distinctly greater than that of the composite coatings containing microcapsules. At the same time, no microcapsules were found to separate out

Table 9.3 Time at which microcapsules start to deposit during composite copper-plating.

Core D_k	Silicone resin	Lube oil
10 mA cm^{-2}	A few microcapsules found after 24 min.	A few microcapsules found after 30 min.
20 mA cm^{-2}	Microcapsules start to deposit on the coating after 18 min; amounts increase with time.	Microcapsules start to deposit on the coating after 20 min; amounts increase with time.
30 mA cm^{-2}	Microcapsules start to deposit on the coating after 14 min; 30 min later microcapsules of various sizes dispersed homogeneously in the coating.	Tiny microcapsules start to deposit on the coating after 16 min; 30 min later microcapsules of various sizes dispersed homogeneously in the coating.

Table 9.4 Weight increment of the composite coatings.

D_k [mA cm^{-2}]	Weight increment of cathode [g]		
	Usual copper coatings	Composite coatings containing organosilicone microcapsules	Composite coatings containing lube oil microcapsules
5	0.0252	0.0227	0.0202
10	0.0460	0.0412	0.0403
20	0.1343	0.1474	0.1378
30	0.1786	0.1842	0.1659
40	0.1309	0.1247	0.1314

from the plating solution. When D_k was >20 mA cm^{-2}, the microcapsules would begin to separate out (in accordance with data in Table 9.3), and the weight increment was clearly surpassed in comparison with the copper-plating coating. This may be due to the fact that, although the microcapsules do not carry any positive charge themselves, their functional groups (e.g., hydroxyl) may chelate positively charged copper ions and migrate towards the cathode under the influence of the electric field (when D_k is sufficiently high), and ultimately co-deposit onto the cathode with metal ions. The composite coating has a higher weight increment than that of the normal copper plating coating, this being due to the presence of embedded microcapsules in the composite coating. As D_k rose to 40 mA cm^{-2}, the evolution of hydrogen at the cathode disturbed the co-deposition of the relatively large microcapsules, and this led to low weight increments.

The incorporation of microcapsules into the plating solution may greatly affect the current efficiency and the weight increment of the plating coating, but have only minimal effect on dispersing capability. The dispersibility of the microcapsules was tested using a curved cathode in the co-deposition process, whereupon the microcapsules were found to have no influence on the dispersion capability of normal copper-plating.

9.4
The Electrodeposition Mechanism of Composite Plating Coating Containing Microcapsules

Although many research articles have been published describing the mechanism of electrodeposition containing solid particles, few have been concerned with composite coatings containing liquid microcapsules. In general, this is because it is difficult to prepare suitable microcapsules that are capable of electrolytic co-deposition. Microcapsules with small size and good stability in the plating bath are suitable for electrodeposition with metal ions under low voltage and direct current. Although it is feasible to prepare suitable microcapsules for electrodeposition, these studies are essentially focused on the selection of process conditions affecting electrodeposition. Moreover, the limitations in obtaining the various types of microcapsules also limit the mechanistic study of electrolytic co-deposition.

9.4.1
Effects of Microcapsules on Conductivity of the Bath Solution

Whether incorporation of the microcapsules would affect the conductivity of the bath solution is another important factor for composite electrodeposition. Figure 9.11 shows conductivity curves in acidic copper-plating solution, Watt solution containing 25 vol.% microcapsules encapsulated organosilicone, lube oil, corrosion inhibitor and hydrophobic agent, respectively. Conductivity of the plating solutions clearly changes according to the type of core material inside the microcapsules. The conductivities of the bath solutions containing hydrophobic agent, organosilicone or lube oil-microcapsules were higher than that of the corrosion inhibitor-microcapsules. Therefore, when D_k was >20 mA cm^{-2}, composite coatings containing such microcapsules were easier to obtain from the corresponding composite baths containing those microcapsules. Although the bath containing lube oil-microcapsules possesses high conductivity, it is not easy to co-deposit lube oil-microcapsules compared to the organosilicone or hydrophobic agent-microcapsules, as the conductivity of the bath falls rapidly upon addition of lube oil-microcapsules to the plating solution.

In the composite nickel-plating bath, conductivity clearly exceeded that of the composite copper-plating bath, but decreased much more rapidly compared to that of the copper-plating bath. The higher conductivity of the organosilicone-microcapsules makes them easier to co-deposit. Clearly, the nature of the wall material also has a remarkable effect on conductivity.

9.4.2
Effect of Microcapsules on Differential Capacitance of the Bath Solution

The incorporation of microcapsules into the plating bath also influences the electrical double layer at the metal–solution interface and the differential capacitance. Usually, a surfactant is added to the bath to induce adsorption onto the electrode, and this reflects a distinct decrement of the differential capacitance on the corresponding curves.

Figure 9.12 illustrates the differential capacitance curves of the plating bath containing different microcapsules. Clearly, the bath containing no microcapsules possesses a quite narrow adsorption potential, while the differential capacitance is fairly large. When hydrophobic agent, corrosion inhibitor, organosilicone or lube oil-microcapsules were introduced into the bath, a wide range of adsorption potentials and small differential capacitance values were observed. This suggests that liquid microcapsules may be adsorbed onto the surface of the electrode during the composite copper-plating, so as to become embedded in the deposited metal coatings. Other experimental analyses, including SEM and composition analysis, further

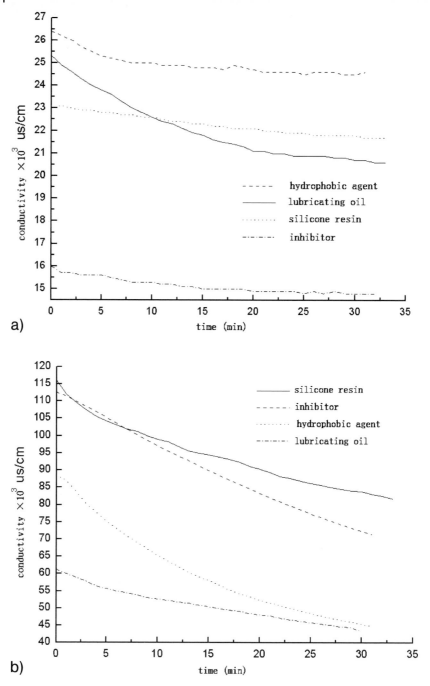

Figure 9.11 Conductivity of copper/nickel plating solution containing different microcapsules. (a) Copper plating solution; (b) nickel plating solution.

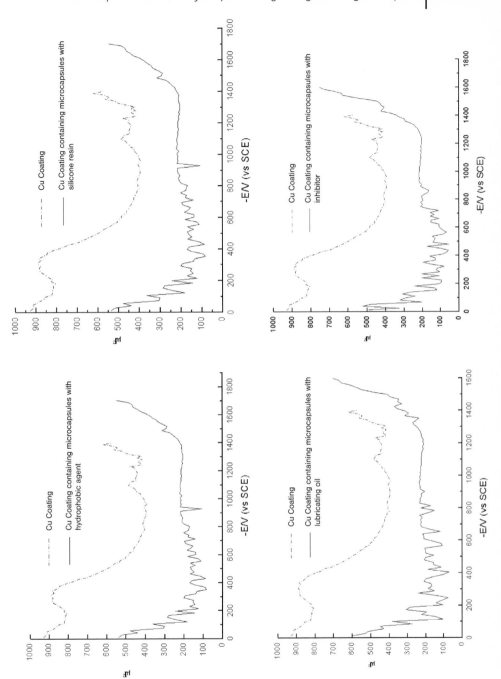

Figure 9.12 Differential capacitance curves of composite copper coating containing different microcapsules.

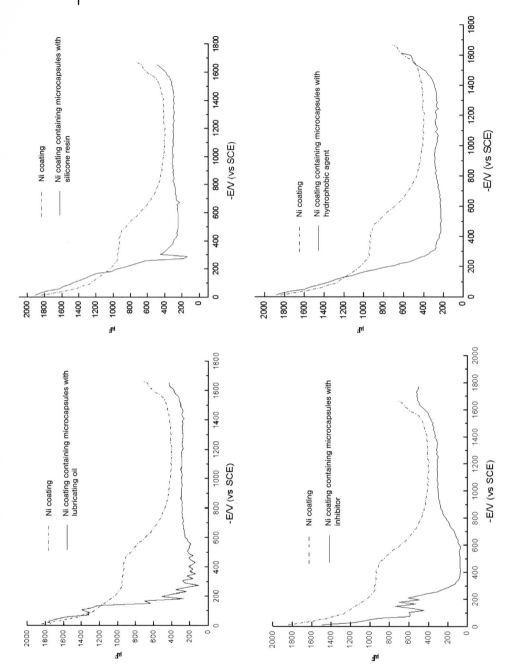

Figure 9.13 Differential capacitance curves of composite nickel coating containing different microcapsules.

confirmed that the resulted plating layer contained embedded microcapsules, and indicated that such adsorption is favored for the co-deposition. The incorporation of microcapsules decreased the differential capacitance, as water molecules with a higher dielectric constant were substituted by microcapsules with a larger volume and a smaller dielectric constant.

The incorporation of microcapsules changes the adsorption potential range and decreases the differential capacitance of the copper-plating bath. Figure 9.13 also shows that the incorporation of microcapsules into the nickel-plating bath induces a wide adsorption potential range and a small differential capacitance value. Such adsorption is also helpful for the co-deposition of the microcapsules with Ni^{2+} in the bath.

9.4.3
Effect of Microcapsules on Cathodic Polarization of the Plating Bath

Figure 9.14 illustrates the cathodic polarization curves of the acidic sulfate copper-plating bath with different amounts of organosilicone-microcapsules. In the solution without microcapsules, the outmost D_k of 12 mA cm^{-2} emerged as the cathodic potential, reduced from -0.36 V to -0.52 V. After adding 8 vol.% of the microcapsules, the outmost D_k appeared at 10 mA cm^{-2}. With continual addition of the microcapsules to 20 vol.%, the cathodic D_k was sustained and increased without any peak value occurring as the cathodic potential decreased. This might be due to elimination of the concentration polarization and to enhancement of the electrochemical control during co-deposition in the presence of a large amount of microcapsules. The cathodic polarization was enhanced as the quantity of microcapsules increased, this also being favorable to obtain copper-plating coatings with fine crystals (see Fig. 9.8, micro-morphology of the copper-plating coating containing mi-

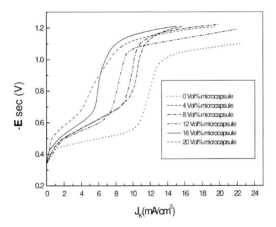

Figure 9.14 Cathodic polarization curves of composite copper coating containing different microcapsules [7]. E sec (V): Electrode potential ($_{SCE}$V); J$_k$(mA cm^{-2}): Electric density (mA cm^{-2}).

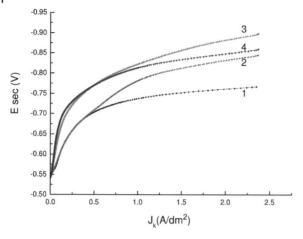

Figure 9.15 Cathodic polarization curves of composite nickel coating containing different microcapsules. (1) Without microcapsules; (2) containing 5% microcapsules; (3) containing 15% microcapsules; (4) containing 25% microcapsules. E sec (V): Electrode potential $(_{SCE}V)$; J_k(A dm^{-2}): Electric density (mA cm^{-2}).

crocapsules). This indicates why polarization is enhanced as the content of microcapsules rises. Combining these data with results of the differential capacitance test indicated that the addition of microcapsules improves adsorption onto the cathode, and that this adsorption tends to accelerate cathodic polarization, leading to the occurrence of tiny crystals in the composite coatings. Certainly, as some microcapsules are added to the bath the volume of the copper-plating bath solution is increased, which in turn reduces the concentration of the metal ions. Although this may be advantageous for accelerating cathodic polarization, the main reason is that the interfacial adsorption brought about by the added microcapsules improved cathodic polarization.

Figure 9.15 illustrates the cathodic polarization curves of the composite nickel-plating bath containing different amounts of microcapsules. The addition of microcapsules enhances polarization and leads to the formation of tiny crystals in the composite coatings.

In view of the fact that the added microcapsules might change the volume of the bath, which in turn would affect the cathodic polarization curve, a corresponding quantity of distilled water was added into the normal nickel-plating bath to determine any effect on cathodic polarization. The data in Figure 9.16 show that a change in bath concentration caused by water addition aggravated the cathodic polarization. However, when the distilled water content reached 5%, the degree of polarization changed only minimally. The data in Figure 9.17 show that the microcapsules aggravated the polarization. Although the addition of distilled water dilutes the bath solution, perhaps leading to a more polarized nickel-plating bath, this has less influence than adding microcapsules, which might aggravate the cathodic polarization.

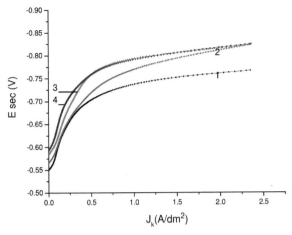

Figure 9.16 Cathodic polarization curves of composite nickel coating containing different contents of distilled water: (1) 0%; (2) 5%; (3) 15%; (4) 25%. E sec (V): Electrode potential $(_{SCE}V)$; J_k(A dm^{-2}): Electric density (mA cm^{-2}).

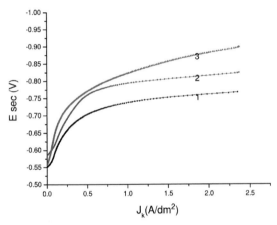

Figure 9.17 Cathodic polarization curves of composite nickel coating containing distilled water and microcapsules. (1) 0% distilled water and without microcapsules; (2) 15% distilled water; (3) 15% microcapsules. E sec (V): Electrode potential $(_{SCE}V)$; J_k(A dm^{-2}): Electric density (mA cm^{-2}).

9.4.4
Mechanism of Electrolytic Co-Deposition of Liquid Microcapsules

Although the mechanism of electrodeposition of solid particles has been well-described in the literature, little attention has been paid to the study of co-deposition behavior of liquid microcapsules. The mechanism of the electrodeposition of liquid microcapsules may also be explained by comparing the mechanism of electrodeposition of solid microparticles.

In referring to Guglielimi's theory of two-step adsorption for the co-deposition of particles, we speculate that, under the influence of an electric field, diffusion and counterflowing, microcapsules in the plating-bath migrate to the electrode. In the first step – the "loose adsorption step" – the microcapsules loaded with ions and solvent molecular film approach the cathode and become adsorbed loosely onto the cathode surface. In the second step – the "strong adsorption step" – the loosely adsorbed microcapsules release the adsorbed ions and solvent film and become irreversibly adsorbed onto the cathode surface. In the subsequent reductive-deposition process of metal ions, the strongly adsorbed microcapsules become embedded into the electroplating coating. Owing to the limitation of the thickness of the electrode–solution interfacial electric double layer, the corresponding particles in the composite co-deposition are 1 μm in diameter. Thus, this theory is clearly inappropriate to explain the electrolytic co-deposition of liquid microcapsules with larger diameters (>3 μm).

The presence of surfactant accelerates adherence of the microcapsules into the electrode surface when they are 3 to 8 μm in size. The effect of the microcapsules on differential capacitance indicates that the added microcapsules might be adsorbed onto the cathode surface, there being a wide adsorption potential range. The microcapsules enter the metal plating coatings by such an adsorption process. Increasing the amount of added microcapsules may cause a weight increment of the microcapsules adsorbed onto the electrode, thereby creating favorable conditions for the incorporation of larger amounts of microcapsules into the co-deposited coatings.

Based on the results of our experiments, smaller microcapsules were found to be more easily embedded in the composite metal-plating coating than were their larger counterparts. This may be explained in terms of the electrochemical changes occurring during co-deposition of the microcapsules. Two types of adsorption are associated with the microcapsules: (i) that the microcapsules themselves are adsorbed onto the electrode, as mentioned above; and (ii) that the microcapsule particles adsorb sufficient positive metal ions from the bath, depending on the strength of the electric field. The adsorbed microcapsules become embedded into the coatings as the metal ions (e.g., Ni^{2+}, Cu^{2+}) in the plating bath are reduced on the cathode; in this way an expected composite plating coating containing the liquid microcapsules is obtained.

It can be concluded from the above-mentioned differential capacitance curve and cathodic polarization curve analyses that the mechanistic model of electrolytic co-deposition of liquid microcapsules is associated with the stable chelation of –OH and –O groups of the wall material (e.g., PVA, gelatin) with metal ions (e.g., Ni^{2+}, Cu^{2+}), and this gives rise to positively charged microcapsules. This in turn helps to accelerate the electrophoretic migration of microcapsules in the plating solution. The liquid microcapsules were also adsorbed onto the electrode due to the presence of a surfactant. Consequently, it is feasible for microcapsules to enter the electrical double layer at the interface and to become embedded in the co-deposited coating.

9.5
Properties of Microcapsules Containing Composite Coatings

The presence of microcapsules in composite coatings may improve their wear resistance, corrosion resistance, water repellence, or weld ability. The present studies have focused on improving these characteristics in copper- and nickel-composite coatings containing lube oil, corrosion inhibitors, or hydrophobic agent-microcapsules by the electrolytic co-deposition of liquid microcapsules with metal ions. The study results provide a theoretical and technical foundation for the wide use of composite coatings containing liquid microcapsules in industrial processes.

9.5.1
Improving the Wear Resistance and Frictional Coefficient of Copper Composite Coatings Containing Lube Oil-Microcapsules

Figure 9.18 shows wearing-induced weight-loss curves of a copper coating and a composite coating containing small (3–8 μm) or large (>10 μm) microcapsules. Weight loss of the normal acidic copper-plating coating was greater under the same frictional conditions when compared to a copper-plating composite coating containing microcapsules, and thus the wear resistance was poor. As weight loss of the copper composite coating was comparatively less until 13 000 frictional movements had been completed, the implication was that the microcapsules in the coatings rupture gradually so that the lube oil in the core are released and spread onto the surface of the coating. This gives rise to self-lubricating properties such that wear resistance is enhanced.

After 13 000 frictional movements, however, weight loss was increased abruptly due to draining of the microcapsules. The slopes of the two weight loss curves tail

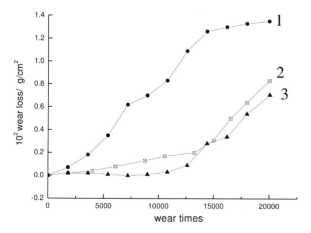

Figure 9.18 Wear loss of copper coating containing microcapsules with lubricating oil. (1) Copper coating; (2) composite copper coating containing microcapsules (diameter >10 μm); (3) composite copper coating containing microcapsules (diameter 3–8 μm).

regions were similar, indicating a decline in self-lubricating capability. These data also revealed that relatively large "holes" appear in coatings containing larger microcapsules following rupture of the latter, and this gives rise to a reduction in supporting force such that the wear resistance becomes inferior. Moreover, the smaller the microcapsule size, the more homogeneous and compact the microcapsules are dispersed, showing better controlled-release properties and providing better wear resistance.

Thus, it seems that the use of smaller microcapsules would be beneficial for co-deposition as these disperse homogeneously in the bath and improve surface properties such as friction resistance of the plating coating. The problem is that these tiny microcapsules encapsulate too-small quantities of core material, and this in turn leads to relatively short periods of controlled release from the microcapsule core. Even microcapsules can lose their functions within a very short period of service!

The morphologies of the composite copper-plating coatings containing antirust or lube oil-microcapsules, after 300 to 20 000 frictional movements, are shown in Figure 9.19. Scratches were not visible on the composite copper-plating coatings after 300 movements, but were clearly visible on the pure copper-plating coating. This shows that breakage of the microcapsules in composite coatings releases oily substances which reduce wear. When the number of frictional movements was increased to 8000, the surface of the composite coatings was changed and the holes that held microcapsules were filled with a mixture of the capsule and resultant products of friction. This showed the wear to have damaged the microcapsules on the composite coatings surface. The presence of the mixture of the lube oil and the products of friction in the microcapsule hole was confirmed by electrochemical impedance spectroscopy (EIS) studies. When the number of frictional movements was increased to 15 000, the weight loss rose sharply due to the presence of fewer releasable lube oil cores, and the wear resistance of the coating declined. It can be seen from the SEM images that after 20 000 movements not only did weight loss become severe but abrasion traces also appeared on the surface, together with the holes and defects which were left when the microcapsules ruptured. These changes proved disadvantageous to the wear- and corrosion-resistance of the coatings. Consequently, the main aspects of any future studies should be to increase the amount of microcapsules of appropriate size in the composite coating, to improve their dispersion homogeneity, and to select lube oil-microcapsules with better lubrication properties.

The data in Figure 9.20 reflect the changes in friction coefficient during the application of friction to composite coatings. It is clear from the figure that the average friction coefficient of copper coating was ca. 0.5, and for the composite coatings was 0.02. The coefficient remained almost unchanged with increased frictional movements for the composite coatings, this being due to the presence of lube oil-microcapsules in the composite coatings lowering the friction coefficient. As the frictional movements increased, the microcapsules in the composite coatings were crushed and released the core lube oils such that the friction coefficient remained stable at ca. 0.02. In other words, the amount of lubricant released was equal to that consumed. However, the coefficient began to increase when frictional movements were increased to a certain value and the difference in friction coef-

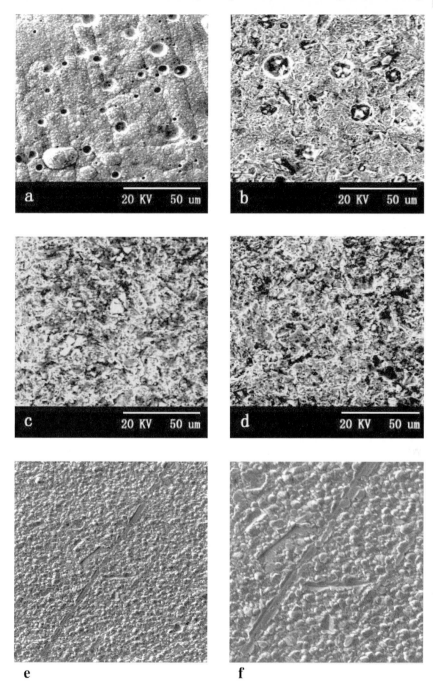

Figure 9.19 Morphology of composite copper plating (a–d) and pure copper-plating (e,f) coatings after different numbers of friction movement. (a) 300 times; (b) 8000 times; (c) 15 000 times; (d) 20 000 times; (e) 300 times; (f) 1000 times.

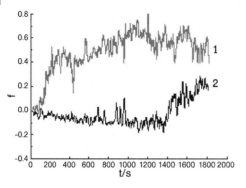

Figure 9.20 Plot of friction coefficient (f) versus time (t) for copper coatings. (1) Normal copper coating; (2) composite copper coating containing microcapsules.

ficient between the composite coating and the pure copper-plating coating was reduced. These data showed that, for the composite coatings, the amount of microcapsules decreased after friction for a certain time, and the friction coefficient began to increase (Fig. 9.20, curve 2).

9.5.2
Improvement of Wear Resistance and Frictional Coefficient of
Copper Composite Coatings Containing Antirust Lube Oil-Microcapsules

Wear loss of nickel-microcapsule composite coatings and normal nickel-plating coatings are plotted against wear time in Figure 9.21. It can be seen from these data that weight loss increased gradually with the increase in friction time for pure nickel plating coatings. During the early movements (up to 2400) there was

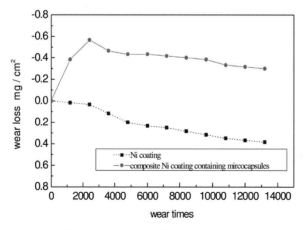

Fgure 9.21 Wear resistance curve of composite nickel coating containing microcapsules.

a weight increment in the composite coating, with a maximum increase of ca. 0.57 mg cm^{-2}. The weight loss began after 2400 movements, but after 13 200 movements the mass of the specimens was greater than that of the original. This may have been due to friction-induced breakage of the microcapsules leading to a gradual release of the encapsulated lube oil, which adhered to solid fragments produced during abrasion. During the later stages of the experiment, however, the samples began to lose weight due to friction, and their mass decreased gradually.

In contrast, the normal nickel-plating coating containing no microcapsules tended to lose weight during the experiment, and weight loss increased with time. This indicated that the release of microcapsules plays a substantial role in self-repair in the case of composite coatings that improve wear resistance.

Figure 9.22 illustrates the morphology of the composite coatings containing microcapsules and normal nickel-plating coatings before and after the wear-resist-

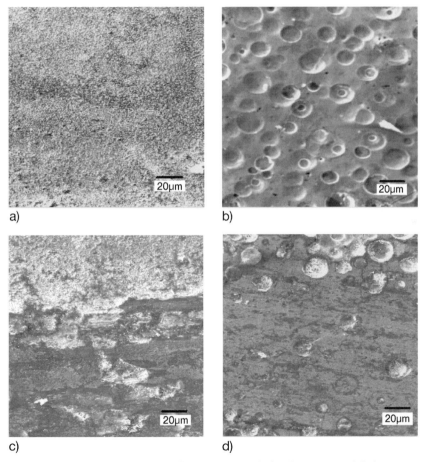

a)

b)

c)

d)

Figure 9.22 Morphology of two types of coating before and after friction. (a) Nickel-plating coating before friction; (b) composite coating before friction; (c) nickel-plating coating after friction; (d) composite coating after friction.

ance experiments. The crystals of normal nickel-plating coatings were tiny and smooth (Fig. 9.22a), while the coating became coarse and showed scratches and wear abrasion (Fig. 9.22c) after 13 200 frictional movements. By contrast, composite coatings containing microcapsules showed an absence of scratches and/or wear abrasion on the surface. The results obtained were, therefore, inconsistent with those observed in the above-mentioned analyses.

Figure 9.23 shows the curves of frictional force for different coatings. The data in Figure 9.23(a) show that for pure nickel plating coatings the curve fluctuates widely, with a frictional force of 1 N. However, the frictional force for composite coatings deposited with 15% microcapsules shows a smooth curve, with a frictional value of 0. This means that the frictional force was decreased substantially in the composite coatings due to the presence of microcapsules, and remained relatively stable. The frictional force will increase only when the bulges of the scraps occurred. Figure

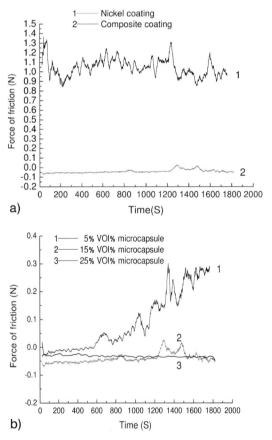

Figure 9.23 Plots of frictional force versus time for composite nickel-plating coatings containing microcapsules. (a) Comparison of friction between different plating coatings; (b) comparison of friction between different composite coatings.

9.23(b) represents the frictional curves of nickel composite coatings deposited with the bath containing 15% and 25% microcapsules, respectively. During continual wear, the frictional forces changed only slightly and were largely equal. In the later stages of the experiment the frictional force of the coating deposited with 15% microcapsules in the bath was raised for a short period, though this may have been due to the presence of scraps left after friction, with two peaks appearing in the curve.

Friction coefficient variations with time for composite nickel-plating coatings containing antirust lube oil-encapsulated microcapsules and normal nickel-plating coating containing no microcapsules are presented in Figure 9.24. The results obtained were similar to those seen with copper plating coatings (Fig. 9.23). The data in Figure 9.24 show that the friction coefficient of the normal nickel-plating coating remained at ca. 0.3, while that of the composite coating containing 15% microcapsules was ca. 0. This was due to the presence of incorporated microcapsules in

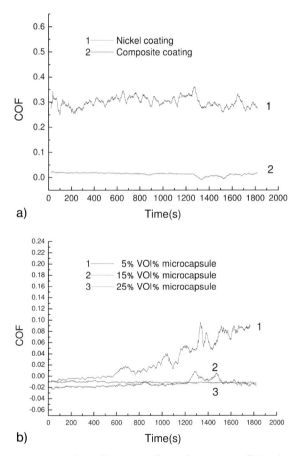

Figure 9.24 Plots of friction coefficient for composite nickel-plating coatings containing microcapsules. (a) Comparison of friction coefficient between different plating coatings; (b) comparison of friction coefficient between different composite coatings.

the composite coatings. Composite coatings with 15% and 25% microcapsules presented similar friction coefficient values.

The above results indicated that the nickel-plating composite coatings containing lube oil-encapsulated microcapsules possessed excellent wear resistance.

9.5.3
Anticorrosion Performance of Microcapsules Containing Composite Coatings

It is well known that pinholes are produced during the electrodeposition of nickel plating coatings, and that these may give rise to corrosion of the metal matrix. If antirust oil-microcapsules are incorporated into the nickel plating coatings by electrolytic co-deposition, the controlled-release effect of the microcapsules will enhance the corrosion resistance of the nickel-plating coating.

Table 9.5 details the time when the corroded points occurred on the normal nickel-plating coatings, nickel-plating coatings coated with antirust oil, and composite nickel plating coatings containing microencapsulated antirust oil in 5% NaCl solution. The corroded points were seen to appear on the composite coating after an immersion time of 40 h, but to appear on normal nickel-plating coatings within 1 h. This explains why microcapsules are capable of reducing the defects of the coatings so as to improve the corrosion resistance of the coating. In addition, no corroded points were found on the normal nickel-plating samples coated with the same antirust oil until 10 h, which indicated that the oil film prevented the surfaces from corroding to a certain extent. However, when the corrosive medium penetrated the oil film, the samples quickly corroded. Composite coatings containing antirust oil-microcapsules remained uncorroded for a longer period, this being due to slow release of antirust oil from the microcapsules that enhanced the corrosion resistance of the plating coatings.

Figure 9.25 illustrates the anodic polarization curves of both the normal nickel-plating coating and a composite coating immersed in 5% NaCl solution. The data in Figure 9.25(a) show the normal nickel-plating coating to be more polarized than the composite coating. An activated current of the normal nickel-plating coating appeared at –0.1 V, and this was increased substantially with increment of the anod-

Table 9.5 Times at which corroded points appeared in 5% NaCl solution.

Sample	Time of corrosion appearance [h]
Nickel composite coating containing antirust oil-microcapsules	40–110
Normal nickel-plating coating	0.5–1
Nickel-plating coating painted with antirust oil	11–15

ic potential. While the composite coating is in passivation between −0.2 and 0.1 V potential range, the anodic activated current is evidently smaller than that of normal nickel-plating coating under the same potential. This implies that the antirust oil-encapsulated microcapsules played an important role in the composite coating, distinctly decreasing the corrosion current in polarization.

Figure 9.25(b) illustrates the polarized curve of the composite nickel-plating coating with or without abrasion (type 600 abrasive paper, 3600 movements). The results obtained revealed that the corrosion current on the surface of the composite coating containing antirust oil-encapsulated microcapsules was smaller after abrasion under the same anodic polarization potential. Abrasion of the composite coating surface ruptured the microcapsules such that the encapsulated oil was released and spread onto the surface. The polarization current was reduced; that is, better wear and corrosion resistance was achieved if the microcapsules were

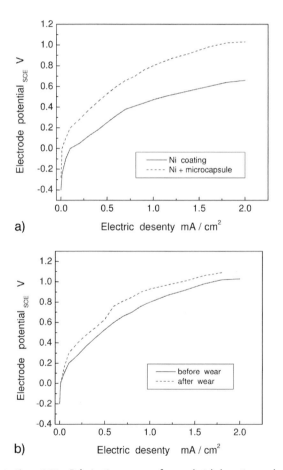

Figure 9.25 Polarization curves of normal nickel coating and composite coatings. (a) Normal nickel coating and composite coatings; (b) composite coating before and after abrasion.

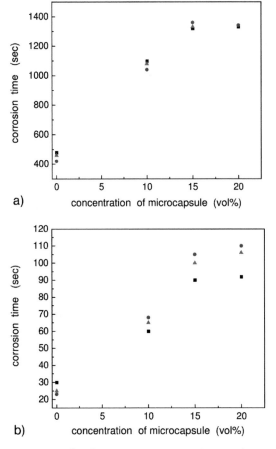

a)

b)

Figure 9.26 Plot of corrosion time versus microcapsule content for composite coatings. (a) In 25% ammonia; (b) in 10% nitric acid.

crushed. Thus, different surface functionalities such as controlled release and self-repairing properties, can be achieved by changing the core material of the microcapsules embedded in the composite coating.

Figure 9.26 illustrates the corrosion time of composite coatings after being drenched with 25% ammonia (Fig. 9.26a) or 10% nitric acid (Fig. 9.26b). Ammonia and nitric acid are both very aggressive to copper coatings, while benzotriazole is a good corrosion inhibiter for copper or copper-alloys. The data in Figure 9.26 show that the copper-plating coating without microcapsules remained uncorroded for only 400–480 s, while the composite coating containing 10% microcapsules showed anticorrosive properties for a period of 1050–1150 s. When the microcapsule content was increased to 15%, the coatings showed anticorrosive behavior for more than 1300 s. However, a further increase in microcapsule loading (e.g., 20%) did not influence anticorrosion time, which remained at ca. 1300 s. Similar results were obtained using a nitric acid medium, whereby the pitting time for pure copper plating was three-fold greater than that for microcapsules containing a composite copper-plating coating. This illustrates, again, that a gradual release of microcapsules improves the corrosion-resistance of a composite copper-plating coating containing corrosion inhibitor-microcapsules.

Figure 9.27 illustrates the anodic polarization curve of composite copper-plating coatings containing corrosion inhibitor-microcapsules in 1 M sulfuric acid solution. It can be observed that the polarization potential of composite coating containing microcapsules is much higher than that of normal copper-plating coating. Likewise, under the same corrosion potential, the corrosion current of the composite coating is smaller due to the presence of a corrosion inhibitor in the microcapsules.

Figure 9.28 illustrates potential versus time plots of copper-plating coatings with or without corrosion inhibitor-microcapsules. The data in Figure 9.28 show the potential of composite coatings to be ca. 200 mV higher than that of the normal plating coatings. While the electrochemical potential of the copper-plating coatings in a sulfuric acid medium fell sharply, from 600 mV to –950 mV, after only 10 min immersion, it changed only slowly for microcapsules containing composite coatings. The potential for composite coatings decreased until it approached –850 mV after ca. 100 min. Subsequently, the electropotential remained almost unchanged, perhaps due to the release of microcapsules in the coating, which affected the electropotential value.

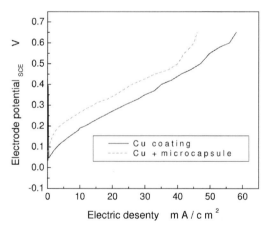

Figure 9.27 Anodic polarization curves of composite coatings containing corrosion inhibitor-microcapsules.

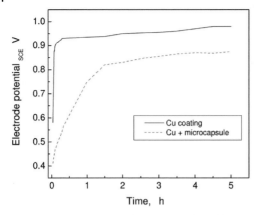

Figure 9.28 Electropotential versus time plots of the composite coatings surface.

9.6
The Controlled-Release Rule of Microcapsules in Coatings

The purpose of incorporating microcapsules into coatings is to improve or repair the surface function of the metal by gradual release of the core materials. In this context, it is very important to study the controlled-release behavior of microcapsules embedded in the composite coating, and of their self-repairing capability for the plating surface.

9.6.1
Controlled Release of Hydrophobic Agent-Microcapsules

Figure 9.29 illustrates the contact angle of water on a composite copper-plating coating containing hydrophobic agent-microcapsules under atmospheric pressure and at room temperature. Generally, release of the core material from microcapsules is affected by environmental conditions such as temperature and permeability of the core. It is clear from Figure 9.29 that water drops spread easily on normal copper-plating coating, showing that the coating itself is wettable. However, in the case of microcapsules containing composites, the encapsulated hydrophobic agent is released gradually with time and enhances the contact angle of water. Thus, it can be concluded that hydrophobic surfaces can be designed by incorporating hydrophobic agent-microcapsules into the composite metal electroplating.

The contact angle of water on the composite copper-plating coating containing hydrophobic agent-encapsulated microcapsules, compared to that of normal copper-plating coating painted with a hydrophobic agent, was plotted against time (Fig. 9.30). In order to measure the contact angle, one drop of distilled water was placed on the composite coating surface, using a 10-mL syringe. The height h and contact length r of the drop were then measured, and the contact angle was calculated according to the following equation:

$$tg\theta = \frac{4rh}{r^2 - 4h^2} \tag{1}$$

The water-resistance behavior of the composite coating was enhanced to a large extent as the hydrophobic agent was released gradually from the microcapsules. However, under the same conditions the normal copper-plating coating painted with hydrophobic agent showed a contact angle of 95–100° (Fig. 9.30). The contact angle of copper-plating coatings with painted hydrophobic agent remained almost unchanged for 150 days of resting, thereby confirming that it possessed stable hydrophobicity. In contrast, the hydrophobic agent-microcapsule-containing coatings

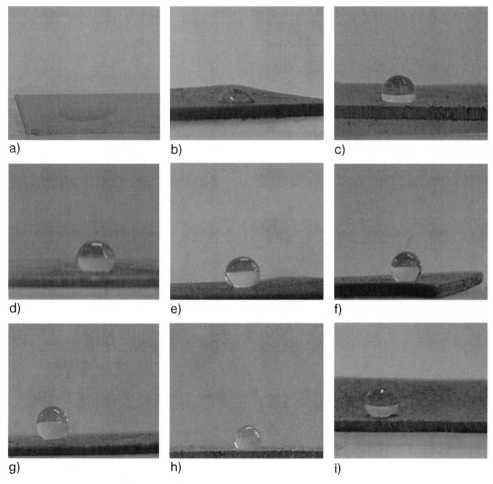

Figure 9.29 Water-repellent behavior of composite coating containing hydrophobic agent-microcapsules. (a) Copper-plating coating. (b) Composite copper-plating coating at various days later: (c) 2 days; (d) 8 days; (e) 12 days; (f) 16 days; (g) 24 days; (h) 32 days; (i) 52 days.

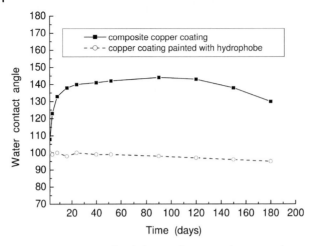

Figure 9.30 Water-repellent behavior of composite copper-plating coating containing hydrophobic agent-microcapsules and that of the copper-plating coating painted with hydrophobic agent under daylight at room temperature.

showed a contact angle of 135–140°. The angle decreased with time, but remained above 130°, even after 150 days. This confirmed that the hydrophobicity of the composite coating was enhanced by its being embedding with hydrophobic agent-microcapsules.

The change of contact angle for composite copper-plating coating containing hydrophobic agent-microcapsules at higher temperature (using a daylight lamp; 80 ± 2 °C) is illustrated in Figure 9.31. The rise in temperature is seen rapidly to increase the water contact angle. For example, the contact angle achieved in 4 h at a

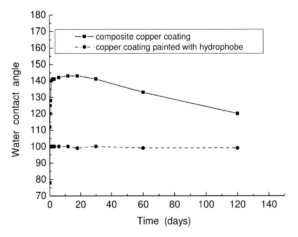

Figure 9.31 Water-repellent behavior of composite copper-plating coating containing hydrophobic agent-microcapsules and that of the copper-plating coating painted with hydrophobic agent under daylight lamp (specimens maintained at 80 ± 2 °C).

higher temperature is equal to that achieved after 10 days at room temperature. Thus, an increase in environmental temperature is beneficial for the release of core material and, thereby, an enhancement of water contact angle.

The water contact angle of the composite coating containing hydrophobic agent-microcapsules under sunlight indicates excellent water resistance (Fig. 9.32). It is worth noting here that the core material is a hydrophobic agent that is gradually re-

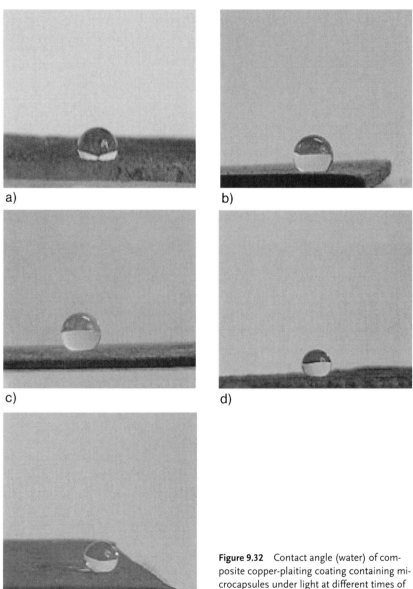

a)

b)

c)

d)

e)

Figure 9.32 Contact angle (water) of composite copper-plaiting coating containing microcapsules under light at different times of illumination. (a) After 2 h; (b) after 4 h; (c) after 16 h; (d) after 24 h; (e) after 32 h.

leased (at ambient temperature) or rapidly released (under a daylight lamp); this leads to a hydrophobic angle of 140° for the composite and of 95° for painted copper-plating coatings. This is due to the fact that the hydrophobic film formed by the hydrophobic agent-paints is continuous, whereas that formed by a gradual release of the hydrophobic agent-microcapsules is discontinuous. It can be seen from the morphology of the composite coatings that the hydrophobic agent-microcapsules dispersed independently in the plating coating. Thus, a hydrophobic film was formed near the microcapsules, whilst a site relatively far away from the microcapsules was not affected by the hydrophobic agent. Hence, the hydrophobic film is not uniform over the entire surface. Furthermore, co-deposition of the microcapsules led to a heterogeneous plating coating (see Fig. 9.7, micro-coarse surface), which was characterized by the fact that the hydrophobic angle of the composite copper-plating coating containing the hydrophobic agent-microcapsules was far larger than that of the hydrophobic agent-painted copper-plating coating.

Besides the above example, composite copper-plating coating containing corrosion inhibitor-microcapsules can improve corrosion resistance and self-repair properties due to the slow release of corrosion inhibitors from the coating. The data in Figure 9.33 show the surface morphology of the composite coating containing microcapsules, and the normal copper-plating coating placed in air at room temperature for 15 days and 45 days, respectively. The surface of the normal copper-

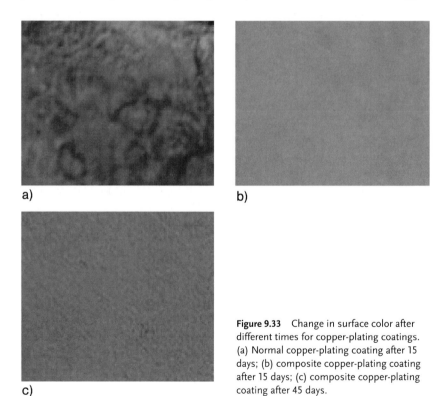

a)

b)

c)

Figure 9.33 Change in surface color after different times for copper-plating coatings. (a) Normal copper-plating coating after 15 days; (b) composite copper-plating coating after 15 days; (c) composite copper-plating coating after 45 days.

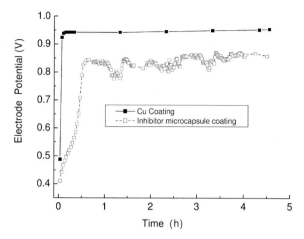

Figure 9.34 Plots of potential versus time for composite coatings containing corrosion inhibitor-microcapsules (solution: 0.1 M sulfuric acid, mercurous sulfate reference electrode).

plating coating was heavily oxidized at room temperature, whereas there was no change in the case of the composite coating. Thus, the slow release of corrosion inhibitor from the microcapsules had prevented the coating from being oxidized.

Figure 9.34 illustrates the time–potential plots of composite coatings containing corrosion inhibitor-microcapsules in 0.1 M sulfuric acid solution, using a mercurous sulfate reference electrode. The potential of the normal copper-plating coating was seen to be essentially stable with an increase in soaking time, but to fluctuate between 30 and 70 mV for the composite coatings; this implied that the corrosion inhibitor was being released gradually. The corrosion inhibitor used, benzotriazole, forms a passive film on the copper plating coatings that protects the surface. The potential was decreased as the sulfuric acid corroded the surface, and a new passive film was formed when the encapsulated corrosion inhibitor was released, making the corrosion potential positive again. This fluctuation in potential was illustrative of self-repair of the composite coatings.

Composite copper-plating coatings containing lube oil-microcapsules showed a similar self-repairing behavior. The data in Figure 9.35 show anodic polarization curves of the composite lube oil-microcapsules containing composite coatings after different times of friction. It is clear from Figure 9.35 that the unworn composite coating (curve 1) bears a smaller cathodic corrosion current than the worn coating (curve 2). This signifies that the corrosion resistance of the worn copper-plating coating is reduced after 300 friction movements, due to scratches occurring on the surface and the integrity of the metal plating being destroyed and accelerating corrosion of the copper coating.

The composite copper-plating coating containing lube oil-microcapsules (particle size 3–8 μm) showed special regularity (curves 3, 4, 5, and 6). First, the anodic corrosion current on the surface of unworn composite copper-plating coating

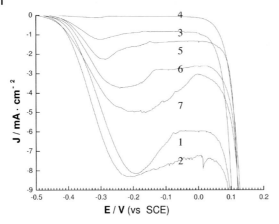

Figure 9.35 Anodic polarization curves of copper-plating coatings after different times of friction movement. (1) Normal copper-plating coating; (2) normal copper-plating coating (300 times); (3) composite coating; (4) composite coating (300 times); (5) composite coating (800 times); (6) composite coating (2000 times); (7) composite coating (diameter >10 μm). E/V (vs. SCE): Electrode potential; J/ma·cm^{-2}: Electric density.

(curve 3) was larger than that of the worn composite coating (curve 4). This showed that the microcapsules in the coating had been crushed after abrasion and that the released oily core substance had spread over the coating surface that isolated the composite coating from the corrosive medium; hence, the anodic corrosion current was decreased. Second, the corrosion current on the anodic polarization curve increased with the frictional times for the composite copper-plating coating. This occurred because most of the microcapsules were broken, and the core and wall materials were lost when the composite copper coating became badly worn. The remaining number of microcapsules was insufficient to fill the gaps and scratches on the surface, and could not produce an isolation barrier against the corrosive medium.

Finally, composite coatings containing smaller microcapsules (2–8 μm; curve 3) possess a corrosion resistance that is quite different from that of coatings containing larger microcapsules (>10 μm; curve 7). This may be attributed to that fact that larger microcapsules tend to break quite easily and are difficult to embed in the coating. Thus, defects such as widespread cavities lead to a deterioration in the integrity of the composite coating and have a detrimental effect on corrosion resistance. Thus, the advantages of using microcapsules with small diameters in electrolytic co-deposition are clear.

9.7
Prospects

The controlled-release properties of microcapsules form the technical basis of the self-repairing properties of composite coatings containing microcapsules, and mechanisms of controlled release will represent the focus of future studies. The release of microcapsular contents may be either rupture-induced or depend on environmental changes. The former effect is caused largely by corrosive media or external frictional forces whereby the core material is released to cover the plating coating surface. The latter effect is caused mainly by factors such as temperature, pressure and pH, whereby the core material is gradually released so as to improve the surface properties of the coating, for example friction resistance, corrosion resistance, and hydrophobic resistance. The release rate corresponds mainly with film thickness, the particle size of the microcapsules, the nature of the wall, the solubility and diffusion coefficients of the core, environmental temperature, and the core content. Thus, it is important to study the influences of environmental factors on the release behavior of microcapsules and to choose appropriate methods to detect the amounts of microcapsules present in composite coatings.

Furthermore, the strength of the wall materials and the different factors associated with electrolytic co-deposition kinetics play additional roles in the development of composite coatings with predetermined properties. Hopefully, future research projects will lead to the practical application of liquid microcapsules to improve the properties of metal plating coatings by using this novel technique of electrolytic co-deposition in a variety of aspects.

Acknowledgments

Information provided in this chapter is based on research investigations conducted within the project supported by the National Natural Science Foundation of China, Grant No. 50171002.

Abbreviations

D_k current density
DTA   differential thermal analysis
EIS   electrochemical impedance spectroscopy
PVA   polyvinyl alcohol
SCE   standard calomel electrode
SEM   scanning electron microscopy
XPS   X-ray photoelectron spectroscopy

References

1. Li-qun Zhu, Chen-min Liu, *Journal of Beijing University of Aeronautics and Astronautics* **2001**, *27(2)*, 133–136.
2. Li-qun Zhu, Qiang Ren, Yan-yan Chang, *Journal of Aeronautical Materials* **2002**, *22(3)*, 30–37.
3. Li-qun Zhu, Wei Zhang, *Acta Phys. Chem. Sin.* **2004**, *20(8)*, 795–800.
4. Li-qun Zhu, Feng Liu, Wie Zhang, *Functional Material* **2004**, *35(4)*, 504–506.
5. Li-qun Zhu, Wei Zhang, Feng Liu, *Material Engineering* **2004**, *39(1)*, 12–15.
6. Li-qun Zhu, Wei Zhang, Feng Liu, *Material Protection* **2003**, *36(3)*, 24–26.
7. Li-qun Zhu, Wei Zhang, Feng Liu, *Journal of Materials Science* **2004**, *39*, 495–499.
8. Li-qun Zhu, Qing-ling Xia, Feng Liu, *Functional Material* **2003**, *34(4)*, 454–457.
9. Yu Chen, Dan-hong Cheng, Xiao-bing Liu, *Plating and Finishing* **2003**, *25(4)*, 1–5.
10. Ke-jun Li, Dan-hong Cheng, Yong-tang Su, *Plating and Finishing* **2005**, *27(1)*, 11–15.
11. I. Masayuki, S. Isao, W. Kunihiro, *Surface Technology* **2003**, *54(9)*, 599–604.
12. I. Masayuki, S. Isao, W. Kunihiro, *Surface Technology* **2003**, *54(3)*, 230–234.

Index